AF361719

Modeling of Soil Erosion and Sediment Transport

Modeling of Soil Erosion and Sediment Transport

Editor

Vlassios Hrissanthou

MDPI • Basel • Beijing • Wuhan • Barcelona • Belgrade • Manchester • Tokyo • Cluj • Tianjin

Editor
Vlassios Hrissanthou
Democritus University of Thrace
Greece

Editorial Office
MDPI
St. Alban-Anlage 66
4052 Basel, Switzerland

This is a reprint of articles from the Special Issue published online in the open access journal *Water* (ISSN 2073-4441) (available at: https://www.mdpi.com/journal/water/special_issues/modeling_soil_erosion).

For citation purposes, cite each article independently as indicated on the article page online and as indicated below:

LastName, A.A.; LastName, B.B.; LastName, C.C. Article Title. *Journal Name* **Year**, *Article Number*, Page Range.

ISBN 978-3-03943-002-4 (Hbk)
ISBN 978-3-03943-003-1 (PDF)

Contents

About the Editor

Vlassios Hrissanthou is Emeritus Professor at the Civil Engineering Department of Democritus University of Thrace (DUTH), Xanthi, Greece. He studied Civil Engineering at the Aristotle University of Thessaloniki (AUTH), Greece, obtaining the Civil Engineer diploma. He then undertook postgraduate and doctoral studies in Hydrology and Hydraulic Structures at the University of Karlsruhe (KIT), Germany. Subsequently, he completed his postdoctoral studies in Hydraulics and Hydraulic Structures at the University of the Armed Forces Munich (UniBw München), Germany. He is the principal investigator or leader of many competitive German, Greek and international research projects. His teaching work includes the following graduate and postgraduate study courses: Fluid Mechanics, Hydraulics, Engineering Hydrology, River Engineering, Hydropower Engineering, Water Resources Management, Open Channel Hydraulics, Hydrology of Groundwater, Advanced Engineering Hydrology, Sediment Transport, Reservoir Design, Time Series Analysis, Selected Chapters of Hydropower Engineering and Hydraulics of Stratified Flows. He has supervised a large number of diploma, post-diploma and doctoral theses. He is the author or co-author of 44 publications in scientific journals, 86 publications in conference proceedings, 11 publications in book chapters, and 44 technical reports. He is also the editor of 3 books and the author of 1 book. His publications are available in English, Greek, and German and have around 470 citations (excluding self-citations). Finally, he has reviewed numerous publications for 43 international scientific journals in addition to serving as a member of the Reviewer Board of the MDPI journal Sustainability.

Preface to "Modeling of Soil Erosion and Sediment Transport"

The theme of the present Special Issue is "Modeling of Soil Erosion and Sediment Transport". Most studies presented in the Special Issue were applied to different basins in Europe, America, and Asia. It is also worth mentioning that most studies are the result of the cooperation between universities and/or research centers in different countries and continents, portraying an optimistic view of the status of international scientific communication. This Special Issue contains 14 articles that can be classified into the following five categories: Category A: "Soil erosion and sediment transport modeling in basins"; Category B: "Inclusion of soil erosion control measures in soil erosion models"; Category C: "Soil erosion and sediment transport modeling in view of reservoir sedimentation"; Category D: "Field measurements of gully erosion"; Category E: "Stream sediment transport modeling". Articles 1–4 belong to Category A: In the first article (Gudino-Elizondo et al.), the Annualized Agricultural Non-Point Source (AnnAGNPS) model is applied to the Los Laureles Canyon coastal watershed (USA–Mexico border) in order to investigate, amongst others, the impact of urbanization on runoff and sediment load. Field measurements are also used for the calibration and validation of the model. In the second article (Lu and Chiang), the SWAT-Twn model, including the Taiwan Universal Soil Loss Equation (TUSLE), is applied to a small mountainous watershed in Taiwan in order to improve sediment simulation and assess the sediment transport functions. Field measurements are also used for the calibration and validation of the model. In the third article (Yin et al.), a process-based, fully distributed soil erosion model, named WRF-Hydro-Sed, is applied to the Goodwin Creek Experimental Watershed in Mississippi (USA) to account for both overland and channel processes. Streamflow and sediment concentration data during rainfall events are used for the calibration and validation of the model. In the fourth article (Al Sayah et al.), the impact of man-made ponds on soil erosion and sediment transport is assessed, especially for limnological basins. The French Claise basin is considered as application example. In concrete terms, the CORINE erosion and SWAT models are applied to the above basin. Additionally, erosion risk zones are distinguished in the basin. Articles 5–7 belong to Category B: In article 5 (Bai et al.), the impact of terraces and vegetation on runoff and sediment routing is investigated. The Pianguanhe basin in the Chinese Loess Plateau is considered as an application example. For the above investigation, a revised time-area method is integrated into the Land Change Model-Modified Universal Soil Loss Equation (LCM-MUSLE). Eight storms in the 1980s and 2010s are selected to calibrate and verify the original LCM-MUSLE model and its revised version. In article 6 (Nabi et al.), the impact of stone bund type structures and vegetation cover change on sediment yield reduction is investigated. The effectiveness of the above erosion control measures is evaluated by applying a semi-distributed Soil and Water Assessment Tool (SWAT) model to various small watersheds in Pakistan. In article 7 (Xin et al.), the effects of bare and residue cover slopes on the infiltration process of the black soil are evaluated under rainfall simulations. A black soil region in northeastern China is considered as an application paradigm. Two articles are classified into Category C: In article 8 (Németová et al.), a physically based model, named EROSION-3D, is applied to the Svacenicky Creek catchment in the western part of the Slovak Republic. Measurements of sediment volume in a small reservoir, at the bottom of the catchment, are available via a bathymetric field survey. Hence, the model performance can be tested through the available sediment volume measurements. In article 9 (Tarar et al.), the HEC-RAS 1D numerical model is used for predicting delta movement in the Tarbela reservoir (Pakistan). Sediment

Rating Curves (SRCs) and Wavelet-Artificial Neural Networks (WA-ANNs) are applied for setting sediment load boundary conditions in HEC-RAS and for finding missing sediment sampling data. Category D is represented by one article: In article 10 of Luffman and Nandi, the relationship between gully erosion and precipitation parameters (duration, total accumulation, intensity) is examined. In concrete terms, the effect of the seasonal precipitation variability on gully erosion is determined in northeast Tennessee (southeastern USA) on the basis of available field measurements of gully erosion. Four articles are classified into Category E: In article 11 (Park et al.), the effect of catchment characteristics on the performance of an already developed model for the estimation of fine sediment dynamics between the water column and sediment bed is tested, using 13 catchments in France, Ireland, Spain, Italy, the Belgian–Dutch border, and USA. Article 12 (Tavelli et al.) presents a new 2D, semi-implicit numerical scheme for the solution of Navier–Stokes equations (momentum conservation), the incompressibility condition (mass conservation), and the mass conservation law for suspended sediment concentration in gravel bed rivers. The above scheme is tested against analytical solutions and performing numerical tests. In article 13 (Kaffas et al.), fuzzy transformation of the total sediment load formula of Yang is conducted. In other words, transformation of the arithmetic coefficients of the Yang formula into fuzzy numbers. A very large set of experimental data, in flumes, is used for fuzzy regression analysis. In article 14 (Wang et al.), a dynamical and numerical simulation of the evolution process of leaked tailings flow from dam failure is presented. At this point, it should be noted that tailings ponds are indispensable facilities in the mine production and operation. The evolution process mentioned above is analyzed at various downstream riverbed slopes and debris blocking dam settings. As application paradigm, a tailings pond dam in Sichuan Province of China is reported.

Vlassios Hrissanthou
Editor

Article

Modelling Runoff and Sediment Loads in a Developing Coastal Watershed of the US-Mexico Border

Napoleon Gudino-Elizondo [1,2,3,*], Trent W. Biggs [2], Ronald L. Bingner [4], Eddy J. Langendoen [4], Thomas Kretzschmar [3], Encarnación V. Taguas [5], Kristine T. Taniguchi-Quan [6], Douglas Liden [7] and Yongping Yuan [8]

[1] Department of Civil and Environmental Engineering, University of California, Irvine, CA 92697, USA
[2] Department of Geography, San Diego State University, 5500 Campanile Dr., San Diego, CA 92182-4493, USA; tbiggs@mail.sdsu.edu
[3] Departamento de Geología, Centro de Investigación Científica y de Educación Superior de Ensenada (CICESE), Carretera Ensenada-Tijuana 3918, Zona Playitas, 22860 Ensenada, B.C., Mexico; tkretzsc@cicese.mx
[4] National Sedimentation Laboratory, Agricultural Research Service, USDA, Oxford, MS 38655, USA; Ron.Bingner@ars.usda.gov (R.L.B.); Eddy.Langendoen@ars.usda.gov (E.J.L.)
[5] Department of Rural Engineering, University of Córdoba, Córdoba, 14071, Spain; ir2tarue@uco.es
[6] Southern California Coastal Water Research Project, 3535 Harbor Boulevard, Suite 110, Costa Mesa, CA 92626, USA; kristinetq@sccwrp.org
[7] USEPA San Diego Border Liaison Office, 610 West Ash St., Suite 905, San Diego, CA 92101, USA; Liden.Douglas@epa.gov
[8] USEPA Office of Research and Development, Research Triangle Park, NC 27711, USA; Yuan.Yongping@epa.gov
* Correspondence: ngudinoe@uci.edu or ngudinoe@gmail.com; Tel.: +52-1646-108870

Received: 22 January 2019; Accepted: 22 April 2019; Published: 16 May 2019

Abstract: Urbanization can increase sheet, rill, gully, and channel erosion. We quantified the sediment budget of the Los Laureles Canyon watershed (LLCW), which is a mixed rural-urbanizing catchment in Northwestern Mexico, using the AnnAGNPS model and field measurements of channel geometry. The model was calibrated with five years of observed runoff and sediment loads and used to evaluate sediment reduction under a mitigation scenario involving paving roads in hotspots of erosion. Calibrated runoff and sediment load had a mean-percent-bias of 28.4 and − 8.1, and root-mean-square errors of 85% and 41% of the mean, respectively. Suspended sediment concentration (SSC) collected at different locations during one storm-event correlated with modeled SSC at those locations, which suggests that the model represented spatial variation in sediment production. Simulated gully erosion represents 16%–37% of hillslope sediment production, and 50% of the hillslope sediment load is produced by only 23% of the watershed area. The model identifies priority locations for sediment control measures, and can be used to identify tradeoffs between sediment control and runoff production. Paving roads in priority areas would reduce total sediment yield by 30%, but may increase peak discharge moderately (1.6%–21%) at the outlet.

Keywords: soil erosion; rainfall-runoff; sediment yield; AnnAGNPS model; urbanization; scenario analysis

---

1. Introduction

Erosion, defined as the detachment, transport, and spatial redistribution of soil particles [1,2], contributes to environmental degradation around the globe [3]. Urbanization can lead to an increase in erosion and the discharge of terrigenous materials into downstream ecosystems, including inland lakes, reservoirs, estuaries, and oceans.

Sheetwash, rill, and gully erosion, hereafter referred to as hillslope erosion, are frequently associated with anthropogenic soil disturbance and are often related to land use change, such as deforestation and urban development [4,5]. Hillslope erosion processes have been well characterized in agricultural settings, but not in urbanized areas where high erosion rates have also been reported [6]. Hillslope erosion rates typically decrease as bare soil in construction sites is replaced by impervious surface and vegetation [7]. Conversely, in developing countries, soil exposure such as vacant lots and unpaved roads can persist for longer periods [8], which increases hillslope erosion rates compared to other urban watersheds with high impervious cover fractions [9] and storm-water management practices [10].

Hillslope erosion and sediment production can be simulated using numerical models that consider the relationship between terrain attributes and climate regimes [11,12]. These models vary in structure, assumptions, and data requirements [13,14]. Erosion modeling is often used to simulate various erosional processes, such as sheet, rill, gully, and channel erosion, to develop sediment budgets and to assess the effect of Best Management Practices (BMPs) on total sediment reduction. The sediment budget is the quantitative tracking of contributing sources, sinks, and spatial redistribution of sediments over a given time scale [15].

Several soil erosion studies have focused on sheet and rill processes [16–19], but ephemeral gullies can also contribute a significant source of sediment at the catchment scale [20–22], especially in arid and semi-arid areas [23]. Such gullies are caused by concentrated overland flow [24] and are commonly cleared by tillage operations [25] or, in urban environments, filled with unconsolidated sediment during grading [9]. These erosional features form due to a complex relationship between terrain and management characteristics such as slope, land cover, soil properties, climate regime, and management activities [26].

The Annualized Agricultural Non-Point Source (AnnAGNPS) model is a simulation tool developed by the USDA-Agricultural Research Service and the Natural Resources Conservation Service (NRCS) to evaluate the effect of land use and management activities on watershed hydrology and sediment transport [12]. AnnAGNPS simulates runoff and sediment generation by tracking their transport through the channel network (AnnAGNPS reaches) at the watershed scale on a daily time step. AnnAGNPS simulates different erosional processes (i.e., sheet, rill, and gullies) as well as channel sources. Sheet and rill erosion are simulated using the Revised Universal Soil Loss Equation (RUSLE). Ephemeral gully erosion is simulated using EGEM (Ephemeral Gully Erosion Model) whose hydrologic routines to calculate peak and total discharge are estimated following the SCS curve number (CN) methodology [27], and gully width and soil erosion calculations are based on the Chemical, Runoff, and Erosion from the Agricultural Management Systems (CREAMS) model [28]. The model simulates colluvial storage of sediment using the Hydro-geomorphic Universal Soil Loss Equation (HUSLE), which calculates a delivery ratio based on particle size distribution and flow transport capacity [29].

The AnnAGNPS model has been tested in small Mediterranean watersheds (< 1.3 km^2) [14,23,30]. Licciardello et al. [30] evaluated AnnAGNPS in a steep catchment under pasture in Eastern Italy. Taguas et al. [23] evaluated the effect of different management activities on total sediment reduction in an agricultural environment in Spain, where ephemeral gullies are a significant contributor to the total sediment production. Gudino-Elizondo et al. [14] reported good performance of AnnAGNPS in simulating ephemeral gullies at the neighborhood scale in an urban watershed. However, AnnAGNPS has not been tested to model hillslope erosion rates in an urbanizing catchment under different soil types and land uses.

This study aims to (1) test the capabilities of AnnAGNPS to simulate runoff and sediment production in an urban watershed in a developing country context and (2) use the model to constrain the sediment budget in order to inform management and policy designed to mitigate sediment loads downstream. This paper addresses the following research questions: (a) How accurately does

AnnAGNPS simulate water and sediment loads in an urban watershed in a developing-country context where ephemeral gullies are likely to be a significant source of sediment? (b) What processes generate sediment in the watershed, and what is the role of soil properties and land use? (c) How does storm size affect the sediment load from different hillslope processes (sheet and rill, and gully erosion)? (d) Where are hot spots of sediment production, and what watershed characteristics control sediment production? and (e) What are the implications of the sediment budget and distribution of hotspots for management designed to mitigate sediment loads?

2. Materials and Methods

2.1. Study Area

The Los Laureles Canyon Watershed (LLCW) is a transboundary urbanizing catchment located in the northwestern part of Tijuana, Mexico, which flows into the Tijuana River Estuarine Reserve, USA (Figure 1). The total catchment area is 11.6 km^2, with 93% in Mexico and 7% in the USA. The climate in LLCW is Mediterranean, with a rainy season during the winter and annual mean rainfall of approximately 240 mm. Most of the erosional storm events occur during the winter. The regional geology (San Diego formation) includes marine and fluvial sediment deposits of conglomerate, sandy conglomerate, and siltstone. Soils are sandy with a wide range of cobble fraction, and are dominated by abrupt slopes (15°, average), which encourages gully formation and results in high erosion rates.

The LLCW is an uplifted and incised marine terrace, where the soil types are controlled in part by the underlying geology. The conglomerate geology in the northern part of the watershed has steep, competent valley walls with relatively flat buttes and mesas. In the central part of the watershed, the sandy conglomerate geology with low cobble fraction has lower slopes and rounded hilltops. The southern part of the watershed is a relatively flat, non-incised conglomerate. A narrow valley floor has Quaternary alluvium, but most of this has been paved or channelized.

Land use in the LLCW is predominantly mixed urban and rural, and was urbanized starting in 2002 with many illegal housing developments ("invasiones"). Gullies form on unpaved roads, which affect civil infrastructure in the upper watershed [31] and downstream ecosystems [32]. Such gullies are filled in with sediment following storms, and this management practice should be taken into account in developing the sediment budget and in soil erosion modeling for the watershed.

Sediment from the LLCW has buried native vegetation in the Tijuana River Estuary, which is located downstream of LLCW in the United States. In response, sedimentation basins were constructed at the outlet in the US in 2004, which costs $3 million USD to clean annually [32]. Stormflow and erosion also threaten human life, which causes damage to roads and houses in Tijuana [31]. The primary sources of sediment from LLCW are gully formation on unpaved roads, channel erosion, and sheet and rill erosion from unoccupied lots in Tijuana [8].

2.2. Field Data Collection and Model Setup

A summary of the data collection activities is reported in Reference [33]. Briefly, a tipping-bucket rain gauge station (RG.HM in Figure 1) was installed in February 2013. A pressure transducer (PT) (Solinst, water level logger) was installed in a concrete channel at the watershed outlet in December 2013 to record the water stage at 5-minute intervals (Figure 1). The stage-discharge relationship was determined using Manning's equation and flow velocity measurements. Manning's roughness coefficient (n) was based on field measurements of discharge in 2016 and 2017, which was used to back-calculate a Manning's n. The discharge measurements were also used to create a stage-discharge relationship for a stream gauge in the US (RG.GC) to complete our observations when the pressure transducer malfunctioned. The PT data were also validated and supplemented using time-lapse photographs of the water stage at the PT station. Suspended Sediment Concentration (SSC) measurements were taken at 10 different locations during a storm on 27 February, 2017 to explore spatial patterns of sediment production within the watershed. Annual sediment load data was collected in two large sediment traps at the watershed outlet. Data on the quantity of sediment removed annually (2006–2012) from the traps were available from the Tijuana River National Estuarine Research Reserve (TRNERR), corrected for trap efficiency, and used for model calibration (Figure 1).

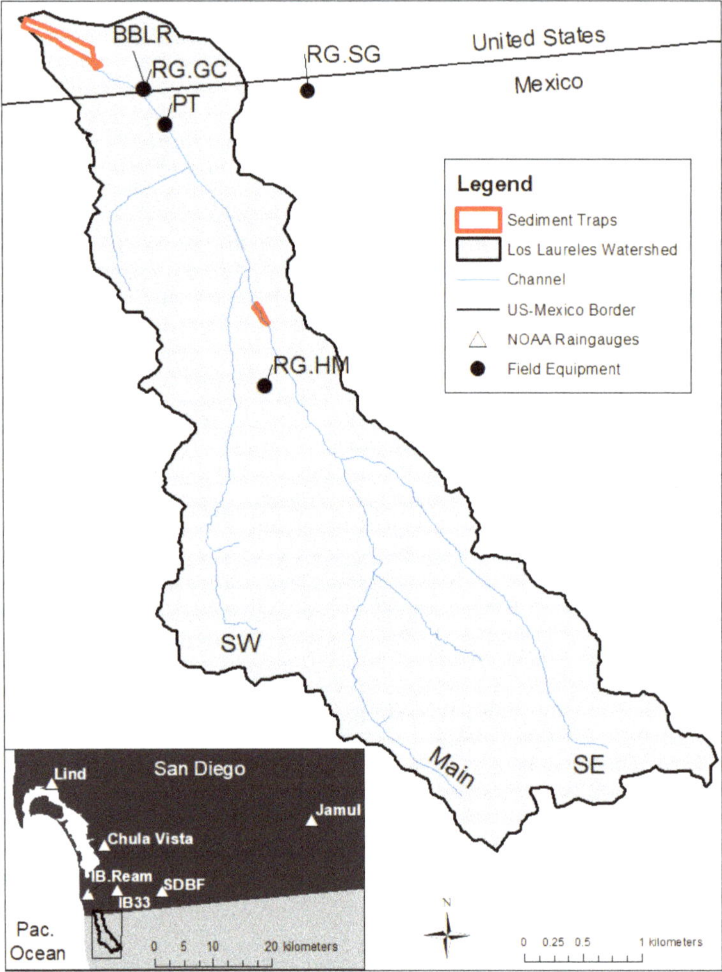

Figure 1. Location of the Los Laureles Canyon Watershed, main channels (SW, Main, and SE), and monitoring stations including sediment traps and rain gauges (RG). Inset shows the geographic locations of nearby rain gauges used in this analysis to span the rainfall time series.

A map of soils and associated parameters needed for the model were not available for the watershed, so soil type and parameters were mapped by modifying an existing geology map [34] and a correlation between geology and soil type that was established using the Soil Survey Geographic Database (SSURGO) from the United States Department of Agriculture. The geology map was modified based on field data collected during September 2015 and visual interpretation of high resolution imagery on Google Earth. First, a seamless cross-border geological map was created using the Instituto Metropolitano de Planeación de Tijuana, Baja California Mexico (IMPLAN) [34] geology map for Mexico and a geology map from the US [35]. The US soils that occurred on geologic types found in LLCW were identified as candidate soils for the study watershed. See Biggs et al. [33] for a full description of field and laboratory data collection.

Three main soil types were identified: (1) Los Flores formation (Lf) is a loamy fine sand, (2) the Chesterton formation (CfB) is conglomerate dominant with a fine sandy loam matrix, and (3) the Carlsbad formation (CbB) is a gravelly loamy sand. Once candidate soils were determined from the US geology and soils maps, the SSURGO soil characteristics were extracted for all horizons for comparison with data collected in the field. Soil samples were then collected from different geology types and analyzed for texture to compare with the SSURGO database and with the observed soil texture in the sediment traps at the watershed outlet (Figure 2). Samples ($N = 25$) were collected from road cuts and other exposed profiles from the near-surface (10–50 cm) and from the subsurface (>50–100 cm). The cobble percentage was determined through point counts along a 1 m transect through each distinct horizon. A bulk sample of sediment smaller than coarse gravel (<32 mm) was collected for texture analysis, analyzed in the laboratory using dry sieving to separate a 2-mm fraction and the pipette method for fines (<2 mm). Soil texture for all soil samples collected in LLCW and near the US-Mexico border was plotted in ternary diagrams and compared to SSURGO surface and subsurface soil texture (Figure 2). For each soil group, the SSURGO soil types that most closely matched the mean texture from the soil samples, were selected and used to update the soils map for LLCW. For some areas, the texture from the samples was similar to SSURGO data (northern part of the watershed). For the CfB soils, the texture from the samples did not match the SSURGO texture (southern part of the watershed), so a new geologic type (CfB.MX) was created, supported by field and laboratory data, to include in the AnnAGNPS model. Lastly, polygons delineating soil types were created by first determining the relationship between soil color, landform, and soil type for soils in the US, and then extrapolating those relationships to map similar soils in the LLCW (Figure 3).

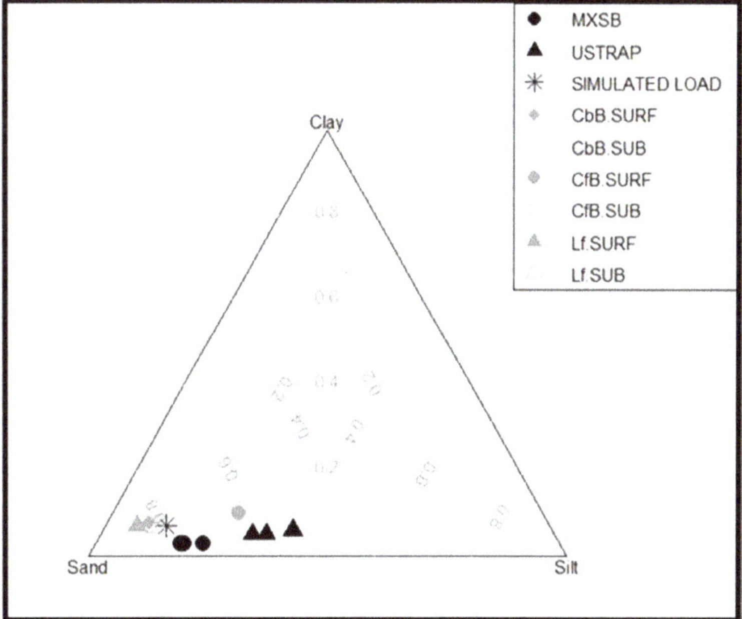

Figure 2. Ternary diagram of the mean grain sizes observed in the surface (SURF) and subsurface (SUB > 50 cm depth) soil layers for each soil type in the watershed, and in the sediment traps in Mexico (MXSB) and at the outlet in USA (USTRAP).

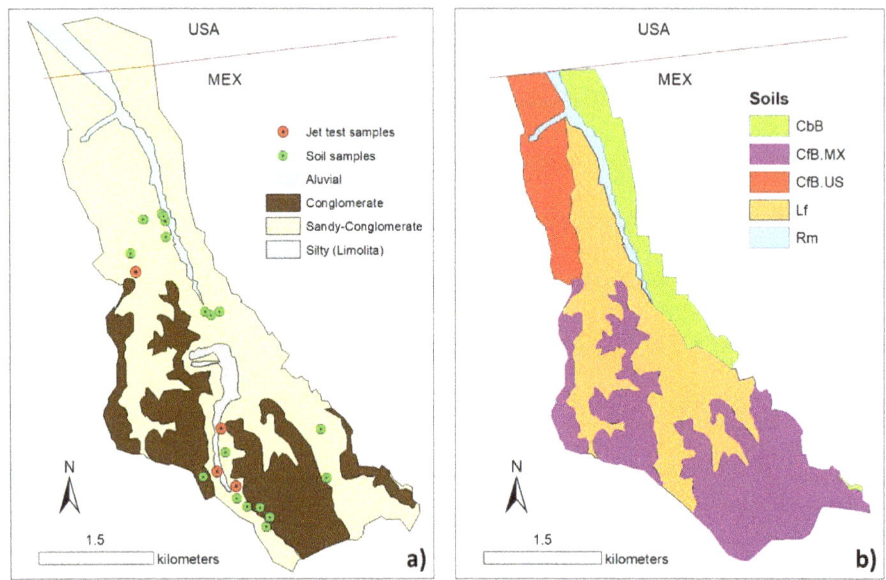

Figure 3. Geology map (**a**) and updated soils map for LLCW (**b**). Las Flores (Lf), fine sandy loam, dominates the central portion of the watershed (orange). CfB.MX represents the Chesterton sandy loam (CfB), but with a cobbly surface horizon. Carlsbad (CbB) and "CfB.US" soils extend south from the US/Mexico border.

The critical shear stress (τ_c) and soil erodibility in the non-cobbly sandy conglomerate (Lf) were taken from Gudino-Elizondo et al. [14], who used a mini-jet erosion test device following the methodology described by Hanson [36]. Values of τ_c for conglomerate soil were initially taken from USGS [37] and were modified during calibration.

The AnnAGNPS model requires daily precipitation data. To extend the simulation period to include the period after the installation of the sediment trap (2004–2017), the precipitation data collected from February 2013 to 2017 at RG.HM were compared with rainfall data from nearby stations in the United States (Figure 1) to select the best rain gauge to use for rainfall data from 2004–2013.

Application of AnnAGNPS can be challenging in a watershed with steep topography and sediment coarser than sand (Figure 4) because the model does not simulate mass wasting processes, only transports sediment up to coarse sand (2 mm), and is designed for mixed-use watersheds in agricultural areas. Mass wasting, including shallow landslides, was observed in the study area, and coarse sediment accounts for ~10–15% of the sediment in the traps at the watershed outlet (unpublished data). However, the valley floor at the base of the steep slopes that are most likely to experience landslides, has been graded and paved for roads on either side of the channel, which limit the transport of coarse material to the channel from landslides or other hillslope processes. Field observations suggest that landslides typically terminate on these flat road segments or other graded areas, and the coarse material that accumulates at the toe of a landslide is periodically cleared mechanically. In this scenario, we assumed that all coarse material is from the channel. The sediment load from channel erosion was taken from Reference [38] and added to the total modelled hillslope erosion. Lastly, the RUSLE equation to estimate sheet and rill erosion was designed for relatively flat agricultural hillslopes, and may not be valid for steep hillslopes (>30%) [39]. In this case, we assumed that the model application was valid in most of the study watershed because 87% of the total watershed area has a slope gradient of less than 30% (mean slope = 15%).

The spatial variability of topography, land cover, soils, and management properties within the catchment area was represented in the model by discretizing the watershed into cells that are relatively similar in slope, soils, and land use. A LIDAR-derived Digital Elevation Model (DEM) (3 m, horizontal resolution) of the LLCW (sponsored by the County of San Diego California, USA) was used as input for a topography-based method (TopAGNPS) [26]. The TopAGNPS method was used to (i) delineate surface flow paths, (ii) subdivide the total catchment area into sub-catchments (cells) along drainage segmentations, and (iii) estimate representative cell parameters, such as slope, area, and soil and management attributes. Cells sizes were based on user-defined values of the Critical Source Area (CSA), which is the minimum drainage area required to form a channel, and the Minimum Source Channel Length (MSCL), which prunes the channel network of channels shorter than the specified MSCL value.

To characterize the hillslope and reach units within the AnnAGNPS model, a CSA of 1 ha and a MSCL of 50 m were assigned, based on field observations, to characterize the hill-slope and reach units within the AnnAGNPS model. LLCW was discretized into 1147 sub-catchments (AnnAGNPS cells) and 462 channels (AnnAGNPS reaches). The cell sizes ranged from 9×10^{-6} to 0.1 km^2 (Figure 4).

TopAGNPS was also used to map Potential Ephemeral Gullies (PEGs) throughout the LLCW following the methodology described by Momm et al. [26]. This method provides an automated estimate of the downstream-most locations of knickpoints (i.e., PEGs), which are used within AnnAGNPS to calculate the length of ephemeral gullies in the landscape. The approach uses improvements on the EGEM described in Gordon et al. [40] and, more recently, revised by Bingner et al. [12]. In the model, the gullies are filled in once per year at the end of each wet season, which corresponds to observed management practices, but may under-represent the filling frequency on larger main roads that are typically filled between each storm event that generates gullies.

The watershed hydrology module of AnnAGNPS uses the SCS Curve Number method [27] to estimate storm event runoff from precipitation in each cell. The storm event water peak discharge and the time-to-peak are determined for the hydrograph at each reach section and at the watershed outlet, following the TR-55 [27] approach that utilizes the time of concentration for each cell and reach, determined with TOPAGNPS, total daily runoff determined from AnnAGNPS, and the storm type entered as an input parameter [12].

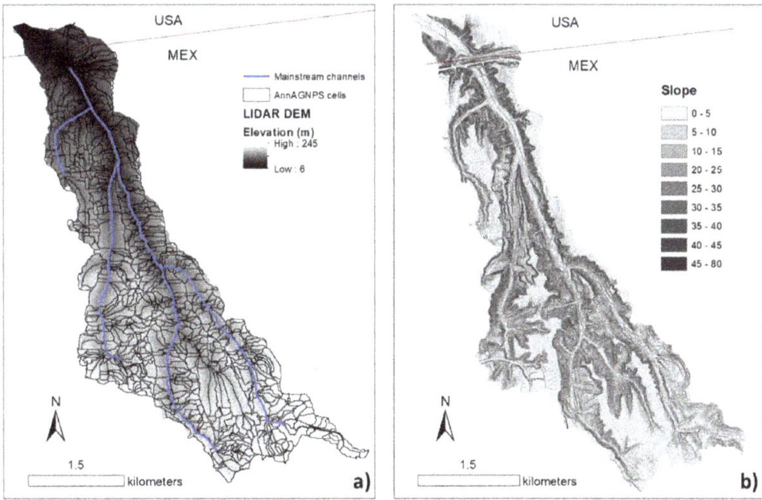

Figure 4. (**a**) AnnAGNPS cells, reaches, mainstream channels, and (**b**) slope gradient at the Los Laureles Canyon watershed (LLCW).

The spatial resolution of the DEM has little impact on AnnAGNPS runoff volume estimates [11,41], but soil erosion and sediment loads can change with DEM resolution, since resolution impacts slope [42]. LIDAR-derived 3-m DEMs should improve the model performance in the study watershed compared with applications that use more-commonly available 10 m or 30 m resolution DEMs. Initial AnnAGNPS reaches did not follow the road network, so the road segments were "burned" into the DEM by lowering the elevation on the DEM cells falling on roads by 1 m.

A land use map was created using Google Earth (11 November 2012, 2017 Digital Globe) imagery based on visual classification of seven categories (rangeland, highway road, paved residential roads, dispersed urban unpaved, unpaved rural roads, unpaved residential roads, and sediment basin). The accuracy of the land use data was validated by comparing land use categories with ground-based photography and field data collection. The land use map was overlain on the AnnAGNPS cells to populate the required hydrologic and management parameters needed for AnnAGNPS. The soils map was used to link the required physical variables from the SSURGO database to the model such as soil texture and erodibility, bulk density, and saturated conductivity. Tillage depth is the depth to an impervious soil layer, which limits the potential depth of gullies, and was determined as the depth of the gullies observed in the field [14]. The main equations solved within AnnAGNPS to estimate soil erosion are listed in Table 1. A detailed description of these equations are given by Bingner et al. [12].

Table 1. Equations and parameters used to simulate soil erosion in the AnnAGNPS model.

Module	Equation
Sediment yield by sheet and rill erosion	$Sy = 0.22 \times Q^{0.68} \times q_p^{0.95} \times KLSCP$
Gully width	$W = 9.0057 \times \left(Q_p \times S\right)^{0.2963}$
Head-cut migration erodibility coefficient	$Kh = 0.0000002 / \sqrt{\tau c}$

Annotation: S_y = sediment yield by sheet and rill erosion (Mg/ha): Q = surface runoff volume (mm). qp = peak rate of surface runoff (mm/s) and K, L, S, C, P are RUSLE factors. W = gully width (m): Q_p = peak discharge at the gully head (m^3/s). S = the average bed slope above the gully head (m/m). K_h = head-cut migration erodibility coefficient (m^3/s/N): τ_c = the critical shear stress (N/m^2).

See Gudino-Elizondo et al. [14] for a detailed description of the usage of these equations in the study watershed.

2.3. Model Calibration and Evaluation

The model performance metrics considered both graphical and statistical analyses to assess the best parameterization based on the coefficient of determination (R^2), Root Mean Square Error (RMSE, Equation (1)), and the percent bias (PBIAS, Equation (2)), which are widely applied in hydrologic and erosion modeling [43].

2.3.1. Runoff

Total and peak runoff measured at PT (outlet) for 14 storm events were used for model calibration. Manning's *n* back-calculated from discharge measurements was consistent with literature values for "ordinary concrete lining" (0.013) [44] and with the channel condition at PT. Simulated total and peak discharge were then compared and evaluated using the RMSE, which is calculated by the equation below.

$$RMSE = \sqrt{\frac{1}{N} \sum_{i=1}^{N} (Observed\_i - Simulated\_i)^2} \tag{1}$$

where *i* is the index of the storm events and '*N*' is the number of events (14).

The selection of the AnnAGNPS model parameters to calibrate the runoff was based on the watershed characteristics, preliminary model runs, and literature values identified for each cell in the watershed [14,45] (Table 2).

Table 2. Final AnnAGNPS runoff parameters values used to calibrate runoff.

	Range Land		Disperse Urban Unpaved		Non-Urban Unpaved	Paved Residential Roads		Unpaved Rural Roads	Unpaved Residential Roads		MX Sediment Basin
Hydrologic soil group	B	D	B	D	D	B	D	D	B	D	D
Percent watershed area	2	30	1	11	5	3	10	1.3	4	32	0.7
Curve number	77	88	88	94	98	98		89	82	89	98
Manning's n of overland flow	0.13		0.07		0.01	0.01		0.07	0.03		0.01

2.3.2. Sediment

Data on sediment removed from the sediment traps at the LLCW outlet (Figure 1) were used for calibration. Both upper and lower traps were excavated in the Spring and Fall of 2005, Winter 2006, and each Fall from 2007–2012 ($N = 7$). The sediment trap efficiency, or the proportion of the total sediment trapped in the sediment basin, was calculated based on Morris and Fan [46] and Urbonas and Stahre [47]. See Biggs et al. [33] for a detailed description.

Preliminary simulations suggested that the channel erosion module of AnnAGNPS resulted in excessive sedimentation in the channels, which was not observed in the field. We, therefore, calculated the simulated sediment load as the total amount of sediment by source (sheet and rill, and gully erosion) that makes it to the stream channel network (Figures 1 and 4), plus channel erosion estimated in previous work [38]. The load was compared to the total sediment load with the total amount of sediment being excavated from the sediment traps for specific dates. The stream channel network was defined as permanent channels within the watershed based on field observations and visual examination of these channels using high resolution imagery. Channel erosion estimates (t/yr) were taken from Taniguchi et al. [38], who calculated channel erosion from the difference between the cross sections observed in 2014 with the cross section under reference (pre-urban) conditions, which was divided by the time since urbanization. Taniguchi et al. [38] estimated that channel erosion accounted for 25% to 40% of total sediment yield to the estuary over 2002–2017. In this scenario, we estimate channel erosion by multiplying the hill-slope erosion estimated by AnnAGNPS by 0.33 and 0.67 to get channel contributions of 25% and 40% of total load, and adding that load from channel erosion to the hill-slope load to get the total load.

The data from the sediment trap, corrected for trap efficiency, were compared with AnnAGNPS simulation results of total load, including both hill-slope and channel erosion. Critical shear stress τ_c and sediment delivery ratio (SDR) were then calibrated to match the observed sediment yield at the LLCW outlet. An initial value of τ_c was set to 1.6 N·m^{-2} for sandy soils based on the average value from nine samples collected on the Lf soil type [14]. Initial values of τ_c for conglomerate soils were taken from USGS [37] dataset for fine cobbles (64 N·m^{-2}) and were modified during calibration to $\tau_c = 32$ N·m^{-2}, which corresponds to very coarse gravel. The parameters used to calibrate sediment yield are presented in Table 3. The SDR for coarse soil formations (CfB and CbB) were calculated internally by the model. The SDR for the Lf type was set to 1 and based on field observations of extensive rill and gully formation, which results in the delivery of most sediment from sheet and rill erosion to the channel network.

Table 3. AnnAGNPS soil erosion parameters used to calibrate the watershed scale model for sediments.

	Lf	CbB	CfB.MX	CfB.US
Critical shear stress (N·m^{-2})	0.1	32	32	32
Tillage depth (cm)	60	60	60	60
Sediment delivery ratio	1	Internally calculated	Internally calculated	Internally calculated
USLE K (t ha hr)/(ha MJ mm)	0.036	0.006	0.048	0.048
Hydrologic soil group	D	B	D	D

The percent bias (PBIAS) was used as a measure of the average tendency of the simulated results relative to the observed data, which indicates over (positive PBIAS) or underestimation (negative PBIAS), respectively [48,49]. The PBIAS was calculated using the equation below.

$$\text{PBIAS} = \frac{\sum_{i=1}^{N}(observed - simulated) \times 100}{\sum_{i=1}^{N} observed} \tag{2}$$

where i is the index of the storm events and 'N' is the number of events (14).

The measurements of sediment accumulation at the outlet provides an aggregate measure of sediment load for the watershed, but does not validate the spatial pattern of sediment load from different soil and land use units in the watershed. Grab samples of water were collected for suspended sediment analysis at 10 sites in the watershed during a large storm (81 mm, total depth) on 27 February 2017. All samples were collected over a 0.5 hour period, which corresponded to a period of maximum runoff. The observed SSC of the storm-water samples were then compared with the simulated AnnAGNPS SSC (SSC = storm event sediment mass/storm event runoff volume) to explore the influence of soil properties and land use on sediment production in the watershed. While SSC at a given location changes during an event, the samples were collected during similar hydrological conditions, and provide a snapshot of the spatial variability of SSC during an event. Table 4 summarizes the data type and parameters set for model calibration and evaluation.

Table 4. Field data collection and time periods for model calibration.

Type of Data	Dates	AnnAGNPS Parameters	Model
Water discharge	14 events (2013–2017)	Storm type, Manning's n	Calibration
Sediment traps	7 excavation periods	SDR, τ_c, tillage depth, USLE-K	Calibration
SSC (grab samples)	1 event (Feb 2017)	None	Evaluation

We used the entire dataset of observations at the outlet, including annual sediment accumulation in the traps (N = 6) and event runoff (N = 14), for model calibration due to the small number of observations. Use of an entire dataset for calibration is consistent with other AnnAGNPS applications in Mediterranean environments such as References [50,51].

2.4. Scenario Analysis

We evaluated the impact of paving roads on runoff and sediment yield using the calibrated model. The simulation paved only those roads in the AnnAGNPS cells that generated 50% of the total sediment yield at the LLCW scale (hotspots) under current conditions. For the scenario analysis, we assumed that the *CN* is the same for all paved roads (*CN* = 98), and that gully sediment yield is zero since gully erosion occurred solely on the dirt road network within the LLCW [9,14]. Composite curve numbers were calculated for unpaved and paved conditions following Gudino-Elizondo et al. [14]. The scenarios were run for 2004–2017 and the impact on sediment and runoff were determined by the change in simulated sediment load, and total and peak runoff between the current conditions and the paving scenario. The change in total and peak discharge were calculated for the largest 14 storm events (event-total precipitation ranging from 28 to 81 mm).

3. Results

3.1. Rainfall Data

Total event rainfall at the rain gauge in LLCW (RG.HM, Figure 1) correlated closely with daily rainfall at nearby stations in the United States (Figure 5). For the events when rainfall data were recorded for the LLCW watershed at RG.HM (2013–2017), the gauge at San Diego Brownfields (SDBF) has the highest correlation coefficient and smallest RMSE out of the stations with good data availability. Rainfall at RG.IIM was higher than that at all other stations for larger events (> 60 mm), but matched the SDBF data well for rainfall between 10 and 50 mm (Figure 4). The SDBF gauge had a higher correlation coefficient and lower error compared to stations closer to LLCW in the Tijuana Estuary (IB3.3). Therefore, SDBF was used to estimate rainfall in LLCW for years when no data was available at RG.HM.

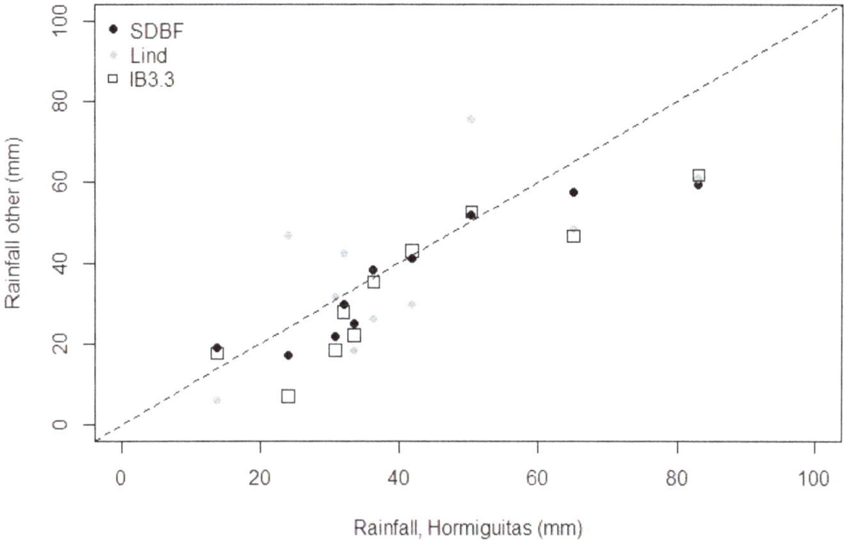

Figure 5. Event-total precipitation at the Hormiguitas rain gauge (RG.HM) versus three other nearby stations (see Figure 1). The dashed line is the 1:1 line. Taken from Biggs et al. [33].

3.2. Rainfall-Runoff Relationships

Event-total rainfall for the 14 events with rainfall (P) and runoff (Q) data ranged from 7 to 83 mm (Table 5). The event-wise runoff coefficients (Q:P) ranged from 0.02 to 0.67. Event-total runoff increased with event-total rainfall and fits a watershed-mean SCS CN of 80–90 (Table 5 and Figure 6). The highest SCS CN occurred for the smallest events and CN generally decreased with the event size (Figure 6). This was consistent with runoff production from surfaces with low infiltration capacity during small events, and from all surfaces, including those with high infiltration capacities, during large events. The largest event (rainfall 81 mm) has a runoff coefficient of 0.51, where most points fell between SCS CN 80 and 90 (Figure 6), which is consistent with literature values for partially urbanized land cover [45]. Thus, no adjustments were needed for the CN as the fit was adequate with the observed storm-wise rainfall-runoff relationships (Figure 6). The 24-hour rainfall distribution used for most of the simulated storms was type II [27] because it is representative of semi-arid regions of South-western USA, and matches the most frequent storm type calculated using rain gauge measurements in the LLCW [33]. Some storms were assigned different storm types based on their rainfall distribution compared to SCS storm types.

The RMSE of the simulated storm-wise runoff was 6.6 mm (89% of mean), and 13 $m^3 \cdot s^{-1}$ (177% of the mean) for peak runoff. The RMSEs were notably influenced by a single large storm of 81 mm total-event precipitation (27 February 2017, Table 5). RMSE without that large storm was 3.6 mm (75% of the mean) for total runoff, and 6.9 $m^3 \cdot s^{-1}$ (105% of the mean) for peak runoff. The AnnAGNPS model was most accurate for medium-sized events (event precipitation between 2 and 20 mm, Figure 7), which are the most frequent events. Therefore, we did not calibrate the model to minimize the error. Peak discharge was generally underestimated for small storms and overestimated for large storms [33].

Table 5. Summary of storm events used for model calibration.

Event	Rainfall (mm)	Peak Discharge (m³/s)		Total Runoff (mm)	
		Observed	Simulated	Observed	Simulated
28 February 2014	12.25	1.13	0.43	0.27	0.74
1 March 2014	7.50	1.54	0.08	0.33	0.20
2 March 2014	7.50	6.14	0.58	1.08	0.90
1 March 2015	23.25	3.36	2.69	1.36	3.91
2 March 2015	9.25	1.43	0.31	0.48	0.56
15 May 2015	22.50	19.46	2.46	5.93	3.62
15 September 2015	30.75	5.27	5.69	6.40	7.27
5 January 2016	22.25	17.72	3.58	3.76	4.79
6 March 2016	6.50	1.03	0.00	0.93	0.01
7 March 2016	23.00	5.07	2.55	4.23	3.74
19 January 2017	13.00	5.37	2.85	2.57	3.51
20 January 2017	28.00	6.86	15.91	18.66	17.24
17 February 2017	33.25	11.16	13.88	7.03	15.31
27 February 2017	81.00	16.69	58.23	42.07	63.12
TOTAL	320	102	109	95	125
RMSE		13		6.6	

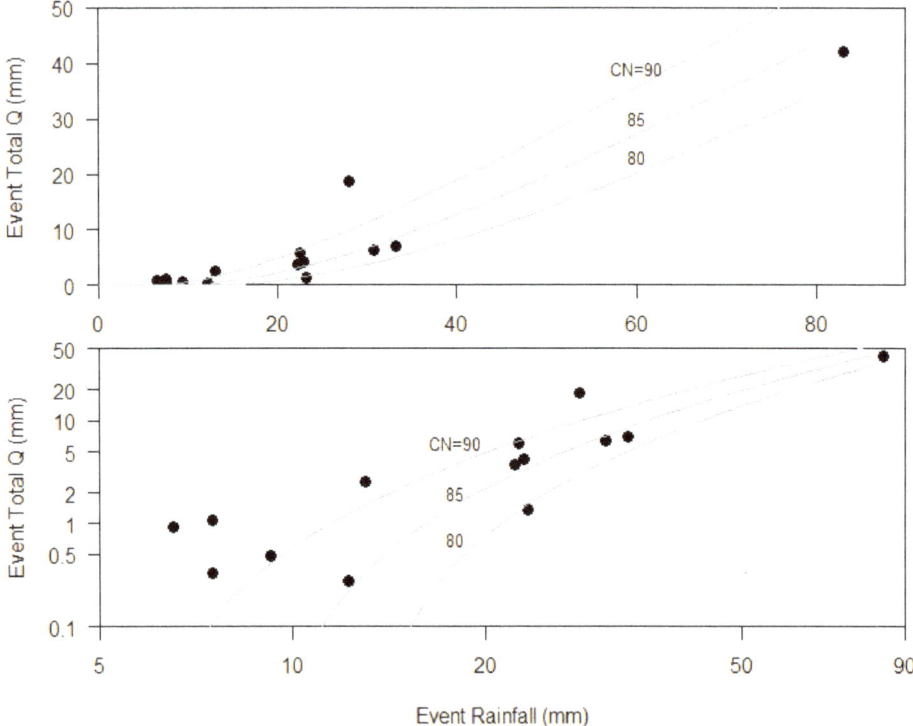

Figure 6. Rainfall-runoff relationship for all observed storm events, with several SCS CN rainfall-runoff relationships, in non-log (top) and log-log (bottom). Taken from Biggs et al. [33].

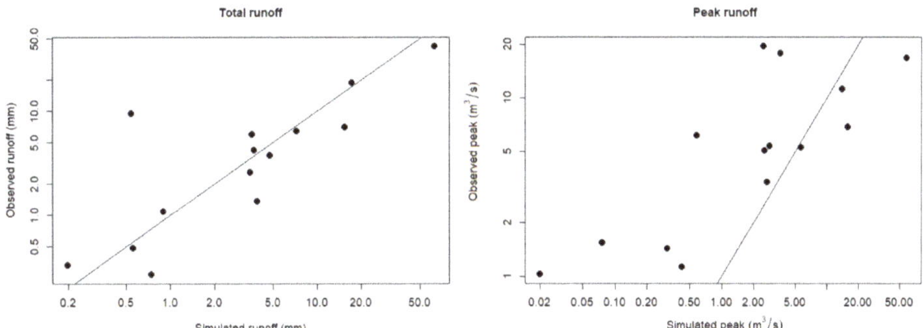

Figure 7. Relationship between observed and simulated total (a, mm per storm) and peak (b, m^3/s) runoff. The solid lines are the 1:1 lines.

3.3. Simulated Sediment Production

The SSC samples were collected from sub-watersheds with different soil characteristics. Fractional covers of soil types in the sub-watersheds draining to each SSC sample location vary from 20% to 100% erodible, non-cobbly soils (Lf soil type) (Figure 8). The observed SSC of the storm-water samples correlated with the AnnAGNPS-simulated SSC (Figure 8b), and modelled SSC correlated with the fraction of the sub-watersheds covered by Lf (Figure 8a,c), which highlight the influence of soil properties on the modelled sediment production. The modelled SSC values were higher than the observed SSC values (mean model = 210.7 g/L, mean observed = 48.7 g/L, RMSE = 203.5 g/L) because the grab samples were not all taken at the time of the peak discharge. No samples were available for areas drained by cobble and gravel soils in the southern and northern parts of the watershed.

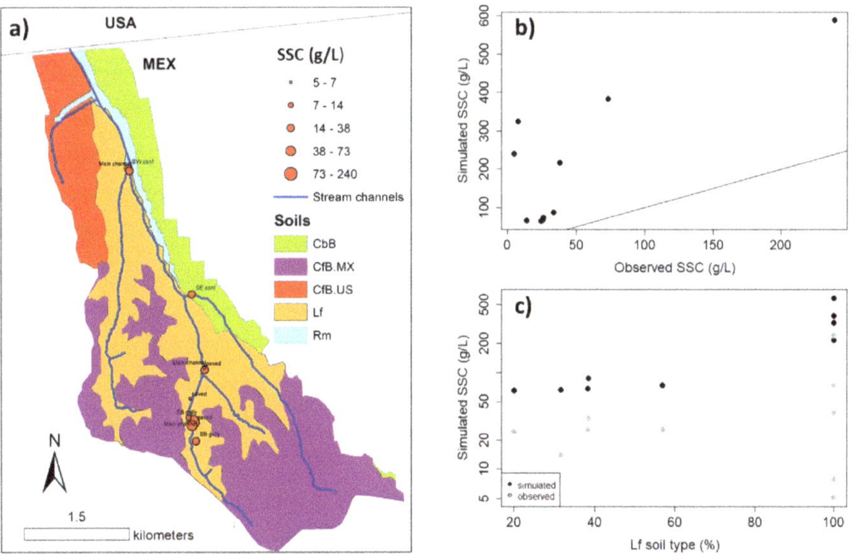

Figure 8. (**a**) Geographic location of storm-water samples, suspended sediment concentration (SSC), and soil types along the LLCW. (**b**) Relationship between observed and simulated SSC (the black line is the 1:1 line) and (**c**) relationship between Las flores (Lf) soil fraction and simulated and observed SSC.

3.4. Sediment Budget

The annual trap efficiency varied from 0.79 to 0.98, and was 0.89 for the cumulative mass removed over 2006–2012 [33]. Uncertainty in sediment trap efficiency calculation, in particular the gradual filling of the traps during the year and subsequent decrease in trap efficiency, may have caused underestimation of sediment load at the traps (Table 6). Total annual sediment accumulation in the traps correlates with annual precipitation at the SDBF station (Figure 9).

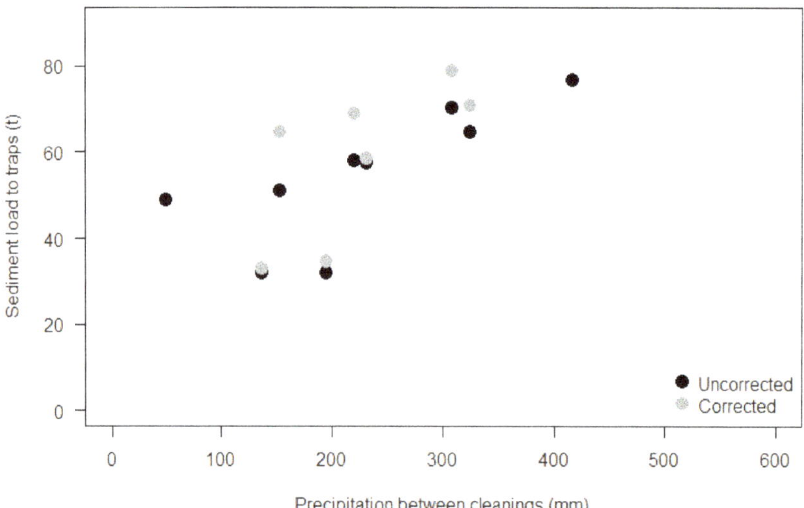

Figure 9. Total sediment removed from the Los Laureles Canyon traps in the United States versus total annual precipitation between removal events from 2005 to 2012. The uncorrected amount of sediment removed is in black and the load corrected for trap efficiency is in grey. Annual precipitation is from the San Diego Brownfield station. Modified from Biggs et al. [33].

Table 6. Simulated annual sediment yield and total observed (in tons) at the watershed outlet by the erosion process. The range values for channel contribution and total yield assumes channel erosion is 25% (minimum value) and 40% (maximum) of the total simulated results from AnnAGNPS.

Year	Rainfall (mm)	Sheet and Rill (t)	Gully (t)	Channel (t)	Total (t)	Observed (t)	Ratio of Simulated to Observed
2006	193	8483	8143	5487–11140	22113–27766	34642	0.64–0.80
2007	136	15257	6022	7022–14257	28301–35536	33079	0.85–1.07
2008	154	10204	5518	5189–10534	20911–26257	64580	0.32–0.40
2009	218	38058	13555	17032–34580	68645–86193	68949	1.00–1.25
2010	298	48347	20669	22775–46240	91791–115256	78935	1.16–1.46
2011	323	51752	29273	26738–54287	107763–135311	70965	1.51–1.90
2012	234	16992	10797	9170–18619	36960–46408	58513	0.63–0.80
Mean	222	27013	13425	13345–27094	53783–67532	58523	0.88–1.10

Total modelled sediment load correlates with the sediment load observed at the sediment trap (Figure 10), with the following errors: (i) $pBIAS_{25\%} = 8.1$, $RMSE_{25\%} = 24{,}115$ t (41% of the mean) considering a channel erosion contribution of 25% of the total hill-slope sediment production, and (ii) $pBIAS_{40\%} = 15.4$, $RMSE_{40\%} = 32{,}570$ (55% of the mean) considering a channel erosion contribution of 40%. These model efficiencies were moderate-to-good according to other studies [44,49,52] reporting relatively similar values.

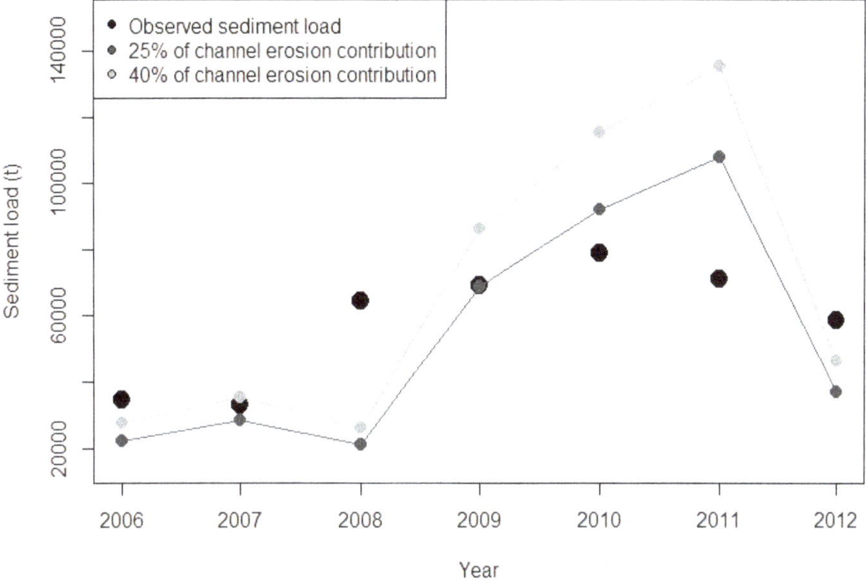

Figure 10. Time series of the relationship between observed and simulated annual sediment load at the LLCW outlet, assuming a channel erosion contribution of 25% and 40% of the total hill-slope sediment production.

The default values of τ_c for conglomerate soil types (CfB and CbB) (τ_c = 64 N·m^{-2}) resulted, on average, in underestimation of the total sediment load at the outlet, so it was changed during calibration to τ_c = 32 N·m^{-2} to fit better with the observed sediment yield in the sediment traps. The τ_c value set in the calibrated model corresponds to very coarse gravel [37], which was consistent with the observed particle sizes of gravelly and cobbled soils in the LLCW [33].

Precipitation correlates with simulated sediment production from sheet and rill erosion, gully erosion, and total sediment yield, while sediment production by sheet and rill erosion correlates more closely with rainfall than gully sediment production does (Figure 11). A minimum precipitation threshold (~25–35 mm) for gully initiation was reported by Gudino-Elizondo et al. [9], which is consistent with the significant contribution by gullies to the total sediment production for those storm events with precipitation greater than 25 mm (Figure 11).

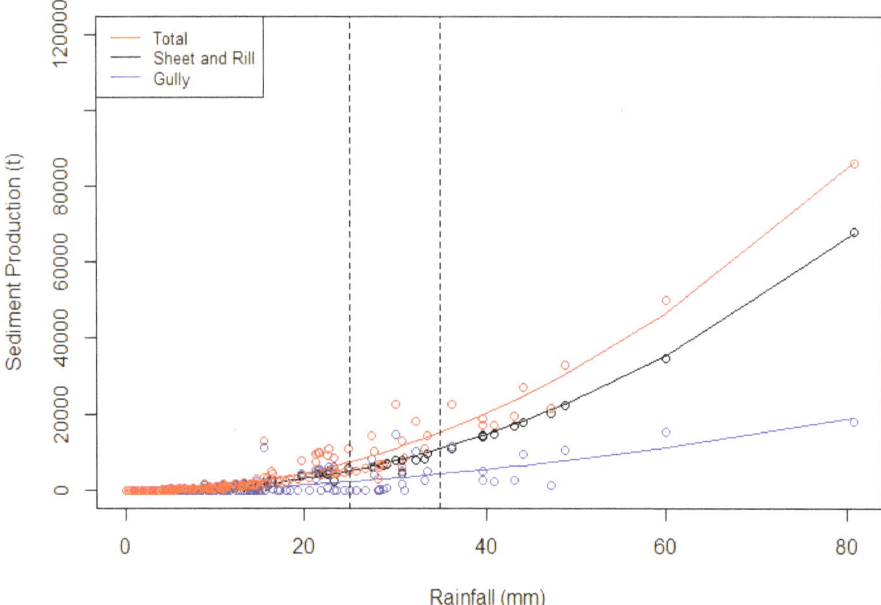

Figure 11. Simulated sediment load by erosion processes in LLCW. The vertical dashed lines indicate the range of the rainfall threshold for gully erosion observed in the field during 2013 to 2018.

Simulated sheet and rill erosion was the dominant erosional processes within the LLCW (Figure 12), which was also reflected in the event-wise rainfall-sediment relationships (Figure 11), especially for larger events. Total sediment load at the sub-watershed scale (AnnAGNPS cells) was dominated by cells characterized by sandy soil types (Lf) on steep slopes, which show evidence of frequent rill and gully formation (Figure 12).

The observed sediment in the trap is finer (higher silt fraction) than both the hill-slope sediment and the AnnAGNPS-simulated sediment load (Figure 2). This suggests that either more sand is being retained in storage on the hill-slopes and in the channel than is simulated by the model, or that silt is preferentially eroded from soils that have a mixture of silt and sand, or that soils with high silt fraction contribute more to the load than is being modelled. The particle size in the Mexico sediment trap is coarser than the US sediment trap, which suggests either retention of sand in the channel downstream of the Mexico sediment trap, or high loads of silt from the sub-watershed outside of the Mexico sediment trap sub-watershed.

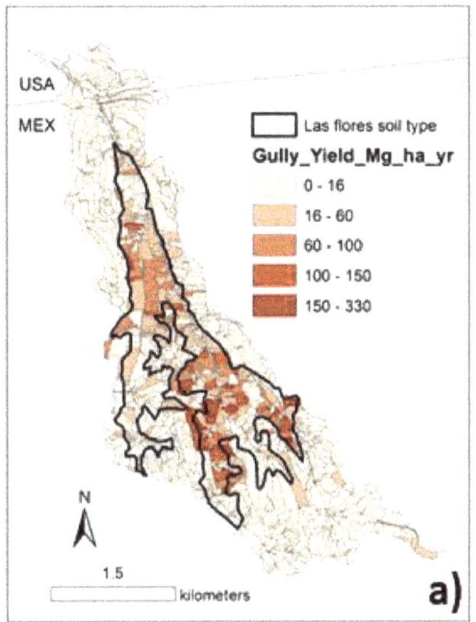

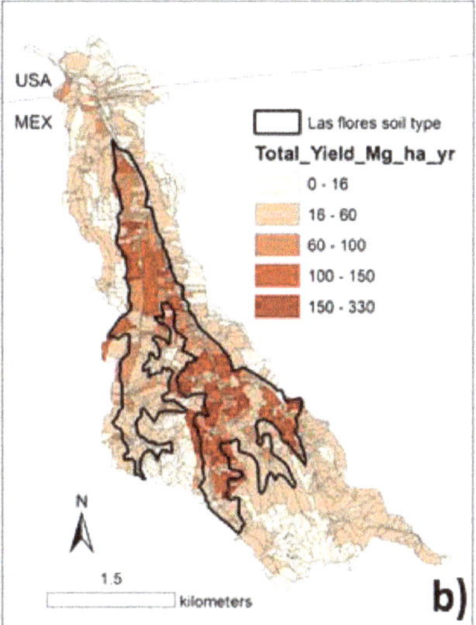

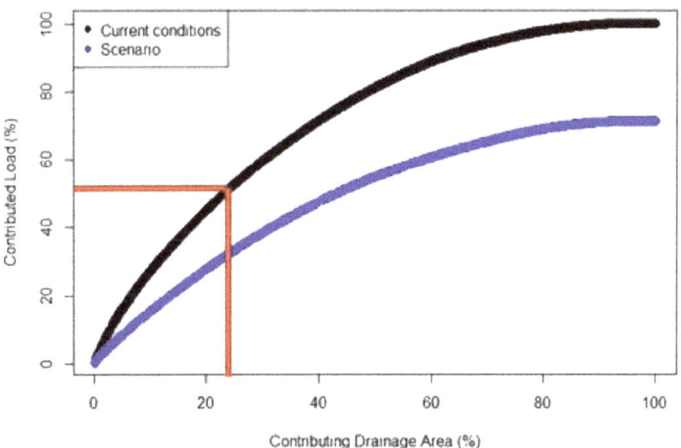

Figure 12. (**a**) Sub-watershed sediment yield by gully erosion. (**b**) Total sediment yield by sub-watershed within the Los Laureles Canyon watershed and (**c**) total sediment production by contributing the drainage area under the current conditions and road-paved scenario.

3.5. Scenario Analysis

Half of the simulated sediment load at the watershed scale is generated by only 23% of the total watershed area under current conditions (red lines in Figure 12). These cells are hotspots of sediment production and, pending validation of erosional severity with additional field observations, could be prioritized for management activities to reduce sediment production at the watershed scale.

The model scenario suggests that, on annual average, paving all the roads in the hotspots would reduce sediment production by 30% (Figure 12). However, storm-wise total runoff increases by an average of 10%, and peak runoff increases from 1.6% to 21% (Table 7). The projected peak discharge increased the most for the medium-sized events (40–49 mm, two-year recurrence interval), and not for the largest event (81 mm, 25-year recurrence interval), which suggests that paving roads in hotspots could be suitable for the study watershed without increasing peak discharge for the largest events. This may be the most responsible factor for flood damage.

4. Discussion

AnnAGNPS simulated total water and sediment load with satisfactory agreement with observed total event runoff and sediment yield in the LLCW. The simulated total runoff and peak discharge were more accurate for medium-sized events (event precipitation between 2 and 20 mm, Figure 7), which are the most frequent events in the region. The model generally underestimated peak discharge for small storms and overestimated peak discharge for large storms, which is partly due to underestimation of the peak rainfall rates of small storms and overestimation of the peak rainfall rates of large storms by the SCS Storm type model [33]. Further research on the impact of paving on peak discharge for a range of storm sizes and sequences is needed.

Table 7. Rainfall and simulated increases in peak and total discharge volume at the outlet under current and scenario (hotspot paving) conditions for the 14 largest storm events.

Date	Rainfall (mm)	Increase in Peak Runoff (%)	Increase in Total Runoff (%)	Decrease in Sediment Load (%)
27 February 2017	80.75	1.63	1.98	29.97
27 February 2004	59.90	10.19	9.82	16.59
7 December 2009	48.80	17.78	15.28	19.32
22 December 2016	47.25	10.34	9.64	37.73
13 December 2012	44.20	21.09	17.68	58.18
17 December 2008	43.20	8.92	8.71	23.84
22 December 2010	40.90	9.13	8.86	35.91
27 October 2004	39.60	7.82	7.40	24.61
19 February 2007	39.60	16.77	14.8	7.36
17 February 2017	31.00	6.07	5.12	16.01
26 February 2011	30.00	7.31	6.36	15.92
3 January 2005	26.70	5.27	5.23	15.86
4 January 2005	24.40	4.50	4.39	26.25
11 January 2005	23.60	3.78	3.58	36.24
Min	23.6	1.63	1.98	7.36
Max	80.75	21.09	17.68	58.18

The AnnAGNPS model satisfactorily simulated ephemeral gully erosion rates in the LLCW at the neighborhood scale [14], which helped identify parameter values for use at the LLCW scale. The model performed well compared with other models applied in semi-arid environments [23,30,49,51,53] which supported its use for runoff and sediment budgets in this watershed. However, uncertainties in soil-resistance-to-erosion parameters, especially critical shear stress for cobbled soils, may affect sediment production by gully erosion. This suggests that more field and laboratory data are necessary to have more accurate sediment yield estimates at the watershed scale.

The SSC from 10 grab samples (Figure 8c) collected during the largest storm event correlated with modelled SSC, which suggests the model represented spatial variations in sediment production within the watershed. Modelled erosion was sensitive to the fraction of highly-erodible Lf soil type, which generated 61% of the total sediment load. Most of the AnnAGNPS cells that contribute significantly to the total sediment load (hotspots) had both highly erodible soils (Lf) and steep slopes (>30%) that encourage gully sediment production (Figure 12). No SSC samples were available for areas drained by cobbled and gravel soils in the northern part of the watershed, so future work should include more grab samples from sub-watersheds draining cobbled soils.

The RMSE of the AnnAGNPS model for sediment load was 41% and 55% of the mean value considering 25% and 40% of channel erosion contribution, respectively. Our observed values of sediment load at the outlet likely underestimate the total load because our method for calculating trap efficiency does not account for a reduction in trap efficiency as the trap fills during the wet season. This underestimate is likely largest during wet years when the trap is full at the end of the season, so future research will explore the impact of reduction in available trap capacity on trap efficiency and estimated sediment load. Channel evolution is also not well characterized by the AnnAGNPS model, which reduces the performance of the model to simulate the observed behavior of the system. Taniguchi et al. [38] noted that urbanization caused extreme channel enlargement in the LLCW, which suggests the necessity to implement and couple a more sophisticated channel evolution model to AnnAGNPS such as the channel evolution computer model (CONCEPTS) [54,55] to better simulate the sediment production at the LLCW scale.

Simulated gully erosion represented approximately 16% and 37% of total sediment production considering 25% and 40% of channel erosion contribution, respectively. This was relatively close to other estimates for human-disturbed watersheds. Bingner et al. [56] reported that ephemeral gullies were the primary source of sediment (73% of the total) in agricultural settings within the Maumee River basin, USA. De Santiesteban et al. [57] found that ephemeral gullies contributed 66% to total soil loss in a small agricultural watershed. Taguas et al. [23] found that contribution of gully erosion to the total soil loss varies substantially depending on the management, on average, from 19% to 46% under spontaneous grass cover and under conventional tillage management. Previous studies reported gully erosion in agricultural and partially urbanized watersheds even though gully erosion can be a significant, and often neglected, portion of the sediment budget. We likely underestimate gully contribution since the model assumes gullies on roads are filled only once per year, while field observations suggest that main roads are repaired several times per year, after every storm that generates gullies. Our estimates of gully contribution are also sensitive to the mapped distribution of fine-textured soils that generate most of the gullies. Future research could refine the soils map and test for the sensitivity of gully filling frequency on the gully contribution.

Our estimate of the contribution of channel erosion to the sediment load (25%–40%) is smaller than other studies, which report channel contributions ranging from 67% [58] to 85% [59] of the total sediment yield in urban areas. The relatively large contribution from hill-slope sources in our study area is likely due to persistent soil exposure and erosion, including vacant lots and unpaved roads, that characterizes urbanization in Tijuana [8] and possibly other cities in developing countries.

In our watershed study, 50% of the sediment production under current conditions is generated only from 23% of the watershed area. Paving all the current unpaved roads in these "hot spots" would reduce sediment production by 30% compared to the current conditions, but it would also increase total discharge by 2%–17% and peak discharge by 2%–21%. The smallest increase in peak was for the largest event, which suggests that the impacts of paving may be small for the events that cause the most flood damage. This is consistent with other studies that document proportionately large impacts of urbanization on the smallest events, and declining impact for larger events [60], even though more complete documentation of the impact of paving for a range of storm sizes under different antecedent moisture conditions is necessary.

This investigation highlights the necessity to implement management activities to mitigate soil erosion such as stabilization of unpaved roads and other management activities (i.e., revegetation, sediment basins, channel stabilization, etc). Future studies should evaluate the uncertainty of the model-estimated parameters as well as implications in scenario analysis [14], which are critical for proper sediment management in the LLCW and potentially in other rural urbanizing watersheds, particularly those in developing countries. Our study highlights the relative importance of various erosion processes, and also key uncertainties for future investigation.

5. Conclusions

Urban development has significant impacts on watershed sediment production in a developing country context. Management activities, especially the practice of filling gullies with poorly-consolidated materials, represent a persistent source of sediment in the watershed. Simulated total runoff correlated well with the observed data whereas simulated peak discharge was best predicted for medium-sized events. Simulated gully erosion contributed significantly to the total sediment load at the watershed scale (20% to 26%, annual average), even though most (40%–50%) of the total load was from sheet and rill erosion. Hotspots of erosion cover 23% of the total catchment area but generate 50% of the total sediment yield, and occur on steep slopes on highly erodible soils. This investigation highlights the necessity to implement management activities to mitigate soil erosion such as asphalt or other stabilization measures on unpaved roads, as well as other management activities (i.e., revegetation, sediment basins, channelization, etc). The scenario analysis showed that paving roads in the hotspots reduced total sediment production by 30%, but may increase peak discharge moderately (2%–21%) at the outlet. Any mitigation activity in the watershed that includes road paving needs to consider the potential impacts on downstream communities and channel erosion. Future studies for improving model calibration, and evaluating more mitigation scenarios are critical for proper sediment management in the LLCW and potentially in other urbanizing watersheds, particularly those in developing countries. Our maps of the spatial distribution of sediment yield are uncertain due to the coarse resolution of land use and soil properties for small sub-watersheds (AnnAGNPS cells), and possible overestimation of sheet and rill erosion on steep slopes. Future research should include more detailed spatial information on soil properties to improve parameters' estimates, which would improve the accuracy of model simulations under current conditions and in various management scenarios.

Author Contributions: Conceptualization, T.W.B., R.L.B., Y.Y., and N.G.-E. Methodology, T.W.B., R.L.B., E.J.L. Y.Y., E.V.T., T.K, and N.G.-E. Software, R.L.B. Validation, T.W.B. and N.G.-E. Formal analysis, T.W.B., R.L.B., and N.G.-E. Investigation, T.W.B. and N.G.-E. Resources, T.W.B., D.L., T.K., and Y.Y. Data curation, T.W.B., R.L.B., K.T.T.-Q., and N.G.-E. Writing—original draft preparation, N.G.-E.; Writing—review and editing, T.W.B., R.L.B., E.V.T., E.J.L., T.K., K.T.T.-Q., and Y.Y. Supervision, T.W.B., T.K., and Y.Y.; Project administration, T.W.B., D.L., and Y.Y. Funding acquisition, D.L. and Y.Y.

Funding: The Consejo Nacional de Ciencia y Tecnología (CONACyT, México; Grant/Award Number: 210925) and the US Environmental Protection Agency (EPA) (Interagency Agreement ID # DW-12-92390601-0) in collaboration with the US Department of Agriculture (USDA, Agreement # 58-6408-4-015), San Diego State University (USA), University of Córdoba (Spain), and the Centro de Investigación Científica y de Educación Superior de Ensenada (CICESE, Mexico) funded this study.

Acknowledgments: We thank Kraemer Stephen and Babendreier Justin from USEPA, and two other anonymous reviewers from *Water* for their review, technical editing, and valuable comments. Thanks to Chris Peregrin and Bronti Patterson for compiling data on sediment removal from the traps in the United States, and to Oscar Romo for initial field visits and allowing installation of a rain gauge at a field station. Special thanks to residents of Los Laureles Canyon, who provided valuable help for data collection.

Conflicts of Interest: The authors declare no conflict of interest. The views expressed in this paper are those of the authors and do not necessarily represent the views or policies of the U.S. Environmental Protection Agency (EPA). Mention of trade names or products does not convey, and should not be interpreted as conveying, official EPA approval, endorsement, or recommendation.

References

1. Govers, G.; Everaert, W.; Poesen, J.; Rauws, G.; De Ploey, J.; Lautridou, J.P. A long flume study of the dynamic factors affecting the resistance of a loamy soil to concentrated flow erosion. *Earth Surf. Process. Landf.* **1990**, *15*, 313–328. [CrossRef]
2. Flanagan, D.C.; Nearing, M.A. *Water Erosion Prediction Project Hillslope Profile and Watershed Model Documentation*; SERL Report No.10; National Soil Erosion Research Laboratory: West Lafayette, IN, USA, 1995.
3. Pimentel, D. Soil erosion and the threat to food security and the environment. *Ecosyst. Health* **2000**, *6*, 221–226. [CrossRef]
4. Bakker, M.M.; Govers, G.; Kosmas, C.; Vanacker, V.; van Oost, K.; Rounsevell, M. Soil erosion as a driver of land-use change. *Agric. Ecosyst. Environ.* **2005**, *105*, 467–481. [CrossRef]
5. Ramos-Scharrón, C.E.; MacDonald, L.H. Runoff and suspended sediment yields from an unpaved road segment, St. John, US Virgin Islands. *Hydrol. Process.* **2007**, *21*, 35–50.
6. Archibold, O.W.; Levesque, L.M.J.; de Boer, D.H.; Aitken, A.E.; Delanoy, L. Gully retreat in a semi-urban catchment in Saskatoon, Saskatchewan. *Appl. Geogr.* **2003**, *23*, 261–279. [CrossRef]

7.	Wolman, M.G. A cycle of sedimentation and erosion in urban river channels. *Geogr. Ann.* **1967**, *49*, 385–395. [CrossRef]

8.	Biggs, T.W.; Atkinson, E.; Powell, R.; Ojeda, L. Land cover following rapid urbanization on the US-Mexico border: Implications for conceptual models of urban watershed processes. *Landsc. Urban. Plan.* **2010**, *96*, 78–87. [CrossRef]

9.	Gudino-Elizondo, N.; Biggs, T.; Castillo, C.; Bingner, R.L.; Langendoen, E.; Taniguchi, K.; Kretzschmar, T.; Yuan, Y.; Liden, D. Measuring ephemeral gully erosion rates and topographical thresholds in an urban watershed using Unmanned Aerial Systems and Structure from Motion photogrammetric techniques. *Land Degrad. Dev.* **2018**, *29*, 1896–1905. [CrossRef]

10.	Emerson, C.H.; Traver, R.G. Multiyear and Seasonal Variation of Infiltration from Storm-Water Best Management Practices. *J. Irrig. Drain. Eng.* **2008**, *134*, 598–605. [CrossRef]

11.	Bisantino, T.; Bingner, R.; Chouaib, W.; Gentile, F.; Trisorio Liuzzi, G. Estimation of Runoff, Peak Discharge and Sediment Load at the Event Scale in a Medium-Size Mediterranean Watershed Using the Annagnps Model. *Land. Degrad. Dev.* **2015**, *26*, 340–455. [CrossRef]

12.	Bingner, R.L.; Theurer, F.D.; Yuan, Y.P. AnnAGNPS Technical Processes. US Department of Agriculture (USDA)—Agricultural Research Service (ARS): Washington, DC, USA, 2015. Available online: www.wcc.nrcs. usda.gov/ftpref/wntsc/HandH/AGNPS/downloads/AnnAGNPS_Technical_Documentation.pdf (accessed on 5 July 2017).

13.	Merritt, W.S.; Letcher, R.A.; Jakeman, A.J. A review of erosion and sediment transport models. *Environ. Model. Softw.* **2003**, *18*, 761–799. [CrossRef]

14.	Gudino-Elizondo, N.; Biggs, T.W.; Bingner, R.L.; Yuan, Y.; Langendoen, E.J.; Taniguchi, K.T.; Kretzschmar, T.; Taguas, E.V.; Liden, D. Modelling Ephemeral Gully Erosion from Unpaved Urban Roads: Equifinality and Implications for Scenario Analysis. *Geosciences* **2018**, *8*, 137. [CrossRef]

15.	Slaymaker, O. The sediment budget as conceptual framework and management tool. *Hydrobiologia* **2003**, *494*, 71–82. [CrossRef]

16.	Gómez, J.A.; Giráldez, J.V.; Vanwalleghem, T. Comments on "Is soil erosion in olive groves as bad as often claimed?" By L. Fleskend and L. Stroosnijder. *Geoderma* **2008**, *147*, 93–95. [CrossRef]

17.	Nouwakpo, S.K.; Williams, C.J.; Al-Hamdan, O.Z.; Weltz, M.A.; Pierson, F.; Nearing, M. A review of concentrated flow erosion processes on rangelands: Fundamental understanding and knowledge gaps. *J. Soil. Water Conserv.* **2016**, *4*, 75–86. [CrossRef]

18.	Phinzi, K.; Ngetar, N.S. The assessment of water-borne erosion at catchment level using GIS-based RUSLE and remote sensing: A review. *Int. Soil Water Conserv. Res.* **2019**. [CrossRef]

19.	Labrière, N.; Locatelli, B.; Laumonier, Y.; Freycon, V.; Bernoux, M. Soil erosion in the humid tropics: A systematic quantitative review. *Agric. Ecosyst. Environ.* **2015**, *203*, 127–139. [CrossRef]

20.	Poesen, J.; Govers, G. Gully erosion in the loam belt of Belgium: Typology and Control Measures. In *Soil Erosion on Agricultural Land*; Boardman, J., Foster, I.D.L., Dearing, J., Eds.; Wiley: Chicester, UK, 1990; pp. 513–530.

21.	Vandaele, K.; Poesen, J.; Govers, G.; van Wesemael, B. Geomorphic threshold conditions for ephemeral gully incision. *Geomorphology* **1996**, *16*, 161–173. [CrossRef]

22.	Martínez-Casasnovas, J.A.; Ramos, M.C.; Ribes-Dasi, M. Soil erosion caused by extreme rainfall events: Mapping and quantification in agricultural plots from very detailed digital elevation models. *Geoderma* **2002**, *105*, 125–140. [CrossRef]

23.	Taguas, E.V.; Yuan, Y.; Bingner, R.L.; Gomez, J.A. Modeling the contribution of ephemeral gully erosion under different soil managements: A case study in an olive orchard microcatchment using the AnnAGNPS model. *CATENA* **2012**, *98*, 1–16. [CrossRef]

24.	Foster, G.R. Understanding Ephemeral Gully Erosion. In *Soil Conservation, Assessing the National Resource Inventory*; The National Academies Press: Washington, DC, USA, 1986; Volume 2, pp. 90–125, ISBN 978-0-309-03675-7.

25.	Poesen, J.; Nachtergaele, J.; Verstraeten, G.; Valentin, C. Gully erosion and environmental change: Importance and research needs. *CATENA* **2003**, *50*, 91–133. [CrossRef]

26.	Momm, H.G.; Bingner, R.L.; Wells, R.R.; Wilcox, D. AGNPS GIS-based tool for watershed-scale identification and mapping of cropland potential ephemeral gullies. *Appl. Eng. Agric.* **2012**, *28*, 17–29. [CrossRef]

27. SCS. *Technical Release 55: Urban Hydrology for Small Watersheds; Natural Resources Conservation Service*; US Department of Agriculture: Washington, DC, USA, 1986. Available online: https://www.nrcs.usda.gov/Internet/FSE_DOCUMENTS/stelprdb1044171.pdf (accessed on 15 November 2017).

28. Knisel, W.G. (Ed.) *CREAMS: A Field-Scale Model for Chemicals, Runoff, and Erosion from Agricultural Management Systems*; Conservation Research Report 26; US Department of Agriculture: Washington, DC, USA, 1980.

29. Theurer, F.D.; Clarke, C.D. Wash load component for sediment yield modeling. In Proceedings of the Fifth Federal Interagency Sedimentation Conference, Las Vegas, NV, USA, 18–21 March 1991; pp. 7-1–7-8.

30. Licciardello, F.; Zema, D.A.; Zimbone, S.M.; Bingner, R.L. Runoff and soil erosion evaluation by the AnnAGNPS Model in a small Mediterranean watershed. *Trans. ASABE* **2007**, *50*, 1585–1593. [CrossRef]

31. Grover, R. Local Perspectives on Environmental Degradation and Community Infrastructure in Los Laureles Canyon, Tijuana. Master's Thesis, San Diego State University, San Diego, CA, USA, 2011.

32. CalEPA, California Environmental Protection Agency. 2018. Available online: https://www.waterboards.ca.gov/sandiego/water_issues/programs/tmdls/TijuanaRiverValley.shtml (accessed on 4 February 2018).

33. Biggs, T.W.; Taniguchi, K.T.; Gudino-Elizondo, N.; Langendoen, E.J.; Yuan, Y.; Bingner, R.L.; Liden, D. Runoff and Sediment Yield on the US-Mexico Border, Los Laureles Canyon. EPA/600/R-18/365; Washington, DC, USA, 2018. Available online: https://cfpub.epa.gov/si/si_public_record_report.cfm?Lab=NERL&dirEntryId=343214 (accessed on 3 January 2019).

34. IMPLAN. Programa Parcial de Mejoramiento Urbano de la Subcuenca Los Laureles. 2004. Available online: https://www.implantijuana.org/informaci%C3%B3n/planes-y-programas/ppmu-ll-2007-2015/ (accessed on 15 May 2017).

35. Kennedy, M.P.; Tan, S.S. *Geologic Map of the San Diego 30′ × 60′ Quadrangle, California; 1:100,000*; California Department of Conservation, California Geological Survey; U.S. Geological Survey: Reston, VA, USA, 2008.

36. Hanson, G.J. Surface erodibility of earthen channels at high stresses part II—developing an in situ testing device. *Trans. ASAE* **1990**, *33*, 132–137. [CrossRef]

37. U.S. GEOLOGICAL SURVEY, Scientific Investigations Report 2008–5093, 2008, Table 7. Available online: https://pubs.usgs.gov/sir/2008/5093/table7.html (accessed on 5 January 2018).

38. Taniguchi, K.T.; Biggs, T.W.; Langendoen, E.J.; Castillo, C.; Gudino-Elizondo, N.; Yuan, Y.; Liden, D. Stream channel erosion in a rapidly urbanizing region of the US–Mexico border: Documenting the importance of channel hardpoints with Structure from-Motion photogrammetry. *Earth Surf. Process. Landf.* **2018**, *43*, 1465–1477. [CrossRef] [PubMed]

39. Liu, B.Y.; Nearing, M.A.; Risse, L.M. Slope gradient effects on soil loss for steep slopes. *Trans. ASAE* **1994**, *37*, 1835–1840. [CrossRef]

40. Gordon, L.M.; Bennett, S.J.; Bingner, R.L.; Theurer, F.D.; Alonso, C.V. Simulating ephemeral gully erosion in AnnAGNPS. *Trans. ASAE* **2007**, *50*, 857–866. [CrossRef]

41. Cochrane, T.A.; Flanagan, D.C. Effect of DEM resolutions in the runoff and soil loss predictions of the WEPP watershed model. *Trans. ASAE* **2005**, *48*, 109–120. [CrossRef]

42. Wang, X.; Lin, Q. Effect of DEM mesh size on AnnAGNPS simulation and slope correction. *Environ. Monit Assess.* **2011**, *179*, 267–277. [CrossRef]

43. Dunne, T.; Leopold, L.B. *Water in Environmental Planning*; W.H. Freeman and Company: New York, NY, USA, 1987; p. 818, ISBN 0-71670079-4.

44. Moriasi, D.N.; Arnold, J.G.; Van Liew, M.W.; Bingner, R.L.; Harem, R.D.; Veith, T.L. Model evaluation guidelines for systematic quantification of accuracy in watershed simulations. *Trans. ASABE* **2007**, *50*, 850–900. [CrossRef]

45. Engman, E.T. Roughness coefficients for routing surface runoff. *J. Irrig. Drain. Eng.* **1986**, *112*, 39–53. [CrossRef]

46. Morris, G.L.; Fan, J. *Reservoir Sedimentation Handbook: Design and Management of Dams, Reservoirs and Watersheds for Sustainable Use*; McGraw-Hill Companies, Inc.: New York, NY, USA, 1998.

47. Urbonas, B.; Stahre, P. *Stormwater Best Management Practices and Detention*; PTR Prentice Hall: Englewood Cliffs, NJ, USA, 1993.

48. Gupta, H.V.; Sorooshian, S.; Yapo, P.O. Status of automatic calibration for hydrologic models: Comparison with multi-level expert calibration. *J. Hydrol. Eng.* **1999**, *4*, 135–143. [CrossRef]

49. Chahor, Y.; Casalí, J.; Giménez, R.; Bingner, R.L.; Campo, M.A.; Goñi, M. Evaluation of the AnnAGNPS model for predicting runoff and sediment yield in a small Mediterranean agricultural watershed in Navarre (Spain). *Agric. Water Manag.* **2014**, *134*, 24–37. [CrossRef]

50. Zema, D.A.; Bingner, R.L.; Denisi, P.; Govers, G.; Licciardello, F.; Zimbone, S.M. Evaluation of runoff, peak flow and sediment yield for events simulated by the AnnAGNPS model in a Belgian agricultural watershed. *Land. Degrad. Dev.* **2012**, *23*, 205–215. [CrossRef]

51. Taguas, E.V.; Ayuso, J.L.; Peña, A.; Yuan, Y.; Pérez, R. Evaluating and modelling the hydrological and erosive behaviour of an olive orchard microcatchment under no-tillage with bare soil in Spain. *Earth Surf. Process. Landf.* **2009**, *34*, 738–751. [CrossRef]

52. Parajuli, P.B.; Nelson, N.O.; Frees, L.D.; Mankin, K.R. Comparison of AnnAGNPS and SWAT model simulation results in USDA-CEAP agricultural watersheds in south-central Kansas. *Hydrol. Process.* **2009**, *23*, 748–763. [CrossRef]

53. Licciardello, F.; Zimbone, S.M. Runoff and erosion modeling by AGNPS in an experimental Mediterranean watershed. In Proceedings of the ASAE Annual Intl. Meeting/CIGR XVth World Congress, St. Joseph, MI, USA, 30 July–1 August 2002.

54. Langendoen, E.J.; Alonso, C.V. Modeling the evolution of incised streams: I. Model formulation and validation of flow and streambed evolution components. *J. Hydraul. Eng.* **2008**, *134*, 749–762. [CrossRef]

55. Langendoen, E.J.; Simon, A. Modeling the evolution of incised streams. II: Streambank erosion. *J. Hydraul. Eng.* **2008**, *134*, 905–915. [CrossRef]

56. Bingner, R.L.; Czajkowski, K.; Palmer, M.; Coss, J.; Davis, S.; Stafford, J.; Widman, N.; Theurer, F.D.; Koltum, G.; Richards, P.; et al. *Upper Auglaize Watershed AGNPS Modeling Project Final Report*; Research Report No. 51; USDA-ARS National Sedimentation Laboratory: Oxford, MS, USA, 2006.

57. De Santisteban, L.M.; Casali, J.; Lopez, J.J. Assessing soil erosion rates in cultivated areas of Navarre (Spain). *Earth Surf. Process. Landf.* **2006**, *31*, 487–506. [CrossRef]

58. Trimble, S.W. Contribution of stream channel erosion to sediment yield from an urbanizing watershed. *Science* **1997**, *278*, 1442–1444. [CrossRef]

59. Cashman, M.J.; Gellis, A.; Gorman-Sanisaca, L.; Noe, G.B.; Cogliandro, V.; Baker, A. Bank-derived material dominates fluvial sediment in a suburban Chesapeake Bay watershed. *River Res. Appl.* **2018**, *34*, 1032–1044. [CrossRef]

60. Du, J.; Qian, L.; Rui, H.; Zuo, T.; Zheng, D.; Xu, Y.; Xu, C.-Y. Assessing the effects of urbanization on annual runoff and flood events using an integrated hydrological modeling system for Qinhuai River basin, China. *J. Hydrol.* **2012**, *464–465*, 127–139. [CrossRef]

 *water* 

Article

Assessment of Sediment Transport Functions with the Modified SWAT-Twn Model for a Taiwanese Small Mountainous Watershed

Chih-Mei Lu and Li-Chi Chiang *

Department of Civil and Disaster Prevention Engineering, National United University, Miaoli City 36063, Taiwan
* Correspondence: lchiang@nuu.edu.tw; Tel.: +886-37-38-2368

Received: 4 July 2019; Accepted: 19 August 2019; Published: 22 August 2019

Abstract: In Taiwan, the steep landscape and highly vulnerable geology make it difficult to predict soil erosion and sediment transportation via variable transport conditions. In this study, we integrated the Taiwan universal soil loss equation (TUSLE) and slope stability conditions in the soil and water assessment tool (SWAT) as the SWAT-Twn model to improve sediment simulation and assess the sediment transport functions in the Chenyulan watershed, a small mountainous catchment. The results showed that the simulation of streamflow was satisfactory for calibration and validation. Before model calibration and validation for sediment, SWAT-Twn with default sediment transport method performed better in sediment simulation than the official SWAT model (version 664). The SWAT-Twn model coupled with the simplified Bagnold equation could estimate sediment export more accurately and significantly reduce the overestimated sediment yield by 65.7%, especially in highly steep areas. Furthermore, five different sediment transport methods (simplified Bagnold equation with/without routing by particle size, Kodoatie equation, Molinas and Wu equation, and Yang sand and gravel equation) were evaluated. It is suggested that modelers who conduct sediment studies in the mountainous watersheds with extreme rainfall conditions should adjust the modified universal soil loss equation (MUSLE) factors and carefully evaluate the sediment transportation equations in SWAT.

Keywords: SWAT; TUSLE; sediment transport; model calibration; mountainous catchment

---

1. Introduction

The amount and occurrence time of sediment export from the watersheds are mainly caused by different types of soil erosion, such as channel bed and bank erosion, overland surface erosion, and glacier erosion [1]. As soil erosion causes shear stress by rainfall, surface runoff brings most sediment yields from the overland eroded soil and then is usually transported to the downstream [2]. Moreover, extreme natural disturbances (i.e., typhoons, earthquakes, landslides, and floods) also play as the trigger for abnormal sediment yields in some regions [3], and further change the characteristics of sediment yields and sediment transports in watersheds [4]. Many studies have indicated that climate change has caused higher rainfall intensity and annual precipitation, leading to increases in sediment yields and soil erosion rates [5].

In order to comprehensively quantify the impacts of rainfall, soil erodibility, land use cover, topography, and support practice to sediment yields, the universal soil loss equation (USLE) [6] and the modified universal soil loss equation (MUSLE) [7] have been developed. These two empirical equations have been used to estimate the soil loss in watersheds worldwide [8]. Some other physically-based erosion models were developed in the past decades, such as those used in ANSWERS (Areal Nonpoint Source Watershed Environmental Response Simulation) [9], EPIC (Environmental Policy Integrated Climate) model [10], WEPP (Water Erosion Prediction Project) [11], and EUROSEM (European Soil

Erosion Model) [12]. The main difference between MUSLE and USLE is that USLE uses rainfall as an indicator of erosive energy, while MULSE uses the amount of runoff to simulate erosion and sediment yield. Sadeghi et al. [8] reviewed 49 papers of MUSLE application worldwide, and presented that most of MUSLE studies were to estimate the sediment yield on a storm basis (73.91%), monthly basis (2.17%), and annual basis (17.39%), while other applications of MUSLE (6.53%) were studied in soil erosion in storm-wise scale, pollutant estimation, and return periods of annual sediment yield. Therefore, MUSLE can better simulate the sediment yields during single storms with a low level of estimation error [13]. Since USLE and MUSLE were developed based on experiment watersheds in the U.S., [8] indicated that MUSLE would have huge errors without the calibration but can present reliable results for sediment yield on a storm basis after calibration, especially when it is applied under appropriate conditions (i.e., rangeland watersheds, similar climatic conditions as the USA) similar to where the original model was developed.

The process of erosion is usually described in three stages, which are detachment, transport and deposition; and four main types, which are sheet, rill, gully and in-stream erosion [14]. Sheet erosion and rill erosion are often classified as overland flow erosion caused by raindrop or overland flow [2,15]. Both types of erosions are usually considered together in erosion modelling. Gully erosion is the removal of soil along the channels of concentrated flow and is controlled by the thresholds related to slope and catchment area [2]. In-stream erosion involves the direct removal of sediment from stream beds and stream banks, especially during high flow periods huge amount of sediment transported through the stream network originates from the stream channel [14]. Erosion/sediment transport models can be categorized into three types: empirical model, conceptual model, and physics-based model [14]. The empirical models, the simplest ones, are based on the statistical observations of experiment areas in response to the characteristics of sediment transport [16], while conceptual models usually represent the catchment as a series of internal storages with parameters values determined through calibration against observed data [17]. Physics-based models integrate some fundamental physical equations, such as the equations of conservation of mass and momentum for flow, and those equations for sediment [14]. Compared to physics-based models and more complex conceptual models, which usually lack of sufficient spatially distributed input data, empirical and conceptual models are suggested to be combined for presenting the event responsiveness and sensitivity to climate variability [14].

The soil and water assessment Tool (SWAT) is a semi-distributed hydrological watershed model, which can simulate water balance, plant growth, and transport of sediment, nutrients, heavy metals and pesticides. The SWAT model has been used worldwide for simulating the impact of climate change and land use change on streamflow, sediment and nutrients exports, and best management practices (BMPs) on watershed responses in different countries [18–20]. Generally, the SWAT model performed well for simulating streamflow, sediment and nutrient exports at various spatio-temporal scales. MUSLE plays an important role of the simulation of continuous sediment loads [21]. Sediment transport is often called total sediment load, which is the sum of bed load and suspended load. The sediment transport is generally modeled through the sediment flow caused by overland flow and the channel erosion [22]. Some studies indicated unsatisfactory sediment simulation of the SWAT model. Addis et al. [23] used the SWAT model to simulate discharge and sediment transport at a small mountainous catchment in Ethiopian plateau, and showed that SWAT performed well for discharge but overestimated daily sediment transport with NSE = 0.07 and −1.76 for calibration and validation, respectively, mainly due to insufficient observed data and the heavy rainfall during simulation period. Bressiani et al. [24] indicated that the SWAT model was suitable for sediment simulation in most areas in Brazil and suggested the change in transmission of sediment should be reflected in the current SWAT version.

As the sediment transport and sediment concentration can affect the nutrient and turbidity in water, the estimation of sediment transport or sediment concentration need to be more reasonable. Arnold et al. [25] advised that SWAT users need to calibrate the discharge and sediment transport or concentration sequentially before simulating the nutrients. However, extreme rainfalls in Taiwan result in serious debris flows and landslides, making it more difficult to simulate the sediment transport in the river. As the runoff and sediment hazards in Taiwan are increasing, Lee et al. [26] indicated that the runoff and sediment yield would increase with the storm events and become more frequently occurred, especially in the small mountainous catchments. Chiu et al. [27] simulated the sediment yield in the Shihmen reservoir in Taiwan, and revealed that natural distributions (i.e., typhoon, storm, and earthquake) have become an important potential source of sediment yield. Therefore, Chang et al. [28] suggested that landslide should be considered in the model to simulate more accurate and reasonable sediment exports. In order to more accurately simulate the sediment transport and sediment yields in a small mountainous watershed in Taiwan, we aim to: (1) integrate Taiwan universal soil loss equation (TUSLE) and the landslide area-volume estimation equation into the SWAT model (version 664) as SWAT-Twn model; (2) examine five sediment transport methods in the SWAT model; (3) provide the calibration and validation experience of sediment transport simulation for the SWAT users.

2. Materials and Methods

2.1. Study Area and Data

The Chenyulan watershed, located in central Taiwan, has an area of 448 km^2 with the elevation ranging from 292 m to 3893 m (Figure 1). The Chenyulan watershed has steep terrain that is 49.74% of the total area with slopes steeper than 60%. Typhoons averagely hit Taiwan for 3–4 times in a year, leading large amounts of flow and sediment in the watersheds. Especially in 1996 and 2009, Typhoon Herb and Typhoon Morakot have brought more than 2000 mm of accumulated rainfall in two days, resulting in 3370 m^3/s and 1860 m^3/s of peak discharge and daily average streamflow, respectively. Moreover, extremely high sediment concentration of 98,499 mg/L was observed at Nemoupu station in the Chenyulan watershed when Typhoon Morakot occurred in 2009. During the study period (2004–2015), several typhoons that influenced the watershed have been recorded (Table 1). Moreover, in 1999, the Chi-Chi earthquake of 7.3 Richter magnitude scale occurred in central Taiwan (Figure 1) and has seriously influenced the watershed landscape to become more fragile and sensitive [29,30].

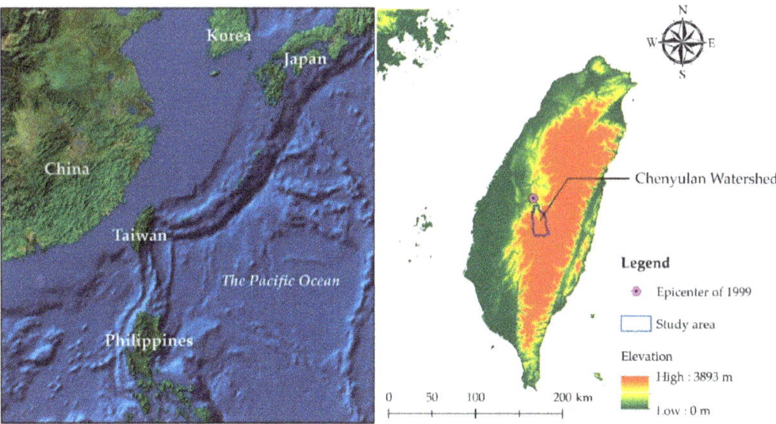

Figure 1. Study area.

Table 1. Top 10 typhoons that influenced the study area during 2004–2015.

Rank	Typhoon	Occurred Period	Accumulated Precipitation (mm)	Weather Station
1	Morakot	5–10 August 2009	1953	C1M440
2	Matmo	21–23 July 2014	1136	C1M440
3	Sinlaku	11–16 September 2008	1072.5	C0H9A0
4	Mindulle	28 June–3 July 2004	774.5	C0H9A0
5	Krosa	4–10 October 2007	771	C0H9A0
6	Haitang	16–20 July 2005	760.5	C0H9A0
7	Saola	30 July–3 August 2012	693.5	C1I310
8	Sepat	16–19 August 2007	660.5	C1M440
9	Soulil	11–13 July 2013	624.5	C0H9A0
10	Jangmi	26–29 September 2008	593	C0H9A0

(Source: Typhoon Database, Central Weather Bureau, Taiwan [31]).

The environmental data required for the SWAT model include: the digital elevation model (DEM), weather, observed streamflow and sediment concentration, land use and soil data. Moreover, the model simulation can be more precisely to represent the characteristics of the watershed by adding detailed information, such as point source pollutants, reservoirs, agriculture management scenarios [32–37]. The DEM data was available from Taiwan Geospatial One Stop (TGOS) [38]. We collected the data (i.e., precipitation, temperature, solar radiation, relative humidity, wind speed) of six weather stations in the watershed during 2003—2015 from the Taiwan Data Bank of Atmospheric and Hydrologic Research (DBAR) [39] (Figure 2). Some precipitation data missing during typhoons were replaced by the typhoon database of Taiwan Central Weather Bureau [31]. The observed daily streamflow and sediment concentration data during 2004–2015 were collected from the annual Hydrological Year Book of Taiwan Republic of China [40]. Most of the gages in the Chenyulan watershed seriously lack in measured data, and the Nemoupu gage is the only station that has complete data from 2004 to 2015 (Figure 2). The measured data during 2004–2009 and 2010–2015 were used for model calibration and validation, respectively. Since the sediment data were not measured continuously for every day, the sediment rating curve was first established to estimate the daily sediment concentration and loads for the model calibration and validation.

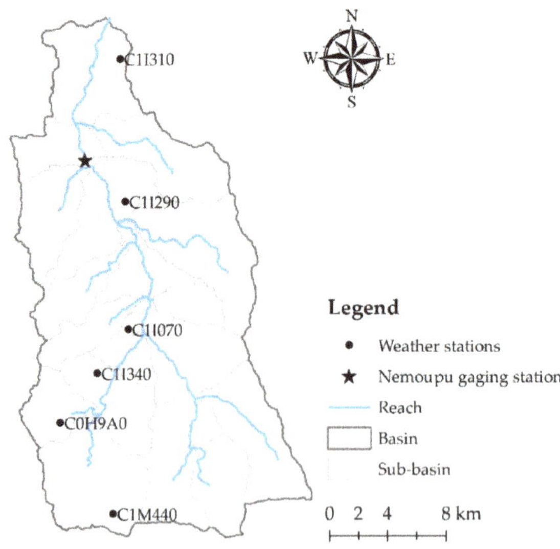

Figure 2. Locations of weather stations, sub-basins, and gaging station.

The land use data were collected form Second Land Use Survey in 2006 from [41] (Figure 3). Forest and agricultural lands are the two major land uses in the watershed, occupying 74.46% and 14.05% of the total area, respectively. The landslide area was set as barren (land use code in SWAT: BARR), accounting for 2.89% of total area. Soil data was collected from [42]. Approximately 82% of the total area has not been surveyed, thus it was assumed to be darkish colluvial soil in this study [30]. The pale colluvial clay accounts for 12% of the study area. The slope was calculated from the DEM data through the GIS spatial analysis tool. The slope was divided into five classes (0–9%, 9–30%, 30–45%, 45–60%, >60%), occupying 3.29%, 12.25%, 15.1%, 19.62%, and 49.74% of the total area, respectively.

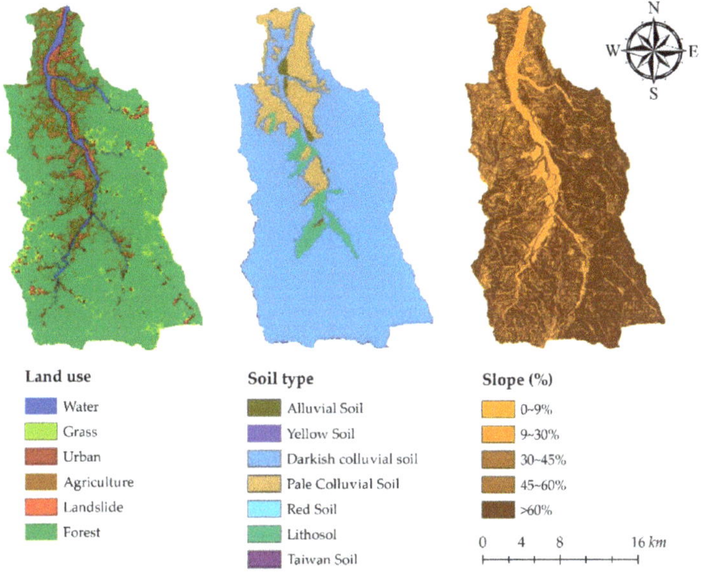

Figure 3. Land use, soil, and slope distribution.

2.2. SWAT Model

2.2.1. Model Description

The soil and water assessment tool (SWAT) model, a semi-distributed watershed-based hydrological model, was developed by the Agricultural Research Service, United States Department of Agriculture in 1994 [43]. The model can simulate the hydrological processes and sediment/nutrient transports at various spatio-temporal scales. In the SWAT model, several sub-basins are first delineated by setting the stream outlets (Figure 2), and further a unique combination of land use, soil and slope forms the basic modeling unit, called the hydrologic response unit (HRU). In this study, a total of 1173 HRUs of 23 sub-basins in the Chenyulan watershed were delineated (Figure 4).

The SWAT model simulates the surface runoff by using the curve number method, developed by the U.S. Soil Conservation Service (SCS). The SCS curve number equation is as follows:

$$Q_{surf} = \frac{\left(R_{day} - I_a\right)^2}{\left(R_{day} - I_a\right) + S} \tag{1}$$

where Q_{surf} is surface runoff (mm); R_{day} is daily precipitation (mm); I_a is the initial loss of surface water storage, interception, and percolation (mm); S is retention parameter (mm), which changes with soil type, land use and slope.

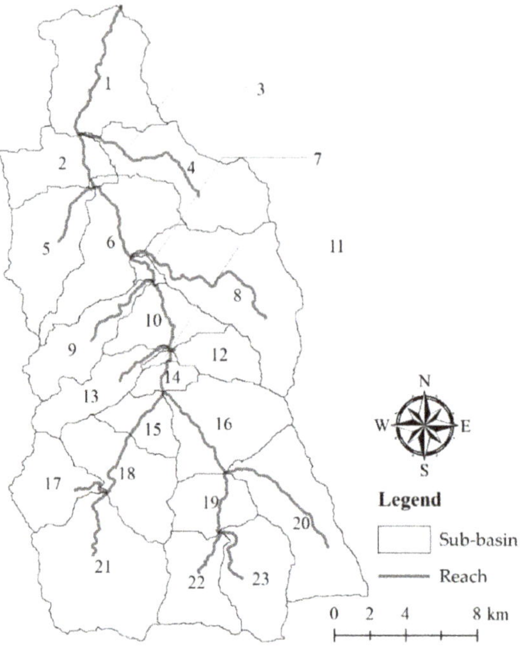

Figure 4. Sub-basin distribution.

In SWAT model, infiltration is calculated by the Green and Ampt infiltration method. For evapotranspiration, three methods of evaporation estimation can be selected, including FAO Penman–Monteith, Hargreaves and Priestly-Taylor method. The hydrologic process is calculated by Equation (2).

$$S_t = S_0 + \sum_{i=1}^{t} R_{day} - Q_{surf} - E_a - W_{seep} - Q_{gw} \tag{2}$$

where S_t is soil water content (mm); S_0 is initial soil water content (mm); R_{day} is the daily precipitation (mm); Q_{surf} is the daily surface runoff (mm); E_a is the daily evapotranspiration (mm); W_{seep} is the daily percolation (mm); Q_{gw} is the daily baseflow (mm). In SWAT, the soil water content is calculated for different layers defined by the users, and 50% of the evaporative demand is extracted from the top 10 mm of soil [44].

SWAT model uses the modified universal soil loss equation (MUSLE) to estimate soil loss at the HRU scale.

$$S = 11.8 \times (Q_{surf} \times q \times A)^{0.56} \times K \times C \times P \times LS \times CFRG \tag{3}$$

where S is soil erosion (t); Q_{surf} is surface runoff (mm/ha); q is peak runoff (m^3/s); A is the area of HRU (ha); K is the soil erodibility factor; C is the USLE land use/cover management factor; P is the USLE support practice factor; LS is the topographic factor; CFRG is the coarse fragment factor, which is calculated as the function of percent rock in the first soil layer. CFRG value equals 1 when there is no rock in the first soil layer. A higher percent of rock will result in smaller CFRG value and less soil loss.

2.2.2. Sediment Transport Methods

There are five sediment transport methods in SWAT model (version 664), which are the simplified Bagnold method with/without routing by particle size, Kodoatie method, Molinas and Wu method, and the Yang sand and gravel method. These methods can be applied for rivers of various river bed materials.

- Simplified Bagnold equation (EQN0 and EQN1):

The Bagnold method is the default method in SWAT model. The maximum sediment concentration is calculated as follows.

$$conc_{sed,ch,mx} = c_{sp} \cdot v_{ch,pk}^{\,spexp} \tag{4}$$

where $conc_{sed,ch,mx}$ is the maximum sediment concentration (t/m^3); c_{sp} is linear coefficient; $v_{ch,pk}$ is peak flow velocity (m/s); spexp is exponential coefficient.

In EQN0, as the default and only one sediment transport method in SWAT 2005 version, the bed load is limited by the channel cover and erodibility factors, and the sediment carried by channel is always near the calculated maximum transport capacity [44]. Moreover, it does not keep track of sediment pools in various particle sizes. Thus, EQN1, additional stream power equation in SWAT 2016 version, has been incorporated with physics-based approach for channel erosion.

- Kodoatie equation (EQN2):

The Kodoatie equation is an optimized sediment transport equation, based on the non-linear sediment equation [45] and the field observation data.

$$conc_{sed,ch,mx} = \left(\frac{a \cdot v_{ch}^{\,b} \cdot y^c \cdot S^d}{Q_{in}} \right) \cdot \left(\frac{W + W_{btm}}{2} \right) \tag{5}$$

where $conc_{sed,ch,mx}$ is the maximum sediment concentration (tons/m^3); v_{ch} is mean flow velocity (m/s); y is mean flow depth (m); S is energy slope (m/m); Q_{in} is water entering the reach (m^3); a, b, c and d are regression coefficients for different bed materials; W is channel width at the water level (m); W_{btm} is bottom width of the channel (m).

- Molinas and Wu equation (EQN3):

Molinas and Wu [46] developed the sediment transport equation based on the universal stream power for rivers of large sand bed. The sediment weight concentration is calculated as follows.

$$C_w = M\Psi^N \tag{6}$$

where C_w is sediment concentration by weight; Ψ is universal stream power; M and N are coefficients. The universal stream power (Ψ) is calculated as follows.

$$\Psi = \left\{ \frac{\left(S_g - 1\right) \cdot g \cdot depth \cdot \omega_{50}}{\left[log_{10}\left(\frac{depth}{D_{50}} \right) \right]} \right\}^{0.5} \tag{7}$$

where S_g is relative density of solid (2.65); g is acceleration of gravity (9.81 m/s^2); depth is flow depth (m); ω_{50} is fall velocity of median size particles (m/s); D_{50} is median sediment size.

- Yang sand and gravel equation (EQN4):

Developed by Yang [47], the sediment weight concentration was calculated based on the sediment size (D_{50}), unit stream power ($V_{ch}S$), shear velocity (V_*), fall velocity (ω_{50}), and kinematic viscosity (υ). The equations are separated for sand and gravel bed material shown as follows.

Sand equation for median size (D_{50}) less than 2 mm,

$$\begin{aligned} log\, C_w &= 5.435 - 0.286\, log\, \frac{\omega_{50} D_{50}}{\upsilon} - 0.457\, log\, \frac{V_*}{\omega_{50}} \\ &+ \left(1.799 - 0.409\, log\, \frac{\omega_{50} D_{50}}{\upsilon} - 0.314\, log\, \frac{V_*}{\omega_{50}} \right) log\left(\frac{V_{ch}S}{\omega_{50}} - \frac{V_{cr}S}{\omega_{50}} \right) \end{aligned} \tag{8}$$

Gravel equation for median size (D_{50}) between 2 mm and 10 mm,

$$\log C_w = 6.681 - 0.633 \log \frac{\omega_{50}D_{50}}{\upsilon} - 4.816 \log \frac{V_*}{\omega_{50}} + \left(2.784 - 0.305 \log \frac{\omega_{50}D_{50}}{\upsilon} - 0.282 \log \frac{V_*}{\omega_{50}}\right) \log\left(\frac{V_{ch}S}{\omega_{50}} - \frac{V_{cr}S}{\omega_{50}}\right) \tag{9}$$

where C_w is weight concentration (ppm); ω_{50} is fall velocity of the median size sediment (m/s); υ is kinematic viscosity (m^2/s); V_* is shear velocity (m/s); V_{ch} is mean channel velocity (m/s); V_{cr} is critical velocity (m/s); S is energy slope (m/m).

2.3. Taiwan Universal Soil Loss Equation (TUSLE)

In order to present the characteristics of Taiwan soil, we applied the Taiwan universal soil loss equation (TUSLE) [48], revised from the universal soil loss equation (USLE) in the SWAT model. The factor adjustments are shown as follows.

- Soil erodibility factor (K)

The adjusted K factors were based on the soil survey in Taiwan conducted by [49,50]. The K values in the study area range from 0.13 to 0.40 (Table 2), and higher value indicates the soil layer is easier to erode.

Table 2. K factor used in the Chenyulan watershed.

Soil Type	Alluvial Soil	Lithosol	Darkish Colluvial Soil	Red Soil	Pale Colluvial Soil	Taiwan Clay	Yellow Soil
K factor	0.4	0.3	0.36	0.13	0.19	0.2	0.29

- Rainfall erosivity factor (R)

The R factor in USLE is calculated by 30-min maximum rainfall intensity and rainfall kinetic energy [6]. Since the R factor of USLE is complicated to calculate, Chen et al. [48] estimated the R factor for TUSLE by the regression equation developed by the U.S. Department of Agriculture, Agriculture Research Service (Equation (10)) [48].

$$R_{eq} = -823.8 + 5.213P_r \tag{10}$$

where R_{eq} is rainfall erosivity factor; P_r is annual precipitation (mm).

In MUSLE, the rainfall erosivity factor is replaced by the runoff factor as the runoff factor could better represent the surface runoff and overland sediment transport characteristics [7]. Thus, in this study, we calculated the runoff factor (R) in MUSLE, instead of the rainfall erosivity factor (R_{eq}).

$$R = 11.8(Q{\cdot}q{\cdot}A)^{0.56} \tag{11}$$

where R is runoff factor; Q is surface runoff (mm/ha); q is peak runoff (m^3/s); A is the area of catchment (ha).

- Cover and management factor (C)

The C factor was estimated by non-linear equation with the normalized difference vegetation index (NDVI) to avoid overestimating the C values of area with low soil erosion rate [51] (Table 3).

$$NDVI \geq 0, \quad C = \left(\frac{1 - NDVI}{2}\right)^{1+NDVI} \tag{12}$$

$$NDVI < 0, \quad \begin{cases} Building\,or\,non-exposed\,ground, \ C = 0.01 \\ Barren, \quad C = 1.0 \end{cases} \tag{13}$$

Table 3. Normalized difference vegetation index (NDVI)-calculated weighted C value for different land uses.

NDVI-Calculated C	Number of Grids			
	Grass	Forest	Agriculture	Landslide
0.01	3914	61,334	0	0
0.1	3049	67,819	8463	707
0.2	8417	153,468	26,007	2362
0.3	4120	59,365	14,141	1590
0.4	4084	56,689	17,422	2793
1	0	0	9205	8026
Weighted C	0.2	0.2	0.4	0.7
SWAT Default C	0.003	0.001	0.2	0.2

- Topographic factor (LS)

Wischmeier and Smith [52] established the LS equation (Equation (14)) as the product of L factor (Equation (15)) and S factor. In the SWAT model, the exponential factor (m) in the L factor equation is defined as Equation (16) [44], while TUSLE adopted the classification suggested by [52] that the exponential factor (m) varies with the slope, where m = 0.5, 0.4, 0.3 and 0.2 is used for the average slope greater than 5%, between 3–5%, between 1–3%, and less than 1%, respectively. Thus, TUSLE can increase the underestimated L factor at flatter slope and reduce the overestimated L factor at a steeper slope (Figure 5a). McCool et al. [53] indicated that Wischmeier and Smith's [52] topographic factor equation could only be suitable for the slope from 0.1% to 18%, and developed S factor equation (Equations (17) and (18)) to more reasonably predict soil loss at steep topography. The comparison of S factor by [52,53] (Figure 5b) showed that the equation would overestimate S factor in areas of steeper slope. Therefore, we combined the L factor equation with m values from TUSLE and the S factor equation by [53] to calculate the LS factor.

$$LS = \left(\frac{X}{22.13}\right)^{m} \left(0.0654 + 4.56\sin\theta + 65.4\sin^2\theta\right) \tag{14}$$

$$L = \left(\frac{X}{22.13}\right)^{m} \tag{15}$$

$$m = 0.6 \times \left(1 - e^{-35.835*\theta}\right) \tag{16}$$

where X = slope length (m), m = exponential factor.

$$S = 10.8\sin\theta + 0.03, \theta < 9\% \tag{17}$$

$$S = \left(\frac{\sin\theta}{0.0896}\right)^{0.6}, \theta \geq 9\% \tag{18}$$

where S is the slope factor of HRUs in MUSLE, θ is the slope of HRUs.

2.4. Calculation of Landslide Volume

The Chenyulan watershed has suffered by several severe natural disasters, such as Typhoon Herb (in 1996), 921 earthquake (in 1999), and Typhoon Morakot (in 2009), resulting in significant landslides, debris flows, change in landslide characteristics of central Taiwan [4]. Many studies have conducted the survey of landslide characteristic changed after the 921 earthquake [26,27,54–57]. Due to the lack of landslide calculation in the SWAT model, we integrated the landslide estimation equation developed by [28] into the SWAT model. The landslide volume was calculated by using the correlation between the landslide area and volume ($R^2 = 0.79$) (Equation (19)).

$$\ln(V) = 0.687 \ln(A) + 2.326 \tag{19}$$

where V is estimated landslide volume (m^3); A is landslide area (m^2).

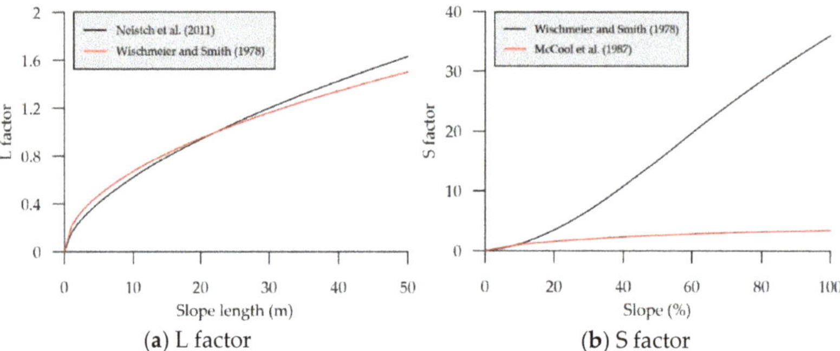

(**a**) L factor   (**b**) S factor

Figure 5. Comparison of L factor and S factor.

It was reported that the percentage of landslide and debris flow in total sediment load would be greater than 60% when the daily cumulative precipitation is higher than 350 mm [58]. Therefore, the landslide volume estimation is triggered only when the daily precipitation exceeds 350 mm. The landslide area is read into the transformed landslide volume estimate equation as follows (Equation (20)).

$$V = e^{2.326} \cdot A^{0.687} \tag{20}$$

In this study, the landslide area was identified by the land use survey map. The landslide volume is calculated by Equation (19), and then added into the sediment yields of HRUs.

2.5. Model Calibration and Validation

We applied the SWAT-CUP (SWAT Calibration Uncertainty Program) [59] for model sensitivity analysis, calibration and validation. In SWAT-CUP, parameter uncertainty can be analyzed by using Sequential Uncertainty Fitting version 2 (SUFI2), Generalized Likelihood Uncertainty Estimation (GLUE), Particle Swarm Optimization (PSO), Parameter Solution (ParaSol), or Marko Chain Monte Carlo (MCMC). Each uncertainty analysis method can run multiple simulations and find the ranges of best parameters for the study project. Parameters at any particular sub-basin, land use, soil, slope, and even HRU can be individually calibrated to reflect the unique spatial characteristics.

Among these uncertainty analysis methods, SUFI2, ParaSol and GLUE are easier to calibrate the parameters [59]. It is suggested that 3000, 7500, 10,000, 100,000, and 45,000 times of simulation are needed for SUFI2, ParaSol, GLUE, PSO, and MCMC, respectively, in order to get satisfactory simulation results [59]. SUFI2 method was selected in this study as it requires the least number of simulations. In sensitivity analysis, *p*-value is used to distinguish whether parameters are sensitive or not. Parameters that have *p*-value smaller than or equal to 0.05 are considered as sensitive parameters for further calibration [59]. After sensitivity analysis, the selection of calibrated ranges and fitted values of the parameters are identified based on the statistical criteria (i.e., R^2, NSE, PBIAS, and RSR), suggested by [60] with the model performance standards (Table 4).

Table 4. Statistical criteria for model performance [60].

Model Performance	NSE	PBIAS (%)			RSR
		Flow	Sediment	Nutrient	
Very Good	$0.75 < \text{NSE} \leq 1.00$	$|\text{PBIAS}| < 10$	$|\text{PBIAS}| < 15$	$|\text{PBIAS}| < 25$	$0.00 \leq \text{RSR} \leq 0.50$
Good	$0.65 < \text{NSE} \leq 0.75$	$10 \leq |\text{PBIAS}| < 15$	$15 \leq |\text{PBIAS}| < 30$	$25 \leq |\text{PBIAS}| < 40$	$0.50 < \text{RSR} \leq 0.60$
Satisfactory	$0.5 < \text{NSE} \leq 0.65$	$15 \leq |\text{PBIAS}| < 25$	$30 \leq |\text{PBIAS}| < 55$	$40 \leq |\text{PBIAS}| < 70$	$0.60 < \text{RSR} \leq 0.70$
Unsatisfactory	$\text{NSE} \leq 0.5$	$|\text{PBIAS}| \geq 25$	$|\text{PBIAS}| \geq 55$	$|\text{PBIAS}| \geq 70$	$\text{RSR} > 0.70$

R^2 is the coefficient of determination, presenting the linear correlation between simulated and observed data. The value of R^2 closer to 1 indicates a higher correlation.

$$R^2 = \frac{\sum_{i=1}^{n}\left(Y_i^{sim} - Y^{mean}\right)^2}{\sum_{i=1}^{n}\left(Y_i^{obs} - Y^{mean}\right)^2} \tag{21}$$

where Y^{sim} is the simulated data; Y^{obs} is observed data; Y^{mean} is the average of observation.

Nash–Sutcliffe efficiency (NSE) presents the residuals of measured data [61], and the value ranges from $-\infty$ to 1. NSE value that equals 1, indicates the simulation is same as the observation, while NSE > 0.5 is acceptable for SWAT model performance [60].

$$\text{NSE} = 1 - \left[\frac{\sum_{i=1}^{n}\left(Y_i^{obs} - Y_i^{sim}\right)^2}{\sum_{i}^{n}\left(Y_i^{obs} - Y^{mean}\right)^2}\right] \tag{22}$$

Percent bias (PBIAS) presents whether the simulated data are overestimated or underestimated. When PBIAS is greater than 0, the simulation is underestimated [62].

$$\text{PBIAS}(\%) = \frac{\sum_{i=1}^{n}\left(Y_i^{obs} - Y_i^{sim}\right)}{\sum_{i=1}^{n} Y_i^{obs}} \times 100 \tag{23}$$

RMSE-observation standard deviation ratio (RSR) is the ratio between root mean square error and standard deviation. The smaller the RSR is, the better simulation performance is [63].

$$\text{RSR} = \frac{\sqrt{\sum_{i=1}^{n}\left(Y_i^{obs} - Y^{sim}\right)^2}}{\sqrt{\sum_{i=1}^{n}\left(Y_i^{obs} - Y^{mean}\right)^2}} \tag{24}$$

2.6. Sediment Load Estimation

In order to calibrate the daily sediment load, we estimated the observed daily sediment load by using the sediment rating curve, which describes the relationship between sediment concentration and water discharge. Since the Chenyulan watershed is a small mountainous watershed in Asia, we adopted the sediment rating curve method suggested by [64], who conducted studies in small mountainous watersheds in Japan.

We collected discontinuous data of sediment concentration and corresponding instant streamflow at the Nemoupu gauging station from 2004 to 2015 with a total of 338 data points. Both streamflow and sediment concentration data were first converted into logarithm and fitted with the linear regression model. A good relationship (r = 0.75) between sediment concentration and streamflow was found (Figure 6). Thus, the sediment rating curve was used for estimating the daily mean sediment concentration with the observed daily streamflow data, and the estimated daily sediment load could be further calculated.

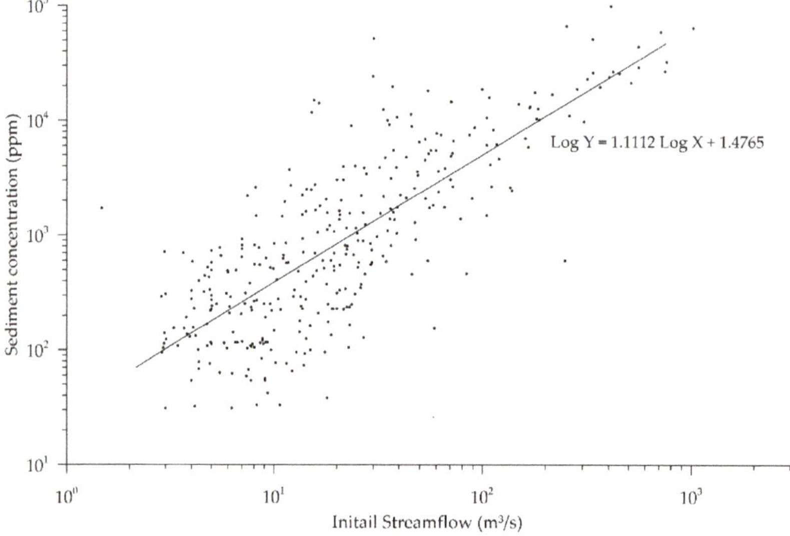

Figure 6. Sediment rating curve.

3. Results

In this study, we compared three SWAT models, which are the official SWAT 2016 (version 664), SWAT-TUSLE and SWAT-Twn. The SWAT-TUSLE was modified with TUSLE which calculates C factor based on NDVI and L factors based on the slope. The SWAT-Twn was the integration of SWAT-TUSLE and landslide volume equation. Since we did not modify the streamflow-related equations in SWAT, the streamflow simulations are the same for all these three models. We first calibrated the official SWAT 2016 model for daily streamflow and compared the performance of sediment load simulations from SWAT 2016, SWAT-TUSLE, and SWAT-Twn.

3.1. Model Calibration and Validation for Streamflow

The observed streamflow from 2004 to 2015 was used for model calibration and validation. The streamflow-related parameters for calibration were referred to the previous study [30]. The parameters include curve number (CN2), plant uptake compensation factor (EPCO), surface runoff lag time (SURLAG), baseflow alpha factor (ALPHA_BF), effective hydraulic conductivity in main channel alluvium (CH_K2), and Manning's "*n*" value for the main channel (CH_N2). In addition, we included Manning's "*n*" value for the tributary channel (CH_N1) and effective hydraulic conductivity in tributary channel alluvium (CH_K1) as they are also sensitive in this study. In order to differentiate the characteristics of the parameters in various land use, slope, and soil, some parameters (i.e., CN2, ALPHA_BF, CH_K1, CH_K2, CH_N1, CH_N2, CH_K1) were individually calibrated for specific land use, slop and soil. Table 5 shows the calibrated ranges and fitted parameter values for daily streamflow simulation. The model did satisfactory and good performance for the calibration and validation, respectively (Figure 7), indicating the fitted streamflow parameters in model could well reflect the runoff characteristics of the Chenyulan watershed.

Table 5. Calibration results for streamflow.

Parameter	Unit	Class [1]	Default	Calibrated Value		
				Fitted	Min.	Max.
		FRST	60	37 (−38.3%)	35 (−41.9%)	57 (−4.7%)
CN2	-	RNGE	69	42 (−38.9%)	39 (−44.1%)	61 (−12.1%)
		AGRL	77	47 (−38.9%)	43 (−44.1%)	68 (−12.1%)
EPCO	-	All	1	0.34	0.10	0.44
SURLAG	-	All	4	20.77	3.24	23.11
ALPHA_BF	1/day	Downstream	0.048	0.31	0.18	0.53
		Sub. mean slope > 60%		0.38	0.06	0.43
CH_K2	mm/h	Downstream	0	543.86	427.89	811.25
		Sub. mean slope > 60%	0	546.43	386.38	744.02
		Head stream	0	515.61	293.20	579.25
CH_N2	-	Sub. mean slope > 60%	0.014	0.13	0.10	0.25
		Head stream		0.37	0.24	0.46
CH_K1	-	Downstream	0	58.88	53.76	161.54
CH_N1	-	Sub. mean slope > 60%	0.014	11.02	2.52	11.08

[1] FRST, RNGE, AGRL denote forest, grassland, agricultural land, respectively; three groups of sub-basin by slope are downstream (sub-basin no. 1–3, 5–7, 9–11, 13–15), sub-basin mean slope that greater than 60% (sub-basin no. 4, 8, 12, 16, 18, 19) and head stream (sub-basin no. 17, 20–23).

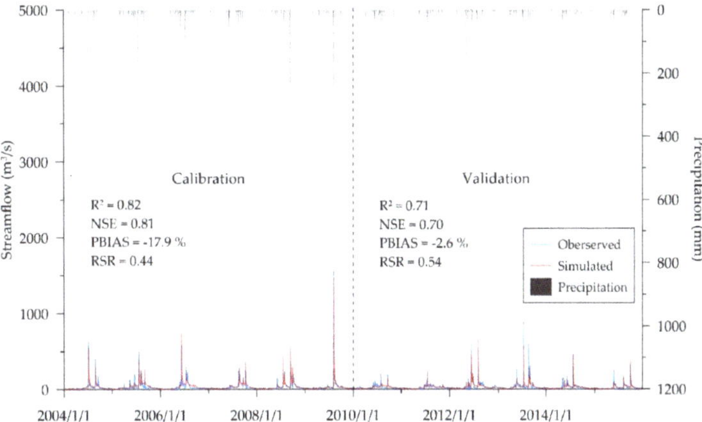

Figure 7. Streamflow calibration and validation results.

3.2. Comparison of SWAT 2016 and Modified SWAT Models

After calibrating the daily streamflow, we compared the uncalibrated simulated sediment yields from SWAT 2016, SWAT-TUSLE, and SWAT-Twn to quantify the impacts of using TUSLE and landslide volume equation on sediment yields at HRU and watershed levels (Tables 6 and 7). It should be noted that we only used the sediment yield data during the streamflow calibration period because the fitted parameter values during validation period are different than those during calibration period. However, the range of parameter values are the same for both calibration and validation periods, and the simulation results can reflect the model uncertainty. Thus, we used the simulated sediment yield data during the calibration period (2004–2009) with calibrated fitted streamflow-related parameters and default sediment-related parameters to demonstrate the difference driven by different models.

Table 6. Sediment yields at hydrologic response unit (HRU) level.

Land Use	Sediment Yield (t/ha)			Difference (%)	
	SWAT 2016	SWAT-TUSLE	SWAT-Twn	SWAT-TUSLE	SWAT-Twn
Urban	315.8	130.3	130.3	−58.7	−58.7
Agricultural land	1201.0	1196.0	1196.0	−0.4	−0.4
Forest	212.6	724.4	724.4	240.7	240.7
Grassland	271.1	933.1	933.1	244.2	244.2
Landslide	5461.6	1949.1	6872.3	−64.3	25.8

Table 7. Sediment yields at watershed level.

Year	Sediment Yield (t/ha)			Difference (%)	
	SWAT 2016	SWAT-TUSLE	SWAT-Twn	SWAT-TUSLE	SWAT-Twn
2004	838	1335	1411	59.2	68.3
2005	756	1295	1320	71.3	74.6
2006	1028	1574	1635	53.1	59.0
2007	671	1133	1164	68.7	73.3
2008	1111	1561	1623	40.5	46.2
2009	746	1336	1382	79.2	85.3
Average	858	1372	1422	59.9	65.7

The major differences between SWAT 2016 and the two other modified SWAT (SWAT-TUSLE and SWAT-Twn) are the LS factor, which has more influence in steep slope areas (slope > 9%), and the C factor, which was calculated by NDVI resulting various C factor values for different land uses. There were 77.12% and 80.29% of urban and agricultural lands located in areas with steeper slope (slope > 9%). Therefore, with unchanged C factor for urban, sediment yield from urban had decreased by approximate 60% due to the modified LS factors in TUSLE (Table 6). However, sediment yields from agricultural lands did not change significantly by modified SWAT models. It was because the C factor calculated by NDVI is doubled than the SWAT default value (Table 3), compensating the decrease in sediment yields by modified LS factor in TUSLE. Besides urban and agricultural lands, significant changes in sediment yields from forest, grassland and landslide were found. Although LS factors could influence the sediment yields, the C factor of forest and grassland which were changed from 0.001–0.003 to 0.2, played an important role in increases in sediment yields.

In the SWAT-Twn model, the landslide volume estimation is activated when the daily precipitation reaches over 350 mm. It should be noted that landslide volume estimation is only applied to the landslide area, not other land uses. It is obvious that sediment yields from landslide area increased significantly when the landslide volume estimation was activated in SWAT-Twn. Moreover, since forest is the main land use occupying 74.46% of the watershed and the NDVI-calculated C factor of forest is greater than SWAT default C factor, the annual sediment yields from the watershed were increased by 59.9% and 65.7% by SWAT-TUSLE and SWAT-Twn, respectively (Table 7). The increase of 5.8% of sediment yield by SWAT-Twn was due to landslide volume estimation at the landslide areas. It shows that landslide volume estimation should be considered as the major contribution to sediment yields.

Before calibrating the sediment, these models overestimated the daily sediment load in terms of great positive PBIAS values (Figure 8). However, SWAT-TUSLE and SWAT-Twn performed better than SWAT 2016 in terms of greater R^2, NSE and smaller RSR, especially the SWAT-Twn performance had better statistical criteria values (R^2 = 0.74, NSE = 0.66, RSR = 0.58). Therefore, we used SWAT-Twn for further sediment calibration and validation.

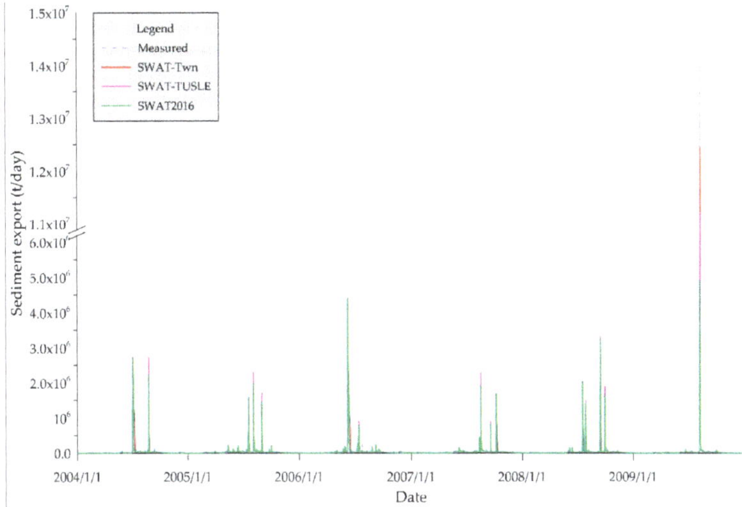

Figure 8. Simulated daily sediment yield by three models with calibrated streamflow.

3.3. Model Calibration and Validation for Sediment

The SWAT-Twn model was calibrated and validated for sediment loads with five different sediment transport methods (i.e., EQN0–4). The calibration (2004–2009) and validation (2010–2015) periods for sediment loads were the same as those for streamflow. First, the sensitivity analyses for sediment parameters for different sediment transport methods were examined (Table 8). A total of eight sediment-related parameters were identified as sensitive parameters. Four of these parameters (i.e., SPCON, SPEXP, PRF_BSN, ADJ_PKR) are estimated on basin level (*.bsn), meaning that the parameter values are fixed for the entire watershed; while the rest of parameters (i.e., CH_COV1, CH_BNK_D50, CH_BED_D50) are estimated on reach level (*.rch), which could be varied by spatial and slope conditions.

Table 8. The sensitivity analysis of sediment-related parameters (p-value < 0.05).

Parameter	File Name	EQN0	EQN1	EQN2	EQN3	EQN4
SPCON	*.bsn	V	V			
SPEXP	*.bsn	V	V			
PRF_BSN	*.bsn	V	V			
ADJ_PKR	*.bsn	V			V	V
CH_COV1	*.rte			V [3]	V [3]	V [3]
CH_BNK_D50 [1]	*.rte					V [3]
CH_BED_D50 [1]	*.rte			V [2]	V [3]	V [3]

[1] unit: μm; [2] the parameter is sensitive for sub-basins with mean slope < 60%; [3] the parameter is sensitive for sub-basins with mean slope > 60%.

The linear parameter (SPCON), exponent parameter (SPEXP) and peak rate adjustment factor (PRF_BSN) are only activated for the simplified Bagnold equation (EQN0 and EQN1). The peak rate adjustment factor (ADJ_PKR) was found to be sensitive to EQN0, EQN3 and EQN4. In order to identify the difference in the reach-level parameters, we separated the watershed by slope of 60% as almost half (49.58%) of the Chenyulan watershed is at a slope greater than 60%.

The channel bank vegetation coefficient for shear stress (CH_COV1) at the sub-basins with mean slope greater than 60% was sensitive for EQN2, EQN3 and EQN4, indicating that the vegetation at steeper slope areas would have great influence on sediment load compared to that at flatter slope areas. The median particle diameter of channel bank (CH_BNK_D50) at steeper slope areas was only sensitive for EQN4, while the median particle diameter of channel bed (CH_BED_D50) were sensitive for EQN1, EQN2, and EQN3. It shows that the median particle diameter of channel bank or bed should be measured for increasing the accuracy of sediment simulation with EQN1-4. Although both EQN0 and EQN1 are simplified Bagnold stream power equations, the EQN0 (default in SWAT 2005 version) does not keep track of sediment pools in various particle sizes, while the EQN1 (additional stream power equation in SWAT 2016) has been incorporated with physics-based approach for channel erosion. Moreover, the simulation of channel erosion with EQN0 is not partitioned between stream bank and stream bed. Thus, both CH_BNK_D50 and CH_BED_D50 are not sensitive for EQN0, while the EQN1 was sensitive with bed erosion (CH_BED_D50).

After identifying the sensitive parameters for those five sediment transport methods, the SWAT-Twn was calibrated and validated separately with each sediment transport method (Table 9 and Figure 9). Generally, the simulation results by EQN0 and EQN1 were better than those by other sediment transport methods, in terms of R^2 and NSE greater than 0.5. It indicates that Bagnold equation is more suitable for the Chenyulan watershed. Moreover, the SWAT-Twn with EQN2, EQN3 or EQN4 was found to underestimate for peak flows and overestimate for low flows (Figure 9). Thus, the overestimated low flows at the most flow period resulted in great negative PBIAS values.

Table 9. Statistical results for sediment calibration and validation.

Statistical Parameter	EQN0		EQN1		EQN2		EQN3		EQN4	
	Cal. [1]	Val. [2]	Cal.	Val.	Cal.	Val.	Cal.	Val.	Cal.	Val.
R^2	0.74	0.55	0.66	0.66	0.24	0.57	0.28	0.58	0.34	0.57
NSE	0.73	0.73	0.65	0.65	0.22	0.22	0.25	0.25	0.31	0.31
PBIAS (%)	−89.8	−81.4	−116.8	−74.5	−164.1	−216.4	−226.5	−168.7	−240.8	−314.1
RSR	0.52	0.52	0.59	0.59	0.88	0.88	0.86	0.86	0.83	0.83

[1] Calibration; [2] Validation.

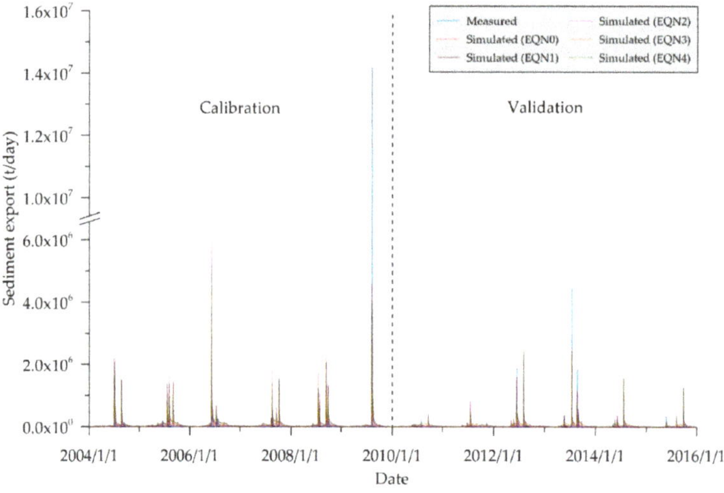

Figure 9. Sediment calibration and validation by different sediment transport methods.

4. Discussion

4.1. Improvement of SWAT-Twn Model for Watershed Sediment Production

In the nearly-flat upland watersheds where surface runoff seldom occurs, sediment transported by surface runoff is usually significantly over predicted in a model [65]. Since the Chenyulan watershed is mountainous with hilly slope, modeling the sediment yield and transport responses with considering unique hydrological responses is important. Compared to SWAT 2016, the major modifications in SWAT-TUSLE are C factor and LS factor. The main land use in the Chenyulan watershed is forest, for which C factor is 0.2 by NDVI calculation and much greater than the SWAT default C factor (0.001). Although the decrease in LS factor would cause less sediment yields, the increase in sediment yield by the adjusted C factor was much greater than the decrease in sediment yield by the adjusted LS factor. Thus, it was found that some sub-basins (i.e., sub-basins no. 5, 8, 9, 12, 13, 16–19) where forest is dominated (>70%) generated more sediment yields (Figure 10 and Table 10). Moreover, some agricultural-dominated (>30%) sub-basins (i.e., sub-basins no. 1, 3, 7, 10, 11 15) are mostly located in central and northern parts of the watershed, where slope is less steep. Thus, the adjusted LS factor has a relatively small impact on sediment yields from agricultural-dominated sub-basins. Great change in sediment yield was found in the south part of the watershed, where both forest and grassland are dominated, as the adjusted C factors of pasture and urban are greater than the SWAT default ones. Since the landslide volume equation is the additional improvement in SWAT-Twn from SWAT-TUSLE, the sediment yields by SWAT-Twn were increased by 2%, 40% and 3% in the sub-basins no. 1, 7, and 17, respectively, where the landslide area is greater than 5% of the sub-basin area.

Figure 10. Sediment yields at sub-basins simulated by three models.

4.2. Calibration with Different Sedmient Transport Methods

Based on the sensitivity analysis (Table 8), the calibrated values are listed in Table 11. For EQN0 and EQN1, the calibrated ranges and fitted values of SPCON, SPEXP and PRF_BSN are similar. Moreover, ADJ_PKR shows similar fitted value (0.56–0.63) and calibrated ranges (0.5–1.5) for EQN0, EQN3, and EQN4, indicating the characteristics of peak flow could be well represented in different sediment transport methods. For the simplified Bagnold method (EQN0 and EQN1), channel erodibility is controlled by the channel erodibility factor (CH_COV1) ranging from 0 to 1. The default value (0) indicates non-erosive channel, while the value of 1 indicates no resistance to erosion. However, CH_COV1, which is conceptually similar to the soil erodibility factor used in the USLE equation, was not sensitive and thus we used the default value for the simulation with EQN0 and EQN1. For other physics-based methods (EQN2-4), the channel erodibility is calculated by shear stress, and the CH_COV1 is defined as channel bank vegetation coefficient for estimating critical shear stress [66]. CH_COV1 is sensitive for EQN2, EQN3, and EQN4, however, the fitted values are quite different. The fitted CH_COV1 values indicate that the channel vegetation is between sparse trees (CH_COV1

= 5.40) and dense trees (CH_COV1 = 19.20), sparse trees, and grassy (CH_COV1 = 1.97) for EQN2, EQN3, and EQN4, respectively. The smaller the CH_COV1 value is, the greater the channel erodibility coefficient is [44]. Interestingly, the reflection of CH_COV1 on the channel erodibility is consistent with the suitability of using the Kodatie model (EQN2) for the stream bed material size ranging from silt to gravel, Molinas and Wu model (EQN3) for primarily sand size particles, and Yang sand and gravel model (EQN4) for primarily sand and gravel size particles.

Table 10. Land use distribution (%) in sub-basins.

Sub-Basin	Area (km^2)	Water	Grassland	Urban	Agricultural Land	Landslide	Forest
1	34.67	10.54	0	3.17	40.43	5.29	40.05
2	16.03	7.15	0	1.98	29.51	1.1	59.2
3	0.75	23.49	0	14.61	35.85	0	26.52
4	30.65	6.36	4.75	0.27	17.29	4.57	66.88
5	24.63	0.15	0	0.12	20.79	1.3	77.02
6	30.04	5.07	0.06	1.03	32.2	3.11	58.91
7	1.84	30.92	0	0	44.87	13.23	11.46
8	43.31	2.04	10.62	0.03	1.14	2.96	83.54
9	17.11	1.89	7.76	0.54	11.3	2.16	75.71
10	14.31	8.79	0.28	2.51	32.45	4.06	52.38
11	0.97	18.34	0	5.25	43.05	1.64	32.2
12	10.3	0	6.63	0	0.88	1.76	91.21
13	14.73	1.27	7.5	0	16.72	3.8	70.71
14	4.76	5.85	0	3.98	23.88	0.86	65.9
15	9.7	5.57	0	1.78	31.11	2.34	59.68
16	27.21	2.82	4.56	0.86	11.92	0.79	79.52
17	16	1.04	12.38	0	4.22	6.53	75.31
18	24.09	3.05	4.16	1.3	9.68	2.02	80.26
19	13.1	2.49	4.21	0	6.22	1.97	85.59
20	33.65	0.01	7.23	0	0.46	3.51	89
21	42.49	1.7	3.29	0.14	3.97	3.02	87.51
22	17.79	0	3.55	0	0.16	1.05	95.16
23	20.68	0	6.27	0	0	0.76	93.01

Table 11. Calibrated sediment-related parameters for different sediment transport methods.

Parameter	EQN0	EQN1	EQN2	EQN3	EQN4
	Range (Fitted)	Range (Fitted)	Range (Fitted)	Range (Fitted)	Range (Fitted)
SPCON	0.01–0.10 (0.012)	0.01–0.10 (0.042)	-	-	-
SPEXP	1.02–1.20 (1.11)	1.00–1.40 (1.25)	-	-	-
PRF_BSN	1.00–1.20 (1.02)	0.80–1.20 (0.95)	-	-	-
ADJ_PKR	0.50–1.20 (0.61)	-	-	0.50–1.20 (0.63)	0.50–1.50 (0.56)
CH_COV1	-	-	7.00-19.00 (10.51)	5.00-15.00 (0.13)	1.00–9.00 (1.70)
CH_BNK_D50 [1]	-	-	-	-	6500–9000 (7956.25)
CH_BED_D50 [1]	-	500–6000 (2411)	100–1800 (605.75)	350–1000 (377.63)	-

[1] unit: μm.

Moreover, due to the lack of the information on channel materials, the calibrated median particle size diameter of channel bank sediment (CH_BNK_D50) or bed sediment (CH_BED_D50) showed various results among different sediment transport methods. For EQN1 and EQN4, the median size of bank and bed sediments were identified as mostly much greater than large aggregate (500 μm), thus the particle size distribution for D_{50} (>2000 μm) assumed by SWAT is 15% of sand, 15% of silt, 5% of clay and 65% of gravel [44]. For EQN2, the calibrated CH_BED_D50 range is between 250 μm and 2000 μm, indicating the channels are characterized as medium to very coarse sand-bed materials [40].

Similarly, for EQN3, the fitted CH_BED_D50 value is between sand (200 μm) and large aggregate (500 μm) [44]. Therefore, for EQN2 and EQN3, the particle size distribution for D_{50} (50 to 2000 μm) assumed by SWAT is 65% of sand, 15% of silt, 15% of clay and 5% of gravel.

In SWAT, the particle size distribution is usually used for estimating the bank and bed erosion, and the percentage of sediment (sand, silt, clay, gravel, small aggregate and large aggregate) that gets deposited, is calculated by the fall velocity of median size particle. The amount of sediment transported out of the reach is calculated as follows: Amount of total load entering the reach at the beginning of the time period minus the amount of total load deposited in the reach plus the amount of sediment due to bank and bed erosion in the reach. The total load in the reach considered includes bed and suspended load coming from the foregoing reaches, as well as suspended load originating from the soil erosion of the surrounding sub-basin. Thus, further improvements (i.e., the assumption of particle size distribution for bank and bed erosion, the calculation of fall velocity for wide range of particle size) need to be done to calculate more reliable and accurate sediment loads in SWAT.

4.3. Selection of a Suitable Sediment Transport Method

Due to the fact that sediment loads are extremely high during heavy rainfall events, the comparison between measured and simulated data in linear-scaled plot was difficult to identify their difference in sediment loads of low values. Therefore, we applied the logarithmic scale for the measured and simulated sediment data with different sediment transport methods and compared their statistical criteria (R^2 and PBIAS) during 2004–2016 (Figure 11 and Table 12). It should be noted that the simulation results are the same in Figures 9 and 11. The simplified Bagnold method (EQN0 and EQN1) performed better for the low sediment loads with smaller PBIAS values. Although the model has been calibrated for sediment by SWAT-CUP, sediment loads are still overestimated for all sediment transport methods. It is because the SWAT-CUP model tends to adjust the parameters to meet the high observed sediment loads, but results in great PBIAS values. Thus, the SWAT model simulated high sediment loads better than the low sediment loads. Interestingly, when the measured and simulated sediment data are plotted in logarithmic scale, the R^2 values increase and PBIAS values decrease (Table 12). The greater improvement in the statistical criteria values shows the higher degree of overestimation when applying the Molinas and Wu model (EQN3) and Yang sand and gravel model (EQN4). By comparing the sediment load data in linear and logarithmic scales, we can better identify the suitable sediment transport method and suggest that further calibration and validation are needed for log-transformed simulated sediment data in the SWAT-CUP model.

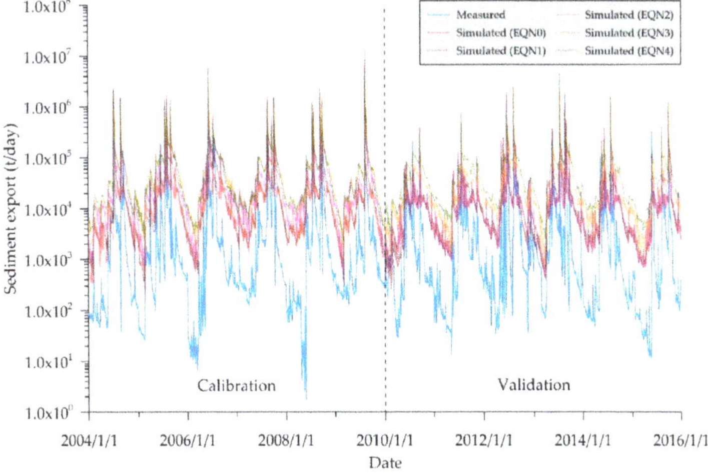

Figure 11. Comparison of simulated and measured sediment loads in logarithmic scale.

Table 12. Statistical results for simulated sediment loads in linear and logarithmic scales.

Method	Plot Scale	R^2	PBIAS (%)
EQN0	Linear	0.71	−86.8
	Logarithmic	0.64	−34.3
EQN1	Linear	0.66	−101.6
	Logarithmic	0.68	−36.4
EQN2	Linear	0.28	−182.8
	Logarithmic	0.68	−48.6
EQN3	Linear	0.31	−205.7
	Logarithmic	0.64	−51.3
EQN4	Linear	0.37	−267.1
	Logarithmic	0.70	−55.4

5. Conclusions

The Chenyulan watershed has suffered from serious landslide and debris flow induced by heavy rainfall and typhoons. In this study, we integrated the TUSLE and landslide volume estimation into the SWAT model as SWAT-Twn. By evaluating the simulated sediment yields from different land uses, the importance of topographic (LS) factor and NDVI-calculated weighted C factor were identified and landslide volume estimation should be taken into concern. The examination of five different sediment transport methods revealed some important issues. First, the level of sensitivity of sediment-related parameters is different for those sediment transport methods, and parameters (i.e., CH_COV1, CH_BNK_D50, CH_BED_D50) that are estimated on each level, are suggested to be calibrated by spatial and slope conditions. Second, it is more accurate to investigate the channel vegetation (CH_COV1) and measure the particle sizes of channel bank and bed sediment (CH_BNK_D50 and CH_BED_D50). The calibrated parameter values by SWAT-CUP for different sediment transport methods may be misleading. Third, the particle size distribution assumed by SWAT is suggested to be an option that can be edited by users. Furthermore, the calculation of fall velocity is suggested to not be only limited for median particle size as it would be biased for channels of wide range of particle sizes. Last but not the least, like the streamflow simulation in SWAT and SWAT-CUP, an option for the user to compare and plot the sediment simulation in logarithmic scale would provide more insights into sediment calibration. In sum, the SWAT-Twn model performed better than SWAT 2016 and SWAT-TUSLE, as TUSLE calculated less sediment at steep area, resulting reasonable sediment export simulation at low flow condition and landslide volume estimation reflected the real situation. Additional improvements in SWAT and SWAT-CUP need to be made to better predict the sediment yields and loads at mountainous watersheds.

Author Contributions: Conceptualization, C.-M.L. and L.-C.C.; Formal analysis, C.-M.L.; Funding acquisition, L.-C.C.; Methodology, C.-M.L.; Resources, L.-C.C.; Software, C.-M.L.; Supervision, L.-C.C.; Validation, L.-C.C.; Visualization, C.-M.L.; Writing—original draft, C.-M.L. and L.-C.C.; Writing—review & editing, L.-C.C.

Funding: This research was funded by the Ministry of Science and Technology, Taiwan (MOST 107-2625-M-239-001).

Acknowledgments: We thank Wen-Cheng Liu from National United University, and Yung-Chieh Wang and Chia-Jeng Chen from National Chung Hsing University for constructive comments that improved the manuscript.

References

1. Garcia-Ruiz, J.M.; Begueria, S.; Nadal-Romero, E.; Gonzalez-Hidalgo, J.C.; Kana-Renault, N.; Sanjuan, Y. A meta-analysis of soil erosion rates across the world. *Geomorphology* **2015**, *239*, 160–173. [CrossRef]
2. Freebairn, D.M.; Loch, R.J.; Silburn, D.M. Chapter 9, Soil erosion conservation for vertisols. In *Developments in Soil Science*; Elsevier: Amsterdam, The Nederlands, 1996; Volume 24, pp. 303–362.

3. Garcia, M.H. *Sediment Engineering*; American Society of Civil Engineers: Reston, VA, USA, 2008.

4. Dadson, S.J.; Hovius, N.; Chen, H.; Dade, W.B.; Hsieh, M.-L.; Willett, S.D.; Hu, J.-C.; Horng, M.-J.; Chen, M.-C.; Stark, C.P.; et al. Links between erosion, runoff variability and seismicity in Taiwan orogen. *Nature* **2003**, *426*, 648–651. [CrossRef] [PubMed]

5. Mullan, D.; Matthews, T.; Vandaele, K.; Barr, I.D.; Swindles, G.T.; Meneely, J.; Boardman, J.; Murphy, C. Climate impacts on soil erosion and muddy flooding at 1.5 versus 2C warming. *Land Degrad. Dev.* **2019**, *30*, 94–108. [CrossRef]

6. Wischmeier, W.H.; Smith, D.D. Rainfall energy and its relationship to soil loss. *Trans. Am. Geophys. Union* **1958**, *39*, 285–291. [CrossRef]

7. Williams, J.R. Sediment yield prediction with Universal Equation using runoff equation. In *Present and Prospective Technology for Predicting Sediment Yields and Sources*; USA Department of Agriculture: New Orleans, LA, USA, 1975; pp. 244–252.

8. Sadeghi, S.H.R.; Gholami, L.; Khaledi Darvishan, A.; Saeidi, P. A review of the application of the MUSLE model worldwide. *Hydrol. Sci. J.* **2014**, *59*, 365–375. [CrossRef]

9. Beasley, D.D.; Huggins, L.F.; Monke, E.J. ANSWERS: A model for watershed planning. *Trans. ASAE* **1980**, *23*, 938–944. [CrossRef]

10. Williams, J.R.; Jones, C.A.; Dyke, P.T. A modeling approach to determining the relationship between erosion and soil productivity. *Trans. ASAE* **1984**, *27*, 129–144. [CrossRef]

11. Nearing, M.A.; Foster, G.R.; Lane, L.J.; Finkner, S.C. A process-based soil erosion model for USDA-Water erosion prediction project technology. *Trans. ASAE* **1989**, *32*, 1587–1593. [CrossRef]

12. Morgan, R.P.C.; Qionton, J.N.; Simuth, R.E.; Govers, G.; Poesen, J.W.A.; Auerswald, K.; Chisci, G.; Torri, D.; Styczen, M.E. The European Soil Erosion Model (EUROSEM): A dynamic approach for predicting sediment transport form fields and small catchments. *Earth Surf. Process. Landf.* **1998**, *23*, 527–544. [CrossRef]

13. Sadeghi, S.H.R.; Mizuyama, T. Applicability of the Modified Universal Soil Loss Equation for prediction of sediment yield in Khanmirza watershed, Iran. *Hydrol. Sci. J.* **2007**, *52*, 1068–1075. [CrossRef]

14. Merrtitt, W.S.; Letcher, R.A.; Jakeman, A.J. A review of erosion and sediment transport models. *Environ. Model. Softw.* **2003**, *18*, 761–799. [CrossRef]

15. Hairsine, P.B.; Rose, C.W. Modeling water erosion due to overland flow using physical principles: 1. Sheet flow. *Water Resour. Res.* **1992**, *28*, 237–243. [CrossRef]

16. Wheater, H.S.; Jakeman, A.J.; Beven, K.J. Progress and directions in rainfall-runoff modeling. In *Modelling Change in Environmental Systems*; Jakeman, A.J., Beck, M.B., McAleer, M.J., Eds.; John Wiley and Sons: Hoboken, NJ, USA, 1993; pp. 101–132.

17. Abbott, M.B.; Bathurst, J.C.; Cunge, J.A.; O'Connell, P.E.; Rasmussen, J. An introduction to the European Hydrological System-Système Hydrologique Europèen SHE. 1. History and philosophy of a physically-based, distributed modelling system. *J. Hydrol.* **1986**, *87*, 45–59. [CrossRef]

18. Storm, D.E.; White, M.J.; Stoodley, S. *Modeling Non-Point Source Component for the Fort Cobb TMDL Final Report*; Oklahoma Department of Environmental Quality: Oklahoma, OK, USA, 2003.

19. Pohlert, T.; Huisman, J.A.; Breuer, L.; Frede, H.G. Modeling of river Dill, Germany. *Adv. Geosci.* **2005**, *5*, 7–12. [CrossRef]

20. Yevenes, M.A.; Mannaerts, C.M. Seasonal and land use impacts on the nitrate budget and export of a mesoscale catchment in Southern Portugal. *Agric. Water Manag.* **2011**, *102*, 54–65. [CrossRef]

21. Arnold, J.G.; Srinivasan, R.; Muttian, R.S.; Williams, J.R. Large area hydrologic modeling and assessment, part I: Model development. *J. Am. Water Resour. Assoc.* **1998**, *34*, 73–89. [CrossRef]

22. Prosser, I.P.; Rustomji, P. Sediment transport capacity relations for overland flow. *Prog. Phys. Geogr.* **2000**, *24*, 179–193. [CrossRef]

23. Addis, H.K.; Strohmeier, S.; Ziadat, F.; Melaku, N.D.; Klik, A. Modeling streamflow and sediment using SWAT in Ethiopian Highlands. *Int. J. Agric. Biol. Eng.* **2016**, *9*, 51–66.

24. Bressiani, D.A.; Gassman, P.W.; Fernandes, J.G.; Garbossa, L.H.P.; Srinivasan, R.; Bonumá, N.B.; Mendiondo, E.M. Review of Soil and Water Assessment Tool (SWAT) applications in Brazil: challenges and prospects. *Int. J. Agric. Biol. Eng.* **2015**, *8*, 9–35.

25. Arnold, J.G.; Moriasi, D.N.; Gassman, P.W.; Abbaspour, K.C.; White, M.J.; Srinivasan, R.; Santhi, C.; Harmel, R.D.; van Griensven, A.; Van Liew, M.W.; et al. SWAT: model use, calibration, and validation. *Am. Soc. Agric. Biol. Eng.* **2012**, *55*, 1491–1508.

26. Lee, T.-Y.; Huang, J.-C.; Lee, J.-Y.; Jien, S.-H.; Zehetner, F.; Kao, S.-S. Magnified sediment export of small mountainous rivers in Taiwan: chain reaction from increased rainfall intensity under global warming. *PLoS ONE* **2015**, *10*, e0138283. [CrossRef]

27. Chiu, Y.-J.; Lee, H.-Y.; Wang, T.-L.; Yu, J.; Lin, Y.-T.; Yuan, Y. Modeling sediment yields and stream stability due to sediment-related disaster in Shihmen Reservoir watershed in Taiwan. *Water* **2019**, *11*, 332. [CrossRef]

28. Chang, C.-T.; Harrison, J.F.; Huang, Y.-C. Modeling typhoon-induced alterations on river sediment transport and turbidity based on dynamic landslide inventories: Gaoping River Basin, Taiwan. *Water* **2015**, *7*, 6910–6930. [CrossRef]

29. Chiang, L.-C.; Lin, Y.-P.; Huang, T.; Schmeller, D.S.; Verburg, P.H.; Liu, Y.-L.; Ding, T.S. Simulation of ecosystem service responses to multiple disturbances from an earthquake and several typhoons. *Landsc. Urban Plan.* **2014**, *122*, 41–55. [CrossRef]

30. Chiang, L.-C.; Chuang, Y.-T.; Han, C.-C. Integrating landscape metrics and hydrologic modeling to assess the impact of natural disturbances on ecohydrological processes in the Chenyulan watershed, Taiwan. *Int. J. Environ. Res. Public Health* **2019**, *122*, 266. [CrossRef]

31. Typhoon Database-Taiwan Central Weather Bureau. Available online: https://rdc28.cwb.gov.tw/TDB (accessed on 21 June 2019).

32. Abbaspour, K.C.; Rouhokahnejad, E.; Vaghefi, S.; Srinivasan, R.; Yang, H.; Klove, B. A continental-scale hydrology and water quality model for Europe: calibration and uncertainty of a high-resolution large-scale SWAT model. *J. Hydrol.* **2015**, *524*, 733–752. [CrossRef]

33. Cibin, R.; Trybula, E.; Chaubey, I.; Brouder, S.M.; Volenec, J.J. Watershed-scale impacts of bioenergy crops on hydrology and water quality using improved SWAT model. *Glob. Chang. Biol. Bioenergy* **2016**, *8*, 837–848. [CrossRef]

34. Liu, R.; Xu, F.; Zhang, P.; Yu, W.; Men, C. Identifying non-point source critical source areas based on multi-factors at a basin scale with SWAT. *J. Hydrol.* **2016**, *533*, 379–388. [CrossRef]

35. Nguyen, H.H.; Reckmagel, F.; Meyer, W.; Frizenschaf, J.; Shrestha, M.K. Modelling the impacts of altered management practices, land use and climate changes on the water quality of the Millbrook catchment-reservoir system in South Australia. *J. Environ. Manag.* **2017**, *202*, 1–11. [CrossRef]

36. Jang, S.S.; Ahn, S.R.; Kim, S.J. Evaluation of executable best management practices in Haean highland agricultural catchment of South Korea using SWAT. *Agric. Water Manag.* **2017**, *180*, 224–234. [CrossRef]

37. Meng, X.; Wang, H.; Shi, C.; Wu, Y.; Ji, X. Establishment and evaluation of the China Meteorological Assimilation Driving Datasets for the SWAT model (CMADS). *Water* **2018**, *10*, 1555. [CrossRef]

38. Taiwan Geospatial One Stop (TGOS). Available online: https://www.tgos.tw (accessed on 21 June 2019).

39. Taiwan Data Bank of Atmospheric and Hydrologic Research (DBAR). Available online: https://www.dbar.pccu.edu.tw (accessed on 21 June 2019).

40. Annual Hydrological Year Book of Taiwan, Republic of China. Available online: https://gweb.wra.gov.tw/wrhygis (accessed on 21 June 2019).

41. Taiwan MAP Service-National Land Surveying and Mapping Center. Available online: https://maps.nlsc.gov.tw (accessed on 21 June 2019).

42. Taiwan Soil Resource and Cultivate Land Cover Map Searching System. Available online: https://farmcloud.tari.gov.tw/SOA (accessed on 21 June 2019).

43. Srinivasan, R.; Arnold, J.G. Integration of a basin-scale water quality model with GIS. *Water Resour. Bull.* **1994**, *30*, 453–462. [CrossRef]

44. Neitsch, S.L.; Arnold, J.G.; Kiniry, J.R.; Williams, J.R. *Soil and Water Assessment Tool Theoretical Documentation*; Texas Water Resources Institute Technical Report No. 406; Texas Water Resources Institute: Temple, TX, USA, 2011.

45. Posada-Garcia, L. Transport of sands in Deeps Rivers. Ph.D. Thesis, Colorado State University, Fort Collins, CO, USA, 1995.

46. Molinas, A.; Wu, B. Transport of sediment in large sand-bed rivers. *J. Hydraul. Res.* **2001**, *39*, 135–146. [CrossRef]

47. Yang, C.T. *Sediment Transport Theory and Practice*; The McGaw-Hill Companies, Inc.: New York, NY, USA, 1996.

48. Chen, S.-C.; Wu, C.-Y.; Wu, Y.-L.; Wang, S.-H. Taiwan Universal Soil Loss Equation (TUSLE) based on revised factors and GIS layers—an example from the Shihmem Reservoir Watershed. *J. Chin. Soil Water Conserv.* **2009**, *40*, 185–197. (In Chinese)

49. Wan, S.-S.; Huang, J.-I. Soil erosion of slope in Taiwan. *J. Chin. Soil Water Conserv.* **1989**, *20*, 127–144. (In Chinese)

50. Hsieh, J.-S.; Wang, M.-G. *Major Soil Maps in Taiwan*; National Chung Hsing University Soil Survey and Testing Center: Taiwan, 1991. (In Chinese)

51. Lin, W.-T. Automated Watershed Delineation for Spatial Information Extraction and Slopeland Yield Evaluation. Ph.D. Thesis, National Chung Hsing University, Taiwan, 2002. (In Chinese).

52. McCool, D.K.; Brown, L.C.; Foster, G.R.; Mutchler, C.K.; Meyer, L.D. Revised slope steepness factor for the Universal Soil Loss Equation. *Trans. ASAE* **1987**, *30*, 1387–1396. [CrossRef]

53. Wischmeier, W.H.; Smith, D.D. Predicting rainfall erosion losses—A guide to conservation planning. In *Agriculture Handbook No. 537*; USA Department of Agriculture: New Orleans, LA, USA, 1978.

54. Hovius, N.; Stark, C.P.; Allen, P.A. Sediment flux from a mountain belt derived by landslides mapping. *Geology* **1997**, *25*, 231–234. [CrossRef]

55. Guzzetti, F.; Ardizzone, F.; Cardinali, M.; Galli, M.; Reichenbach, P.; Rossi, M. Distribution of landslides in the Upper Tiber River basin, central Italy. *Geomorphology* **2008**, *96*, 105–122. [CrossRef]

56. Dong, J.J.; Lee, C.T.; Tung, Y.H.; Liu, C.N.; Lin, K.P.; Lee, Y.J. The role of sediment budget in understanding debris flow susceptibility. *Earth Surf. Process Landf.* **2009**, *34*, 1612–1624. [CrossRef]

57. Chang, C.-W.; Lin, P.-S.; Tsai, C.-L. Estimation of sediment volume of debris flow caused by extreme rainfall in Taiwan. *Eng. Geol.* **2011**, *123*, 83–90. [CrossRef]

58. Liu, G.-F.; Lai, J.-S.; Chen, Y.-C.; Chiu, Y.-J.; Shen, C.-W.; Liang, W.-S.; Chiang, M.-H. *A Study of Flood Control and Sediment Management Due to Climate Change of Jhoushuei River*; Water Resource Agency: Taiwan, 2015. (In Chinese)

59. Abbaspour, K.C. *SWAT-CUP: SWAT Calibration and Uncertainty Programs—A User Manual*; Swiss; EAWAG: Dübendorf, Switzerland, 2015.

60. Moriasi, D.N.; Arnold, J.G.; Van Liew, M.W.; Bingner, R.L.; Haemel, R.D.; Veith, T.L. Model evaluation guidelines for systematic qualification of accuracy in watershed simulation. *Trans. ASABE* **2007**, *50*, 282–290. [CrossRef]

61. Nash, J.E.; Sutcliffe, J.V. River flow forecasting through conceptual models 1. A discussion of principles. *J. Hydrol.* **1970**, *10*, 282–290. [CrossRef]

62. Gupta, H.V.; Sorooshian, S.; Yapo, P.O. Status of automatic calibration for hydrologic models: Comparison with miltilevel expert calibration. *J. Hydrol. Eng.* **1999**, *4*, 135–143. [CrossRef]

63. Singh, J.H.; Knaoo, H.V.; Arnold, J.G.; Demissie, M. Hydrologic modeling of the Iroquois River watershed using HSPF and SWAT. *J. Am. Water Resour. Assoc.* **2004**, *41*, 361–375.

64. Sadeghi, S.H.R.; Mizuyama, T.; Miyata, S.; Gomi, T.; Kosugi, K.; Fukushima, T.; Mizugaki, S.; Onda, Y. Development, evaluation and interpretation of sediment rating curves for a Japanese small mountainous reforested watershed. *Geoderma* **2008**, *144*, 198–211. [CrossRef]

65. Mitchell, J.K.; Banasik, K.; Hirschi, M.C.; Cooke, R.A.C.; Kalita, P. There is not always surface runoff and sediment transport. In Proceedings of the International Symposium on Soil Erosion Research for the 21st Century, Honolulu, HI, USA, 3–5 January 2001.

66. Julian, J.P.; Torres, R. Hydraulic erosion of cohesive river banks. *Geomorphology* **2006**, *76*, 193–206. [CrossRef]

Article

A Process-Based, Fully Distributed Soil Erosion and Sediment Transport Model for WRF-Hydro

Dongxiao Yin [1], Z. George Xue [1,2,3,*], David J. Gochis [4], Wei Yu [5], Mirce Morales [6] and Arezoo Rafieeinasab [4]

[1]   Department of Oceanography and Coastal Sciences, Louisiana State University,
       Baton Rouge, LA 70803, USA; dyin2@lsu.edu
[2]   Center for Computation and Technology, Louisiana State University, Baton Rouge, LA 70803, USA
[3]   Coastal Studies Institute, Louisiana State University, Baton Rouge, LA 70803, USA
[4]   National Center for Atmospheric Research, Research Applications Laboratory, Boulder, CO 80305, USA;
       gochis@ucar.edu (D.J.G.); arezoo@ucar.edu (A.R.)
[5]   Weather Tech Services, LLC, Longmont, CO 80503, USA; weiyutyy@gmail.com
[6]   School of Engineering, National Autonomous University of Mexico, Mexico City CP 04510, Mexico;
       mirce.morales@comunidad.unam.mx
*   Correspondence: zxue@lsu.edu; Tel.: +1-225-578-1118; Fax: +1-225-578-6513

Received: 18 April 2020; Accepted: 24 June 2020; Published: 26 June 2020

Abstract: A soil erosion and sediment transport model (WRF-Hydro-Sed) is introduced to WRF-Hydro. As a process-based, fully distributed soil erosion model, WRF-Hydro-Sed accounts for both overland and channel processes. Model performance is evaluated using observed rain gauge, streamflow, and sediment concentration data during rainfall events in the Goodwin Creek Experimental Watershed in Mississippi, USA. Both streamflow and sediment yield can be calibrated and validated successfully at a watershed scale during rainfall events. Further discussion reveals the model's uncertainty and the applicability of calibrated hydro- and sediment parameters to different events. While an intensive calibration over multiple events can improve the model's performance to a certain degree compared with single event-based calibration, it might not be an optimal strategy to carry out considering the tremendous computational resources needed.

Keywords: WRF-Hydro; CASC2D-SED; Goodwin Creek Experimental Watershed; NLDAS-2; calibration

---

1. Introduction

Eroded by different forcing agents, soil is being lost at a rate that is orders-of-magnitude greater than its replenishment [1]. Among all the erosive agents, water is the most prevalent and usually dominates. Moreover, due to the current climate change, the frequency, and the intensity of extreme rainfall events are projected to increase, which will lead to more intensive erosion [2]. With this in mind, a lot of soil erosion models have been developed to mainly simulate water-induced erosion. Based on the numerical algorithms applied, these models can be classified as conceptual, empirical, and process-based models [3]. The latter, with detailed representation of physical processes, is becoming the mainstream in both academia and industry [4].

Depending on how processes and parameters are described, process-based models can be further grouped into semi-distributed and fully distributed models. Semi-distributed models, such as CREAMS (Chemical Runoff and Erosion from Agricultural Management Systems model [5]), WEPP (Watershed Erosion Prediction Project [6]), EUROSEM (EUROpean Soil Erosion Model [7]), KINEROS (KINematic EROsion Simulation [8]), and THREW (TsingHua Representative Elementary Watershed [9]) break down the model domain (a watershed or a catchment) into a series of basic elements such as hillslopes, planes and channels, over which the algorithms are applied and physical parameters are represented [10]. However, since hydrological parameters are lumped over the basic elements, semi-distributed models are not able to fully account for spatial heterogeneity. Fully distributed models divide a study domain into grid cells with certain sizes. The spatially distributed input data of fully distributed models are usually generated by the Geographical Information System (GIS). In this way, the physical heterogeneity is better represented. With the advances of scientific computation, several fully distributed, process-based models have been developed over the past three decades. Notable examples are LISEM (LImburg Soil Erosion Model [11]), SHESED [12] and CASC2D-SED (CASCade 2 dimensional sediment model [13]). Recently, a physically based hydrological and soil erosion model has been developed by coupling the Soil Conservation Service model with a 2D fully Dynamic Wave model and a Hillslope Erosion model by Juez et al. [2] The model is applied over a Mediterranean watershed to simulate the rainfall-runoff and soil erosion process during two rainfall events with satisfactory results generated. An efficient approach has been applied on the model for calibration process, which has largely reduced the computational cost. The model has demonstrated its potential applicability to large and long-term scale hydro-sedimentary process studies under climate change [2]. In addition, the distributed model usually simulates the sedimentary processes in 2D mode. In the study of the replenishment of sediments in a water-worked channel using the 2D shallow water equation model coupled with Exner equation, Juez points out that the 2D distributed model can better resolve the bidimensional water and sediment flux compared to the 1D model [14]. Moreover, the 2D model is more computationally efficient than the complicated 3D model while still meeting research and engineering requirements [14].

In this study, we present a newly developed, fully distributed, process-based soil erosion and sediment transport model. Our model is built on WRF-Hydro, which simulates the hydrological cycle and provides hydraulic parameters for soil erosion and transport. WRF-Hydro was developed as the hydrological modeling extension package of WRF at the National Center for Atmospheric Research in Boulder, Colorado [15]. Compared to other hydrological models such as the Variable Infiltration Capacity model (VIC, [16]) and the Soil and Water Assessment Tool (SWAT, [17]) model, WRF-Hydro's advantage is its capability of simulating multi-processes at multi-scales while considering the spatial distribution of hydrological variables. Besides, WRF-Hydro takes advantage of various available meteorological and terrain datasets and has been fully coupled with meteorological and climate models such as WRF. WRF-Hydro's performance has been evaluated by its applications in flooding [18], water resource management [19], water budget estimation [20], decadal scale hydroclimatic change [21], and others. Currently, an instance of WRF-Hydro is running operationally as the National Oceanic and Atmospheric Administration's (NOAA) National Water Model, which provides streamflow forecasts on 2.7 million river reaches of the contiguous United States. However, until this study, WRF-Hydro does not include a sediment module, which limits its capability in water quality-related forecasts and studies.

In this paper, the architecture and algorithms of the WRF-Hydro-Sed model, the study area, and datasets used as well as the calibration method are presented in Section 2. Results of model calibration and validation are detailed in Section 3. Section 4 discusses the model's uncertainty, the validity of applying calibrated parameters to different events, the relationship between landscape patterns and soil erosion as well as the limitation of current model regarding long-term simulation under climate change. A conclusion is given in Section 5.

2. Materials and Methods

2.1. WRF-Hydro

The structure and schematic representation of WRF-Hydro-Sed are shown in Figures 1 and 2, respectively. WRF-Hydro is an integrated modeling platform that incorporates a land surface model (LSM), grid aggregation/disaggregation, subsurface flow routing, surface overland flow routing, channel routing, lake and reservoir routing, and a conceptual base flow model. In this study the soil erosion and sediment transport processes are built on WRF-Hydro's overland flow and channel flow routings.

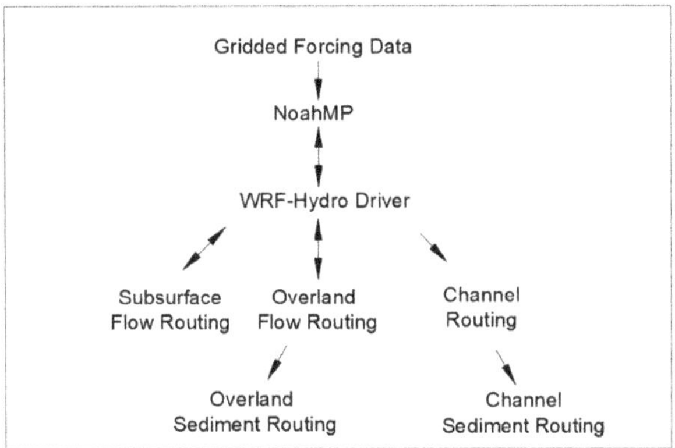

Figure 1. Schematic structure of WRF-Hydro-Sed.

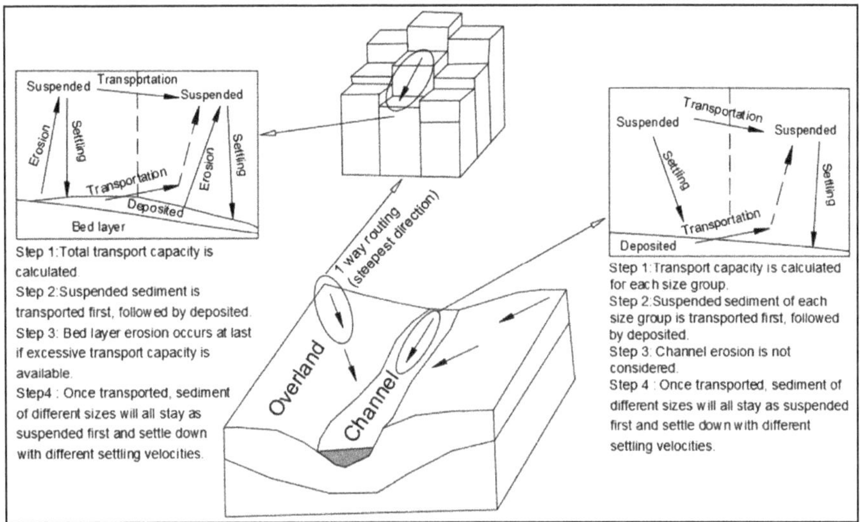

Figure 2. Schematic representation of WRF-Hydro-Sed.

2.1.1. Overland Flow Routing

In WRF-Hydro, the surface runoff is generated through infiltration-excess and saturation-excess of a supersaturated soil column. It can be routed either two-way or one-way (along the largest gradient of slope), depending on the routing method specified in the model name list. As the physics and algorithms of both routing options are identical, only the equations used for two-way routing are presented here. The overland flow is assumed to be fully unsteady and non-uniform. The diffusive wave formulation, which is a simplification of Saint-Venant equations for a shallow water wave, is applied:

Continuity Equation:

$$\frac{\partial h}{\partial t} + \frac{\partial q_x}{\partial x} + \frac{\partial q_y}{\partial y} = i_e \tag{1}$$

Momentum Equation:

$$\begin{aligned}
S_{fx} &= S_{ox} - \frac{\partial h}{\partial x} \\
S_{fy} &= S_{oy} - \frac{\partial h}{\partial y}
\end{aligned} \tag{2}$$

where h is the surface flow depth (L), q_x, q_y are the unit discharge in x and y direction, respectively ($L^2\,T^{-1}$) (in 1-way routing only discharge in the steepest direction is calculated), and i_e is the infiltration excess ($L\,T^{-1}$). S_{fx}, S_{fy} are the friction slope in x and y directions, S_{ox}, S_{oy} are the bed slope in x and y directions, and $\frac{\partial h}{\partial x}$, $\frac{\partial h}{\partial y}$ are the gradient of surface flow depth in x and y directions.

Manning's equation is used to calculate q_x and q_y in order to solve Equation (1),

$$\begin{aligned}
q &= \alpha h^{\beta}, \ \beta = 5/3 \\
\alpha &= \frac{S_f^{1/2}}{n}
\end{aligned} \tag{3}$$

where q is the unit discharge in x or y direction ($L^2\,T^{-1}$), h is the surface flow depth (L), S_f is the friction slope in x or y direction, n is the Manning roughness coefficient of land surface.

Since WRF-Hydro's performance is steady with the one-way routing in a parallel mode, we use the one-way overland routing method to simulate overland flow and provide the hydraulic parameters needed to drive the sediment model.

2.1.2. Channel Flow Routing

Once overland flow gets into the channel network, the water will be routed as channel flow. Currently, WRF-Hydro provides three channel routing options: Muskingum, Muskingum-Cunge, and Diffusive Wave Routing. As the first two options are usually applied for vector-based reaches, we use the third option for gridded channel routing. An explicit, one-dimensional, variable time-stepping diffusive wave formulation is used as follows:

Continuity Equation:

$$\frac{\partial A}{\partial t} + \frac{\partial Q}{\partial x} = q_{lat} \tag{4}$$

Momentum Equation:

$$\frac{\partial Z}{\partial x} = -S_f \tag{5}$$

where A is the cross-section area (L^2), t is the time (T), Q is the flow discharge, which is the product of cross-section area and mean flow velocity perpendicular to cross-section area ($L^3\,T^{-1}$), q_{lat} is the unit discharge of the lateral flow ($L^2\,T^{-1}$), Z is the water surface elevation, which is the sum of bed elevation and water depth (L), S_f is the friction slope.

The friction slope S_f is solved as follows:

$$S_f = \left(\frac{Q}{K}\right)^2 \tag{6}$$

$$K = \frac{C_m}{n} A R^{\frac{2}{3}} \tag{7}$$

where K is the flow conveyance coefficient ($L^3\,T^{-1}$), C_m is the dimensional constant (1.0 for SI units), n is the Manning roughness coefficient, R is the hydraulic radius (L).

2.2. Soil Erosion and Sediment Transport Model

The sediment algorithm is adapted from the CASC2D-SED model, which is a fully distributed, physically based hydrological and upland erosion model developed at Colorado State University. CASC2D-SED is capable of reproducing hydrograph and sediment graph at different grid scales [22]. The code is converted from C to Fortran, and then incorporated into WRF-Hydro as an independent module.

On watershed scale, the annual gross erosion includes upland erosion, gully erosion as well as local bank erosion [13]. In the model, three sediment size groups—sand, silt, and clay—are included. Each sediment group can be presented in three states: suspended, deposited, or in bed layer. The bed layer represents the original soil layer and thus serves as the source of the sediment to the model domain. Once eroded from the bed layer, the sediment becomes suspended. After settling down, it goes into deposited.

As shown in Figure 2, sediment is first eroded from and transported through overland area by surface runoff. At the beginning of each time step, the total transport capacity is calculated with the revised USLE [23] over upland area. With the transport capacity, the suspended sediment is transported first, followed by the deposited part. At last, if there is still transport capacity left (excessive transport capacity), it will be used to erode the bed layer. During these processes, both the wash load and bed load transportation are considered. The different treatment of transport of the fine materials as wash load and coarse materials as bed load are achieved mainly by assigning different settling velocities to different sediment size groups.

In the channel, the sediment is carried by streamflow through the river network delineated in the model, meanwhile settling and resuspension processes are calculated. According to the Engelund and Hansen (1967) equation [24], the transport capacity is calculated for each sediment group separately in the channel. With the transport capacity, the suspended part of each sediment group is transported first by advective process. Then, bed material will be transported with excessive capacity. Channel erosion is not considered by the model yet. The different treatment of fine materials and coarse materials is accomplished by the differences in transport capacity as well as settling velocity between sediment groups.

Once sediment is transported from the "source grid (where the sediment comes from)" to the "sink grid (where the sediment goes to)", all the transported materials will stay in suspension first and then settle down at assigned settling velocities. By the end of each computational step, the deposited sediment is added to the original layer and the net accretion/erosion is updated.

2.2.1. Overland Sediment Routing

The revised USLE [23] is applied to predict soil loss from upland area (sheet and rill), which is caused by rainfall and associated overland flow. In a single grid, the transport capacity is calculated as follows:

$$T_{ovrl} = 58390 \times S_0^{1.664} \times q^{2.035} \times K \times C \times P \times dx \times dt \tag{8}$$

where T_{ovrl} is the overland transport capacity (L^3), here in the model m^3, q is the unit flow discharge ($L^2\,T^{-1}$), in this study m^2/s, K is the soil erodibility factor, which is in t/acre, C is the dimensionless cropping-management factor, P is the conservation practice factor, which is dimensionless, dx is the grid size (L), which is m in this study, and dt is the time step (T), which is in second. In this study, dx and dt are the same as those used in overland routing. S_0 is the water surface slope. The overland sediment transport capacity is used to transport suspended sediment first, and then the deposited sediment if any. After the suspended and deposited sediment in the source grid are transported, if there is still transport capacity left, the remaining capacity will be used to erode the original bed layer.

2.2.2. Channel Sediment Routing

The Engelund and Hansen (1967) [24] equation is applied to calculate the sediment transport capacity for each sediment size group in channels. For each size group, the suspended part is transported by advective process first and the transport capacity will be subtracted by the amount of the suspended sediment transported. Then, the excessive capacity is used to transport deposited material. The amount of the deposited material that can be transported is limited by excessive transport capacity as well as advective processes. After the transport of suspended material and deposited material, if there is still any transport capacity left, it won't be used as channel erosion is not considered in the current version of the model.

$$T_i = \frac{Q \times C_i \times d_t}{2.65} \tag{9}$$

$$C_i = 0.05 \times \left(\frac{G}{G-1}\right) \times \frac{V \times S_f}{\sqrt{(G-1) \times g \times d_i}} \times \sqrt{\frac{R_h \times S_f}{(G-1) \times d_i}} \tag{10}$$

where T_i is the sediment transport capacity in the channel for sediment type i (L^3), Q is the flow discharge ($L^3 \, T^{-1}$), C_i is the sediment concentration of type i by weight, d_t is the time step of channel routing (T), G is the specific gravity of sediment and was set to 2.65 in this study, V is the depth-averaged velocity in channel ($L \, T^{-1}$), S_f is the friction slope, R_h is the hydraulic radius (L), g is the gravity acceleration ($L \, T^{-2}$), d_i is the diameter of sediment type i (L). Calculated transport capacity is used to transport the suspended sediment first and then the previously deposited sediment.

2.3. Study Site and Data Availability

We applied the WRF-Hydro-Sed model on the Goodwin Creek Experimental Watershed (GCEW), Mississippi, USA to assess its performance (Figure 3). GCEW is located at northwest Mississippi, close to Batesville. The watershed has a drainage area of 21.3 km^2 with an outlet located at its southwest corner (89°54′50″ W, 34°13′55″ N). The Goodwin Creek is a tributary of the Long Creek, which flows into the Yocona River, one of the main rivers of the Yazoo River Basin [13]. The weather is hot and humid in summer and mild in winter, with an average annual rainfall of 1440 mm and a mean annual runoff of 145 mm during 1982 to 1992 [9]. Within the watershed, the elevation ranges from 68 m to 130 m above sea level. Around 50 percent of the watershed has a slope less than 0.02 and 15 percent has a slope larger than 0.03 [9]. The channels in the watershed extend mainly from northeast to southwest with an average slope of 0.004. Based on the State Soil Geographic Database [25], soils within the watershed are mainly silt loam and sandy loam, with the former one dominating. According to the 24-category Land Use Categories by the U.S. Geological Survey (USGS), the most common land cover types in this watershed are "dryland cropland and pasture", "irrigated cropland and pasture", and "deciduous broadleaf forest".

GCEW was originally established in 1977 and has been operated by the U.S. Department of Agriculture (USDA) National Sedimentation Laboratory (NSL) to study the influence of land use and upland erosion on sedimentary process and channel stability and to test numerical models. It is highly instrumented, with 32 standard recording rain gauges distributed uniformly within the watershed, 14 stream gages and supercritical flow structures located along the channel to collect discharge and sediment concentration data. In addition, periodic surveys are conducted to track land use conditions, channel geometry, and channel migration. In this study, rainfall data from 16 rain gauges and the streamflow and sediment concentration data at the outlet (MSGC1) was collected to calibrate and validate the model (Figure 3). The data interval is ~15 min.

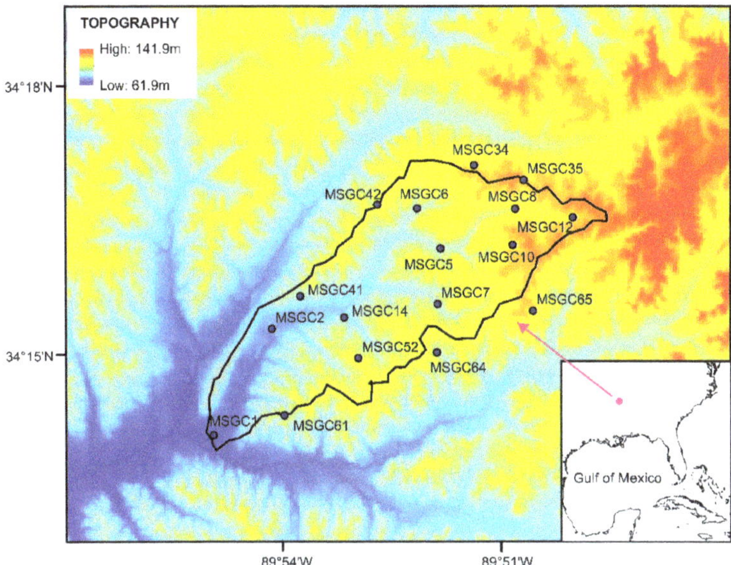

Figure 3. Topography of Goodwin Creek Experimental Watershed (GCEW) and gage location (DEM data source: National Hydrography Dataset Plus (Version 2), [26]).

2.4. Model Setup

WRF-Hydro supports the coupling between hydrological components and atmospheric models. However, as the focus of this study is to introduce the sediment module instead of investigating the interaction between atmosphere and hydrological processes, WRF-Hydro is used in a "one-way coupled" mode.

The Noah land surface model with multi-parameterization options (Noah-MP) [27] is responsible for simulating land surface physics in the model (Figure 1). In this study, Noah-MP has a domain of 14 km × 11 km with a spatial resolution of 1 km. To drive the land surface model, static land surface physiographic data was generated using the WRF Pre-processing System (WPS). The land use information was interpolated from the 24-category USGS land use database derived from the 1-km Advanced Very High Resolution Radiometer (AVHRR) satellite images [28].

The meteorological forcing data is from the North American Land Data Assimilation System Phase 2 Forcing Dataset (NLDAS-2 hereafter, [29]). NLDAS-2 contains incoming longwave and shortwave radiation, near surface wind, specific humidity, air temperature, surface pressure as well as rainfall intensity with hourly temporal resolution. In this study, the 1/8-degree NLDAS-2 was regridded to 1 km to match the Noah-MP grid. Given its coarse resolution, the rainfall intensity of NLDAS-2 might not be able to fully represent the condition of such a small watershed. Therefore, we replaced NLDAS-2's original rainfall data field with interpolated precipitation from the records at the 16 rain gauges using the inverse distance weighting interpolation method.

With a disaggregation factor of 20, the hydrological physical processes are simulated at the spatial resolution of 50 m. The subsurface routing, overland routing and channel routing of WRF-Hydro are all activated. The time step of the overland routing and channel routing is six seconds. Since the sediment processes are driven by the overland flow and channel flow, the setup of the sediment model highly depends on that of the hydrological model. The sediment model is calculated on the same grid as the hydrological model (50 m), with a time step of six seconds as well. Model tests and simulation were carried out on Louisiana Optical Network Initiative (LONI)'s QB2 and NCAR's Cheyenne super computers. In a serial mode, it took 2.5 h to conduct a 2-year simulation of streamflow with sediment module deactivated, and 30 min to finish a 24 h sediment simulation.

2.5. Model Calibration

In this study, the streamflow, sediment concentration, sediment flux and sediment yield were selected as the major variables to calibrate and validate the hydro- and sediment model. A stepwise calibration was performed where first the hydro-parameters and then the sedimentation parameters were calibrated. The streamflow was calibrated first manually through trial and error, and then automatically using the NCAR developed calibration toolbox, which is based on the Dynamically Dimensioned Search (DDS) calibration methodology [30]. Once finished, the calibrated hydro-parameters were used to drive the sediment module and the sediment parameters were then calibrated manually.

2.5.1. Streamflow Calibration

Streamflow was calibrated for a 3-year rainfall event on 17 October 1981 (calibration event hereafter), which started at 21:19 and lasted for approximately five hours. The average rainfall intensity is 14.7 mm/h and total rainfall is 74.4 mm. Calibrated hydro-parameters were selected based on sensitivity analysis and previous studies [23,31,32].

The calibration was carried out through two ways: automated calibration using the NCAR developed calibration toolbox and manual calibration based on trial and error. The reason for calibrating the model in two ways is that the automated calibration tools, which are usually based on standard objective metrics, may weigh more on timing error while weighing less on amplitude error, or the other way around, and thus may result in unreasonable results [33]. A manual evaluation, if executed in a rational way, can take the advantage of both visual inspection and standard metrics. While laborious and highly dependent on researchers' experience, the manual calibration may produce a better result.

Before calibration, a two-year run was performed starting from 1 January 1981 to let the model reach an equilibrium state before the calibration rainfall event. The results from each calibration were statistically evaluated using the correlation coefficient, Root Mean Square Error (RMSE), Nash-Sutcliffe coefficient (NSE) and Kling-Gupta efficiency (KGE). A detailed description of relevant equations is provided in the Supplementary Materials.

The Dynamically Dimensioned Search tool [30] was used for automated calibration. The objective function is the weighted *NSE* and *logNSE*:

$$ObjFn = 0 - \left(\frac{NSE}{2} + \frac{\log NSE}{2} \right) \tag{11}$$

$$NSE = 1 - \frac{\sum_{t=1}^{T} (O_t - P_t)^2}{\sum_{t=1}^{T} \left(O_t - \overline{O_t} \right)^2} \tag{12}$$

where *NSE* is the Nash-Sutcliffe coefficient, O_t is the observed streamflow at time t, P_t is the modeled streamflow at time t, $\overline{O_t}$ is the average of observed streamflow.

Table 1 summarizes the hydro-parameters calibrated by automated calibration with their lower and upper limits and default values. These parameters are mostly related to the land surface model. The hydro-parameters were adjusted within a reasonable bound of values through a 300-iteration automated run. With 32 processors, it took 10 h to finish the automated calibration.

For manual calibration, the refkdt and RETDEPRTFAC in Table 1 were selected as they were identified as the most sensitive parameters by the automated calibration and previous studies [23,31,32]. In addition, channel parameters including channel bottom width (Bw) in meters, channel side slope (Chsslp) and Manning roughness coefficient (MannN) were also calibrated.

Table 1. Calibrated hydro-parameters using the NCAR developed calibration tool.

Parameter	Values			Description
	Min	Max	Initial	
bexp	0.4	1.9	1.0	Multiplier on pore size distribution index
smcmax	0.8	1.2	1.0	Multiplier on saturation soil moisture content
dksat	0.2	10	1.0	Multiplier on saturated soil hydraulic conductivity
refkdt	0.1	4	1.0	Surface runoff parameter
slope	0	1	0.3	Linear scaling of "openness" of bottom drainage boundary
RETDEPRTFAC	0.1	20,000	1.0	Multiplier on maximum retention depth before flow is routed as overland flow.
LKSATFAC	10	10,000	1000	Multiplier on saturated hydraulic conductivity in lateral flow direction
cwpvt	0.5	2	1.0	Multiplier on canopy wind parameter
vcmx25	0.6	1.4	1.0	Multiplier on maximum rate of carboxylation at 25 °C (umol CO_2/m^2/s)

2.5.2. Sediment Calibration

Sediment concentration, sediment flux, and sediment yield were calibrated manually via a series of sensitivity tests following previous studies [9,13,23,34]. Calibrated sediment parameters can be categorized into two groups: soil-type related and land-use-type related. Soil erodibility factor K is soil-type related, while cropping-management factor C and conservation practice factor P are considered to be land-use determined. The calibrated sediment parameters are shown in Table 2.

Table 2. Calibrated sediment parameters and values for sediment model.

Land Use Category	C	P	Soil Type	K (t/Acre)
Dryland cropland and pasture	0.01	0.1	Silt loam	0.2
Irrigated cropland and pasture	0.03	0.1	Sandy loam	0.1
Deciduous broadleaf forest	0.001	0.1		

3. Results

3.1. Streamflow Calibration

The best results from the automated and manual calibration at the outlet of GCEW are shown in Figure 4 and the related values of statistical metrics are in Table 3. Overall, both automatically and manually calibrated hydrographs exhibit a satisfactory performance with a high correlation coefficient (>0.90), NSE (>0.70) and a low RMSE (<8 m^3/s). However, total amount of runoff is overestimated by both manual and automated calibration. While the automatically calibrated total volume of runoff is closer to observation, manually calibrated hydrograph fits better with the observed one over the high flow part of the measured hydrograph around the peaking time. As GCEW is self-drained watershed, we assume that large river discharge corresponds to large overland runoff flux. With the exponential relationship between overland runoff flux and transport capacity (Equation (8)), the sediment concentration and sediment flux during high flow periods should be much larger than during low flow periods. In this case, during a single rainfall event, the sediment concentration, flux, and total yield ought to be dominated by high flow period of the runoff event. (The observed sediment concentration and sediment flux graphs shown in Figure 5a,b justified this assumption as the main sediment event only lasted two hours, which corresponds to the high flow duration in the observed hydrograph shown in Figure 4). In this case, although the total volume from the automated calibration is closer to observation than the manually calibrated one, we chose to use the manually calibrated hydro-parameters to drive the sediment model since the high flow part of the manually calibrated hydrograph is much closer to observation. The values of manually calibrated hydro-parameters are shown in Table 4.

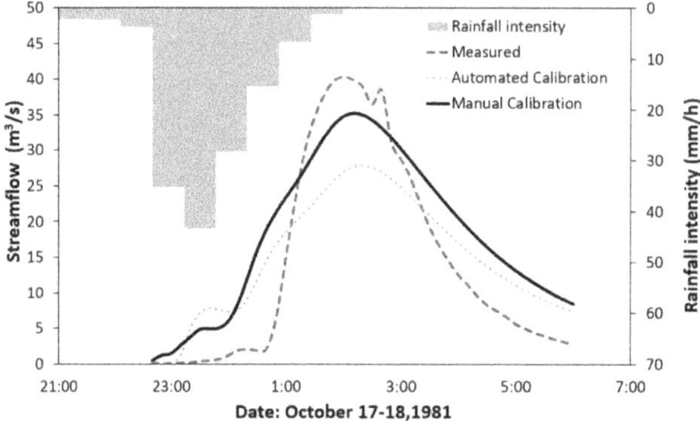

Figure 4. Measured (dash line), manually calibrated (solid line) and automatically calibrated (dot line) streamflow at MSGC01 for calibration event. Time series of rainfall intensity is also shown.

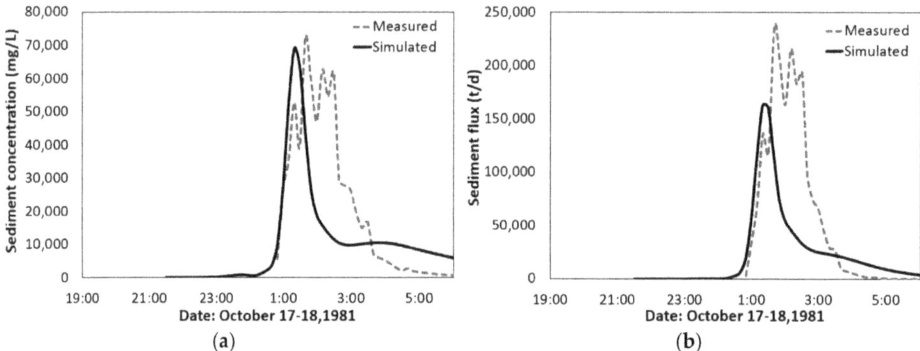

Figure 5. Measured (dash line) and simulated (solid line) sediment concentration (**a**), and sediment flux (**b**) at MSGC01 for calibration event.

Table 3. Performance of the streamflow simulation for calibration event.

Location	Calibration Methods	Correlation Coefficient	RMSE (m^3/s)	NSE	KGE
MSGC01	Automated	0.93	7.21	0.74	0.57
(Outlet)	Manual	0.94	6.72	0.77	0.63

Table 4. Values of manually calibrated hydro-parameters.

Parameter		Value	Parameter		Value
refkdt		0.7	Chsslp	1st order channel	0.35
RETDEPRTFAC		1.0		2nd order channel	0.25
Bw	1st order channel	3.0	MannN	1st order channel	0.35
	2nd order channel	5.0		2nd order channel	0.15

3.2. Sediment Calibration

Sediment concentration and sediment flux results are shown in Table 5 and Figure 5. Overall, the sediment concentration and sediment flux show good agreement with measurements at the outlet, indicating that the model can well reproduce the sediment erosion and transport processes on a watershed scale. Compared with streamflow, sediment concentration and sediment flux are less well predicted, with the NSE value of 0.25 and 0.28, respectively. For both sediment concentration and sediment flux, the simulated rising limbs fit well with the measured ones, while the peak values are underestimated. The falling limbs recess earlier but last longer. The long recession should inherit from that of the streamflow, and essentially the overland flow. The total sediment yield simulated during this event is around 9600 tons, which is 40 percent lower than the measurement. This error is close to the one reported by [13], which is considered to be acceptable in hydrological engineering.

Table 5. Performance of the simulation of sediment concentration and sediment flux for calibration event.

Location	Variable	Correlation Coefficient	RMSE	NSE	KGE
MSGC01	Sediment concentration	0.57	19,692.64 mg/L	0.25	0.44
(Outlet)	Sediment flux	0.62	65,321.10 t/d	0.28	0.30

We notice that the model simulated only one peak for sediment flux and sediment concentration, while the observation exhibited multiple peaks (Figure 5). Such a mismatch can be attributed to the uncertainty of observation or limitation of model algorithm. Due to the large fluctuation of sediment concentration during a rainfall event, the point-wise observation data might not be able to detect the real condition of sediment concentration. Thus, the multiple peaks shown in Figure 5 might not be the real case. If we assume that the observed multiple peaks are the real condition, the multiple peaks or rapid fluctuation of the sediment concentration and sediment flux might be attributed to local erosion of channel or bank collapse, which are currently not represented in the model.

3.3. Model Validation

Using calibrated parameters, we validated the model performance on the rainfall event of 28 August 1982. The return period of the event is one year. It started at 23:30 and lasted 4.5 h with an average rainfall intensity of 10.4 mm/h. The simulation was initialized on 1 January 1981 to assure the model reached a relatively stable condition before the rainfall event. Figure 6 shows the simulated and measured hydrographs at the outlet. The model was able to reproduce the measured hydrograph. The statistical results shown in Table 6 indicate satisfactory simulation skill of the model with a high NSE (0.86). We notice that the simulated rising limb starts earlier than the measurement and the falling limb drops slowly and lasts longer. The simulated peak is within 20 min of the measured one and the simulated peak value ($27~\text{m}^3/\text{s}$) is 26 percent lower than the measurement ($36.4~\text{m}^3/\text{s}$). Simulated water discharge at the outlet is within 10 percent of the measurement.

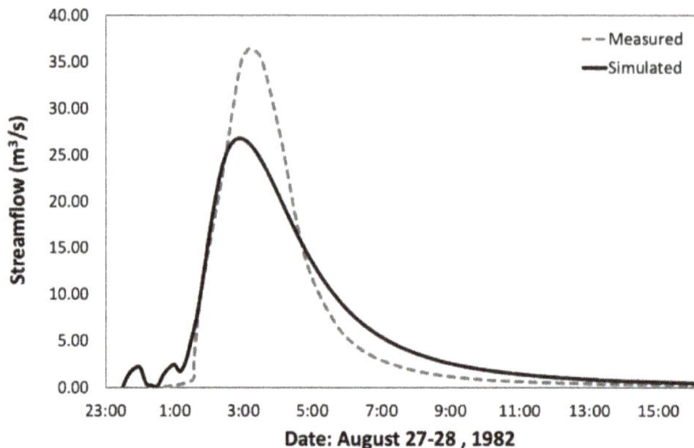

Figure 6. Measured (dash line) and simulated (solid line) streamflow at MSGC01 for validation event.

The comparison between simulated and measured sediment concentration and sediment flux is shown in Figure 7. The model can catch the shape of the observed sediment graphs but with an overall overestimation for both sediment concentration and flux. Sediment flux is better simulated than sediment concentration with NSE and KGE 0.09 (vs. −2.26 for concentration) and 0.34 (vs. −0.15 for concentration), respectively. A comparable model performance in sediment concentration simulation is also reported by Elliot [35], which was partially attributed to the fact that simulation of sediment concentration involves an error in both sediment simulation and runoff simulation. Due to the absence of observation data before 2:00, 28 August 1982, it is impossible to calculate the sediment yield during the entire event. Yet it is reasonable to assume that the real sediment yield is larger than the estimation based on the available data (9100 t). Thus, we consider that our model-simulated sediment yield (14,950 t) is acceptable.

Table 6. Performance of the simulation of streamflow, sediment concentration, and sediment flux for validation event.

Location	Variable	Correlation Coefficient	RMSE	NSE	KGE
MSGC01 (Outlet)	Streamflow	0.97	5.09 m³/s	0.86	0.70
	Sediment concentration	0.43	30,541.63 mg/L	−2.26	−0.15
	Sediment flux	0.85	44,922.10 t/d	0.09	0.34

In spite of the acceptable performance of the model in simulating the sediment yield for both events, the model tends to overestimate the sediment yield for validation events while underestimating it for calibration events. This difference in model behavior should be partly due to the difference in initial soil conditions of the two events. According to Rojas [23], the initial soil condition could be much wetter during the validation event, since a series of preceding rainfall events brought >110 mm rain in the previous month. Yet there was only one preceding rainfall event bringing ~15 mm rainfall before the calibration event. In this case, the soil erodibility factor calibrated for the calibration event might be too large for the validation event, as soil erodibility tends to be smaller in wet conditions than in dry conditions [36]. Thus, the sediment yield during the validation event is overestimated by the model.

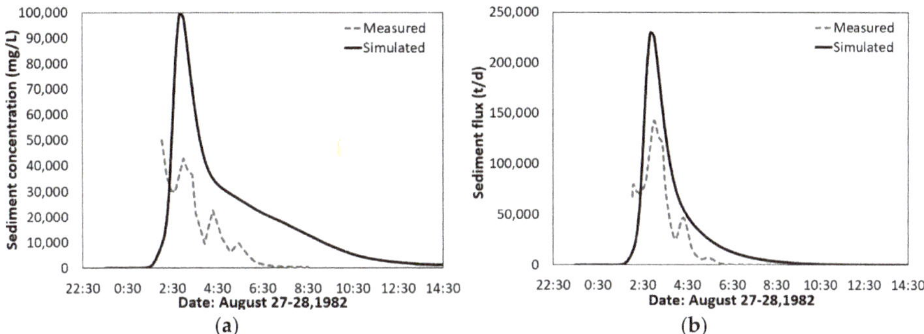

Figure 7. Measured (dash line) and simulated (solid line) sediment concentration (**a**), and sediment flux (**b**) at MSGC01 for validation event.

4. Discussion

4.1. Unceratinty Quantification

For the hydrologic simulation, uncertainty can result from different sources such as forcing data, model structure, model parameter, and initial and boundary condition uncertainties [37]. Each of these sources of uncertainty could be investigated and reduced through different treatments such as by providing more accurate forcing data to reduce the forcing uncertainty or by calibrating the model parameters to limit the parameter uncertainty. Data assimilation techniques are also used commonly to reduce the uncertainty due to the initial and boundary conditions in the models and in some cases to address the parameter uncertainty. Although investigating the model structure uncertainty is possible via the multi-parameterization scheme of NoahMP as the Land Surface Model used in WRF-Hydro as well as through different routing options available in WRF-Hydro, such a study was beyond the scope of this paper. We investigated the impact of the forcing on the model simulation as well as the parameter uncertainty, a subset of sources of uncertainty, of which the findings are summarized in the following subsections.

4.1.1. Forcing Uncertainty

Poor representation of meteorological forcing or errors may propagate into the hydrologic simulation and affect the result in a nonlinear way. Similar to previous WRF-Hydro studies [20,21], the NLDAS-2 dataset was used in this study as meteorological forcing. As explained in Section 2.4, in order to better represent the rainfall intensity, NLDAS-2's rainfall field was substituted with interpolated rain gauge observation. However, how well the model can perform when driven by the original NLDAS-2 dataset needs to be investigated. We performed a sensitivity experiment on the calibration event. The streamflow was simulated with default parameters (un-calibrated) and driven by the original NLDAS-2 (1/8 degree resolution) and updated rain gauge data (1 km resolution), respectively. The comparison between simulated and observed hydrographs is shown in Figure 8.

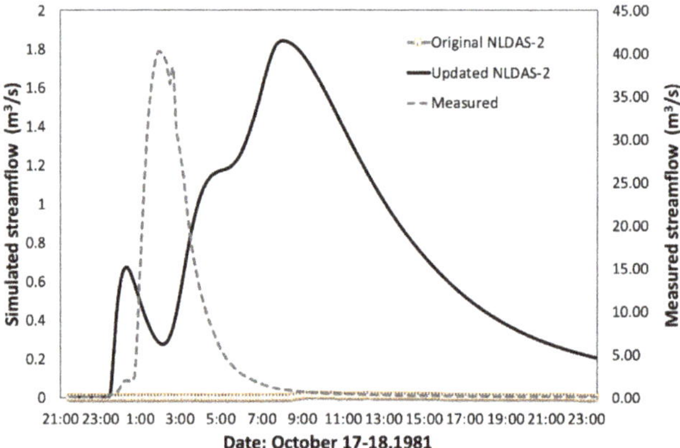

Figure 8. Measured (dash line) and simulated hydrographs (solid line for updated NLDAS-2, dash-dotted line for original NLDAS-2) at MSGC01 for calibration event.

Without calibration, using the original NLDAS-2 forcing data, the simulated hydrograph (Figure 8, dash-dotted line) fails to reproduce the rainfall event. Once the rainfall field is updated with rain gauge data and used to drive the model, a significant improvement is achieved in terms of amplitude of simulated hydrograph (Figure 8, solid line). However, timing error still exists, which will be largely minimized through calibration, as shown in Figure 4. A similar improvement has also been reported previously [37]. Thus, rainfall data with compatible resolution is recommended when applying WRF-Hydro-Sed to a relatively small watershed under local storm events in order to generate satisfactory results. This comparison could serve as a good example of how sensitive the model response is to the forcing dataset and in this case, precipitation. Ideally, one would like to force the model with an ensemble of the different forcing datasets to cover a range of forcing uncertainty in the model simulations.

4.1.2. Parameter Uncertainty

Model parameters are another source of uncertainty in the model simulation. This uncertainty is reduced to some degree through the calibration process. In this part, based on the single event calibration, we assess the goodness of calibration and prediction uncertainty using P-factor adapted from the Sequential Uncertainty Fitting method (SUFI-2) following previous studies [38–40]. P-factor is defined as the percentage of the measured data bracketed by the 95% prediction uncertainty (95 PPU), which represents the degree of uncertainties considered by the model parameters. The 95 PPU is calculated based on the cumulative distribution of the model outputs from different experiments corresponding to different model parameters. Here, the model output is the streamflow simulation and the model experiments are different calibration iterations having different model parameters. It is believed that the streamflow measurements reflect all the uncertainty in the model and inputs [41]. A P-factor of 100% indicates full coverage of observation in the 95 PPU, indicating that all uncertainty is explained by the model parameter uncertainty [38].

Based on the single event calibration we conducted, the P-factor is 95%, indicating that most of the measurements were bracketed by the model parameter uncertainty. Figure 9 shows the ensemble of simulated hydrographs (from the calibration process), with the 95 PPU band (red) against the observation and the best simulation during the calibration event. In this case, it can be concluded that the simulations based on single event calibration has generated a large coverage that covers the observation except for the overestimation over the beginning of the rising limb.

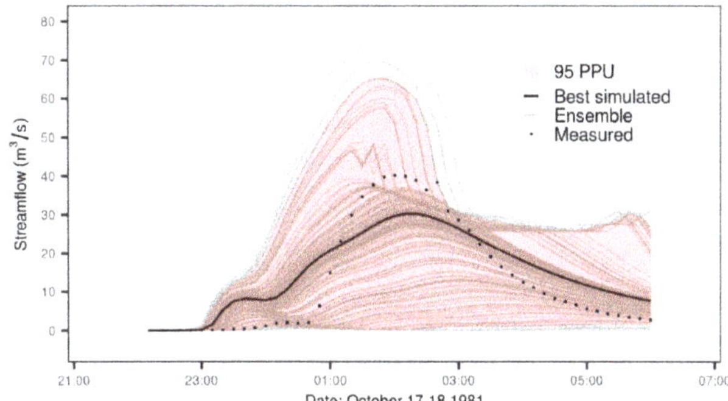

Figure 9. The best-simulated streamflow with 95 PPU and observations during the calibration event in 17–18 October 1981.

4.2. Applicability of Calibrated Parameters

Fine scale, grid- and process-based sediment models are usually used to simulate the sediment processes for a single rainfall event (e.g., [13,22,35,42]), instead of continuously simulating soil erosion over a long time scale. Part of the reason for this is that soil erosion at the watershed scale is thought to be controlled mainly by a few rainfall events [3]. In addition, continuous calculation of soil erosion with process-based models requires a large quantity of computational time as well as huge amounts of observation data, which are usually not available. In this case, such models are usually calibrated on one event and the calibrated parameters are then applied to another event.

In this study, WRF-Hydro-Sed was calibrated for the rainfall event of 17 October 1981 and then verified by the validation event with calibrated parameters. As mentioned in Section 3.3, in spite of the difference in initial conditions, with a reasonable spin-up period, the calibrated hydro-parameters can be transferred to the validation event and generate satisfactory hydrographs with high NSE values (0.86). For sediment simulation, although simulated sediment concentration and sediment flux exhibit larger bias (Table 6), which are well acknowledged by researchers in sediment modeling as a challenge, the simulated sediment yield at the outlet is acceptable, which validated the model's satisfactory performance with calibration.

However, variability in land use character and soil condition, as well temporal and spatial distribution of rainfall between different rainfall events, which haven not/cannot fully have been/be considered in our model, can restrict the application of the calibrated parameters based on a single event calibration to other events. In order to evaluate how well the model can perform over different rainfall events with calibrated parameters based on one single event, we applied calibrated hydro- and sediment parameters to the year of 1982 to conduct a one-year simulation. The year 1982 was selected mainly because it covers various rainfall events with different intensities and rainfall totals (Figure 10).

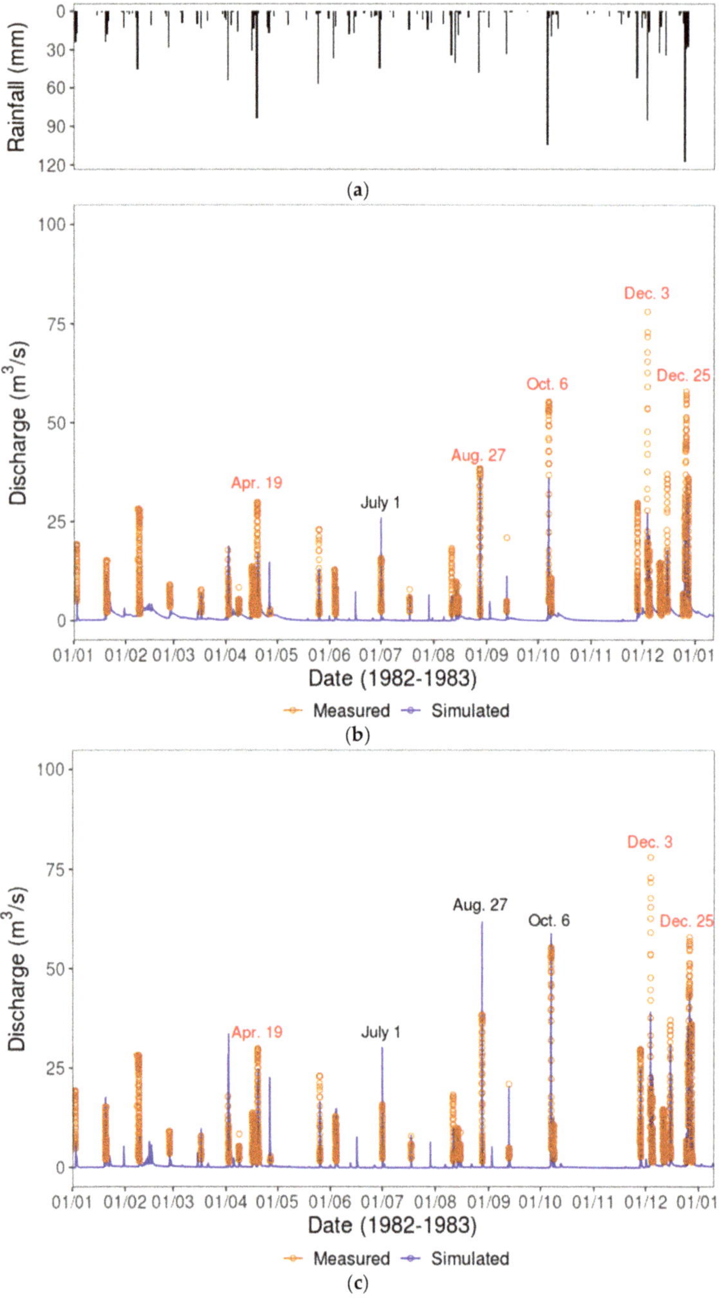

Figure 10. (**a**) Daily rainfall (black) of 1982. (**b**) Measured (dark orange) and simulated (blue, based on calibrated hydro-parameters) hydrographs at MSGC01 for all the rainfall events of 1982. (**c**) Same as (**b**) but based on recalibrated hydro-parameters. Several rainfall events are annotated with red, indicating underestimation and black represents overestimation.

4.2.1. Hydro-Parameters

Figure 10b shows the simulated and observed streamflow during 30 rainfall events of 1982, with the calibrated hydro-parameters from Section 3.3. Overall, with the calibrated hydro-parameters based on one single event (calibrated hydro-parameters hereafter), the model can reproduce all the streamflow events in the year with an NSE value of 0.43. However, simulation underestimates the streamflow mainly during the heaviest rainfall events of the year, i.e., rainfall events of 19 April, 6 October, 3 and 25 December, while overestimation can be found during less intense rainfall events such as that of 1 July. This indicates that the calibrated hydro-parameters based on one single event might favor the calibration event itself, while they are less suitable to fully reproduce hydrographs over events that have much different rainfall characteristics. In this case, we recalibrated the streamflow on the 30 rainfall events of 1982 (recalibrated hydro-parameters hereafter) with the observed streamflow to investigate how much the model performance can be improved through the multiple events recalibration. In addition, the calibrated hydro-parameters can be better evaluated by comparing them to the recalibrated hydro-parameters in terms of model performance improvement.

The multiple events recalibration was conducted automatically with a 150-iterations run using the NCAR developed calibration tool. The recalibrated hydrograph against the observation is shown in Figure 10c. The NSE value is 0.51, which is 1.19 times better than that using calibrated hydro-parameters. However, the multiple events recalibration consumed more than 21 times the computational hours than the single event calibration (6840 versus 320 computational hours) to achieve such an improvement. In addition, streamflow due to three rainfall events (19 April, 3 and 25 December) is underestimated (Figure 10b) and is still subject to underestimation after recalibration (Figure 10c). Streamflow during the rainfall event of July 1 is overestimated both in Figure 10b and after recalibration in Figure 10c. Meanwhile, simulated streamflow during 27 August and 6 October changed from being underestimated with single event calibration to being overestimated under recalibration. This implies that for the event-based simulation, it might not be practical to find a set of parameters that can be suitable for all events. Multiple events calibration can be used to improve the model's performance to a certain degree, yet it requires a substantially higher computational cost than the single event calibration. With this regard, intensive calibration over a long time scale might not be an optimal strategy if computational cost is a major concern and model performance based on a single event calibration is acceptable.

4.2.2. Sediment Parameters

To evaluate the applicability of calibrated sediment parameters based on one single event to other events, we applied them to simulate the sediment processes for the year of 1982. With 20 processors, 168 h were used to finish the simulation. Based on the available observation data of the sediment, the simulated sediment yield is compared against the observation for 17 sediment events. The characteristics of rainfall events, the simulated and the observed sediment yield during those events are shown in Table 7. It is noted that the sediment event of 3–4 June, 3–4 December, and 24~28 December includes 3, 2, and 3 rainfall events, respectively, as the sedimentary processes are correlated during such rainfall events.

For all of the 17 sediment events simulated, the minimum and maximum ratios between the observed sediment yield and the simulated sediment yield are 0.13 and 5.47, respectively (Table 7). This proves that with the calibrated sediment parameters based on one single event, simulated sediment yield for other different events can be expected to be at least within the same magnitude of the measured one. Furthermore, for 8 out of 17 sediment events, the simulated sediment yield is within 50–150% of the measurements, which corresponds to at most 50% under- or over-estimation. For 12 out of 17 events, the simulated sediment yield is within 33–300% of the measured sediment yield, in response to 200% under- or over-estimation at most, which is generally acceptable in sediment simulations. The coefficient of determination R^2 between the simulated and the measured sediment yield is 0.57, which also indicates the acceptable performance of the model [43]. In addition, the simulated total sediment yield (228,698 t) of all the events is only 11% higher than that of the observation

(203,387 t), which implies that it is also promising to use the model to estimate annual soil erosion on a watershed scale.

In spite of the overall acceptable performance of the model in simulating sediment yield, substantial over- and under-estimation can be found during events of 17 April, 25 May, 3–4 June, 11 August and 10–11 December. Considering the exponential relationship between overland runoff and sediment transport capacity, the bias of the simulated sediment yield can partly be attributed to the under- or over-estimation of the streamflow. In addition, model bias can also be attributed to the absence of a channel and bank erosion algorithm in the current model. As the sediment yield may be sourced from not only upland erosion, but also from channel and bank erosion, model bias may occur as a consequence of the model's failing to account for the sediment contribution from the channel bed and bank. With this regard, future development of the model should include the bank and channel erosion to further improve the model performance.

Table 7. Rainfall intensity, duration, and return period, simulated and observed sediment yield for 17 sediment events during 1982. Sediment events of 3–4 June, 3–4 December, and 24–28 December include 3, 2 and 3 rainfall events, respectively. Return period < 1 represents normal rainfall event.

Date	Rainfall Intensity (mm/h)	Duration (h)	Return Period (years)	Sediment Yield (t)		
				Measured	Simulated	Measured/Simulated
2–3 January	4.1	5		1566	1738	0.90
8 February	5.3	6.5		7903	3711	2.13
8 April	3.6	4	<1	133	374	0.36
17 April	3.8	4.5		969	177	5.47
19 April	4.2	9		6777	11,650	0.58
25 May	10.6	2.5		2684	19,893	0.13
3–4 June	8.4	8	1	5334	33,685	0.16
	1.9	1.5	<1			
	6.0	2				
11 August	5.6	6		4403	805	5.47
14 August	5.5	5	<1	2054	4641	0.44
15 August	2.9	6.5		180	168	1.07
6–7 October	9.0	11.5	7	34,996	57,769	0.61
8 October	6.2	3	<1	3019	4071	0.74
27 November	3.7	14		8923	8612	1.04
3–4 December	4.1	21	~1	48,380	25,932	1.87
	3.1	5	<1			
10–11 December	1.5	29	<1	5984	1596	3.75
15 December	3.9	8.5		9842	9569	1.03
24–28 December	5.8	1.5	<1	60,240	44,307	1.36
	4.3	34	6			
	4.9	7.5	<1			
Sum				203,387	228,698	0.89

4.3. Landscape Pattern Index and Sediment Yield

A landscape pattern index is used to describe the type and spatial arrangement of the landscape by considering different features such as size, shape, and connectivity [44]. A series of work has been conducted to investigate the relationship between soil erosion and landscape patterns (e.g., [45–47]). In this study, we introduce a landscape pattern index following Zhou [47]. The index (SI, Equation (13)) considers the soil erodibility factor (K), the cropping-management factor (C) and the topography factor slope (α, °).

$$SI = C \times K \times sin\alpha \tag{13}$$

In this study, C and K values are calibrated over GCEW manually. Topographic slope is calculated using ArcGIS on the digital elevation grid of the study area with a resolution of 50 × 50 m. Then the SI is calculated over GCEW following Equation (13). Furthermore, we analyzed the relationship between *SI* and the magnitude of the soil erosion simulated by our model during the calibration event. Figure 11a–d shows the spatial distribution of *C*, *K*, slope and *SI* over GCEW. Figure 11d exhibits the net erosion and deposition over the study area during the calibration event.

To investigate the relationship between *SI* (Figure 11d) and net erosion/deposition (Figure 11e), we conducted a correlation analysis using the Spatial Analyst Tools of ArcGIS. With a correlation coefficient of −0.017, landscape pattern and soil erosion show no significant linear relation. This indicates that during a storm event with a duration of hours over GCEW, the landscape pattern might not be the dominating factor controlling the spatial distribution of soil erosion.

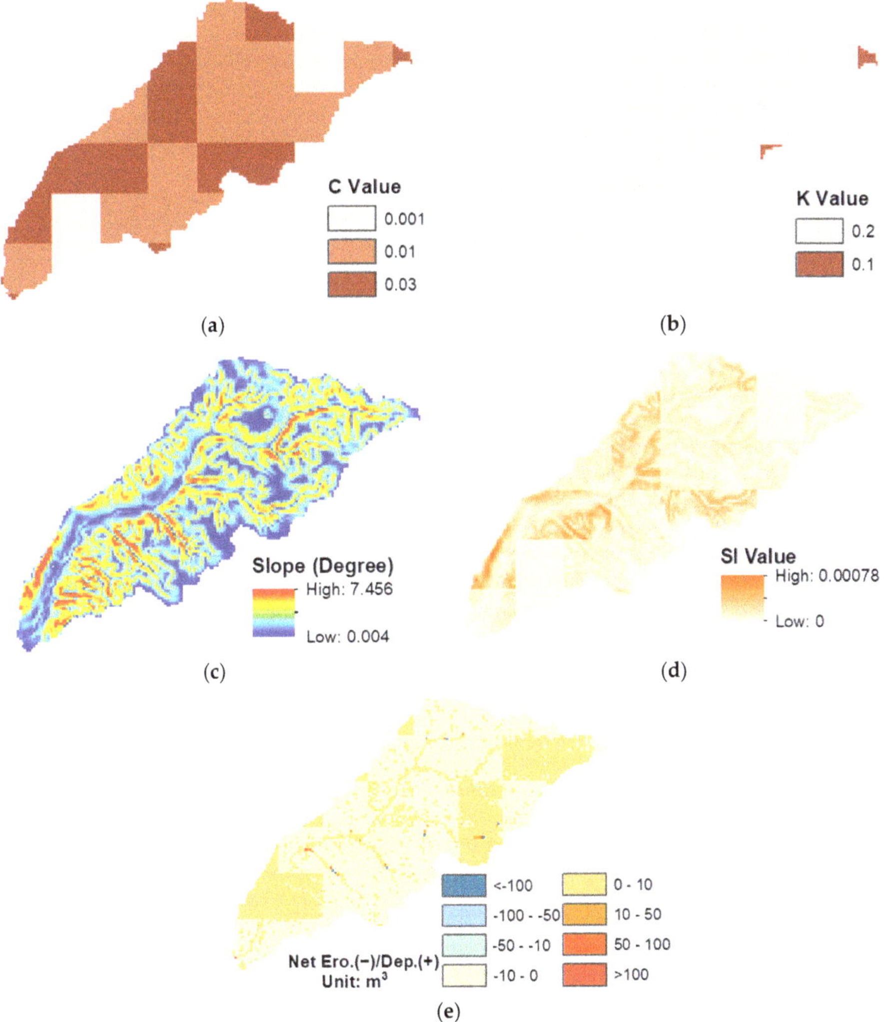

Figure 11. *C* value (**a**), *K* value (**b**), slope (**c**), and *SI* (**d**) over GCEW, and Net Erosion (−)/Deposition (+) (**e**) during calibration event over GCEW.

4.4. Climate Change Scenarios

Under climate change, the frequency, intensity as well as the incidence of extreme rainfall events are subject to change [48]. Meanwhile, temperature, surface runoff, and land use will also be influenced by the changing climate. The shift in extreme rainfall characteristics can influence the soil erosion directly by changing the erosive capacity [49]. The variance of soil moisture, temperature, and land use can affect soil erosion via affecting soil properties such as erodibility. In addition, these factors will interact in a nonlinear way to regulate soil erosion and related sediment transport processes. One of the best ways to project the rainfall-runoff and soil erosion dynamics under climate change is via numerical models that can address the complicated hydro-sedimentary dynamics. Such studies have been conducted by several scientists [35,44,45].

Over GCEW, the decrease in the cultivated land is proved to reduce the fine sediment source. In addition, such land use change can also decrease the peak flow and runoff volume, resulting in the reduction of sediment supply and transport in channels [50]. In this study, the current sediment model is built on WRF-Hydro in the expectation that it will be used to nowcast/hindcast the streamflow and soil erosion during rainfall events over GCEW. In addition, unlike the WRF-Hydro, WRF-Hydro-SED, for now, only supports simulation in serial mode. Thus, long-term simulation considering climate change scenarios is not feasible to by carried out using the current model. Future model development by parallelizing the model code, introducing a morphological evolution algorithm, and considering the rainfall and land use evolution, as well as the complex interrelation between those factors under climate change scenarios, are expected to alleviate such limitations.

5. Conclusions

In this study, by adapting the sediment algorithm from CASC2D-Sed, we introduced a sediment module into the WRF-Hydro platform, allowing for the development of a fully distributed, process-based soil erosion and sediment transport model (WRF-Hydro-Sed). The model's performance was evaluated via a comparison with the observed streamflow and sediment concentration data at the Goodwin Creek Experimental Watershed during rainfall events.

WRF-Hydro-Sed is able to generate satisfactory results of streamflow and sediment yield during rainfall events. The streamflow can be calibrated successfully based on a single rainfall event with the adjustment of a few hydro-parameters including refkdt (the parameter that controls runoff–infiltration partition) and channel geometries. With the single event calibrated hydro-parameters, the model can also perform satisfactorily in simulating the hydrograph during a validation event. Based on calibrated hydro-parameters, sediment concentration, sediment flux, as well as sediment yield can also be calibrated successfully at watershed scale by adjusting sediment parameters related to land use and soil category. Satisfactory results are also generated for a validation event using calibrated sediment parameters. The model's performance in simulating sediment yield is better than sediment concentration and flux.

The model's performance in streamflow simulation is sensitive to forcing data. The original NLDAS-2, given its 1/8 degree coarse resolution, may not be an optimal choice to provide rainfall forcing for simulation over a relatively small watershed like the Goodwin Creek under local storm events. High resolution meteorological forcing data is recommended for application of the WRF-Hydro-Sed on a small watershed.

Calibrated hydro-parameters based on a single event can be applied to different rainfall events to reproduce the hydrograph. While it might not be practical to have a set of parameters that can be suitable for any rainfall event, an intensive calibration based on multiple events can improve the model's performance to a certain degree, but with extensive computational efforts. In this case, intensive calibration over a long time scale might not be an optimal strategy if computational cost is a major concern and if the model performance based on a single event calibration is acceptable. With the calibrated sediment parameters based on a single event, the sediment yield over different events can be simulated within the same magnitude observed. Moreover, the model shows promising potential in simulating annual soil erosion on a watershed scale. While simulated sediment yield is considered acceptable for 71% of the events (12 out of 17), substantial bias can be found during certain events mainly due to the bias transferred from the streamflow simulation. Future development of the model by including the bank and channel erosion algorithm is expected to further improve model performance.

Supplementary Materials: The following are available online at http://www.mdpi.com/2073-4441/12/6/1840/s1, the equations used to calculate correlation coefficient, RMSE, NSE and KGE.

Author Contributions: Conceptualization, Z.G.X. and D.Y.; methodology, D.Y.; software, D.Y. and W.Y.; validation, D.Y.; formal analysis, D.Y. and M.M.; investigation, D.Y. and M.M.; resources, Z.G.X. and D.J.G.; data curation, D.Y.; writing—original draft preparation, D.Y.; writing—review and editing, D.Y., Z.G.X. and A.R.; visualization, D.Y. and Z.G.X.; supervision, Z.G.X.; project administration, Z.G.X. and D.J.G.; funding acquisition, Z.G.X. and D.J.G. All authors have read and agreed to the published version of the manuscript.

Funding: This project was funded by The Water Institute of the Gulf under project "Project Louisiana Rivers' Sediment Flux to the Coastal Ocean using a Coupled Atmospheric-Hydrological Model" (award number RCEGR260003-01-00). This project was paid for (in part) with federal funding from the Department of the Treasury through the Louisiana Coastal Protection and Restoration Authority's Center of Excellence Research Grants Program under the Resources and Ecosystems Sustainability, Tourist Opportunities, and Revived Economies of the Gulf Coast States Act of 2012 (RESTORE Act). The statements, findings, conclusions, and recommendations are those of the author(s) and do not necessarily reflect the views of the Department of the Treasury, CPRA or The Water Institute of the Gulf. Dongxiao Yin is also supported by the Economic Development Assistantship program at LSU. Support from NCAR's Advanced Study Program's Graduate Student (GVP) Fellowship is appreciated. This research is also partially funded by USGS The Cooperative Ecosystem Studies Units (award# G20AC00099).

Acknowledgments: Louisiana Optical Network Initiative (LONI), and Cheyenne (doi: 10.5065/D6RX99HX) provided by NCAR's Computational and Information Systems Laboratory, sponsored by the National Science Foundation are acknowledged for high-performance computing support. We acknowledge the NCAR WRF-Hydro developing team for help with model setup. Glenn V. Wilson and Jacob Ferguson are thanked for providing us with the data of GCEW.

Conflicts of Interest: The authors declare no conflict of interest.

References

1. Amundson, R.; Berhe, A.A.; Hopmans, J.W.; Olson, C.; Sztein, A.E.; Sparks, D.L. Soil and human security in the 21st century. *Science* **2015**, *348*. [CrossRef]
2. Juez, C.; Tena, A.; Fernández-Pato, J.; Batalla, R.J.; García-Navarro, P. Application of a distributed 2D overland flow model for rainfall/runoff and erosion simulation in a Mediterranean watershed. *Geogr. Res. Lett.* **2018**, *44*, 615–640. [CrossRef]
3. Aksoy, H.; Kavvas, M.L. A review of hillslope and watershed scale erosion and sediment transport models. *Catena* **2005**, *64*, 247–271. [CrossRef]
4. Fatichi, S.; Vivoni, E.R.; Ogden, F.L.; Ivanov, V.Y.; Mirus, B.; Gochis, D.; Downer, C.W.; Camporese, M.; Davison, J.H.; Ebel, B.; et al. An overview of current applications, challenges, and future trends in distributed process-based models in hydrology. *J. Hydrol.* **2016**, *537*, 45–60. [CrossRef]
5. Knisel, W.G. *CREAMS A Field-Scale Model for Chemicals, Runoff, and Erosion from Agricultural Management Systems*; Science and Education Administration: Washington, WA, USA, 1982.
6. Nearing, M.A.; Foster, G.R.; Lane, L.J.; Finkner, S.C. A Process-based soil erosion model for USDA-water erosion prediction project technology. *Trans. ASAE* **1989**, *32*, 1587–1593. [CrossRef]
7. Morgan, R.P.C.; Quinton, J.N.; Smith, R.E.; Govers, G.; Poesen, J.W.A.; Auerswald, K.; Chisci, G.; Torri, D.; Styczen, M.E. The European soil erosion model (EUROSEM): A dynamic approach for predicting sediment transport from fields and small catchments. *Earth Surf. Process. Landf.* **1998**, *23*, 527–544. [CrossRef]
8. Smith, R.E. A kinematic model for surface mine sediment yield. *Trans. Am. Soc. Agric. Eng.* **1981**, *24*, 1508–1514. [CrossRef]
9. Patil, S.; Sivapalan, M.; Hassan, M.A.; Ye, S.; Harman, C.J.; Xu, X. A network model for prediction and diagnosis of sediment dynamics at the watershed scale. *J. Geophys. Res. Earth Surf.* **2012**, *117*, 1–17. [CrossRef]
10. Merritt, W.S.; Letcher, R.A.; Jakeman, A.J. A review of erosion and sediment transport models. *Environ. Model. Softw.* **2003**, *18*, 761–799. [CrossRef]

11. De Roo, A.P.J.; Wesseling, C.G.; Ritsema, C.J. LISEM: A Single-event physically based hydrological and soil erosion model for drainage basins. I: Theory, input and output. *Hydrol. Process.* **1996**, *10*, 1107–1117. [CrossRef]

12. Wicks, J.M.; Bathurst, J.C. SHESED: A physically based, distributed erosion and sediment yield component for the SHE hydrological modelling system. *J. Hydrol.* **1996**, *175*, 213–238. [CrossRef]

13. Johnson, B.E.; Julien, P.Y.; Molnar, D.K.; Watson, C.C. The two-dimensional upland erosion model CASC2D-SED. *J. Am. Water Resour. Assoc.* **2000**, *36*, 31–42. [CrossRef]

14. Juez, C.; Battisacco, E.; Schleiss, A.J.; Franca, M.J. Assessment of the performance of numerical modeling in reproducing a replenishment of sediments in a water-worked channel. *Adv. Water Resour.* **2016**, *92*, 10–22. [CrossRef]

15. Gochis, D.J.; Yu, W.; Yates, D. The WRF-Hydro model technical description and user's guide, Version 3.0. *NCAR Tech. Doc.* **2015**, 120. [CrossRef]

16. Liang, X.; Lettenmaier, D.P.; Wood, E.F.; Burges, S.J. A simple hydrologically based model of land surface water and energy fluxes for general circulation models. *J. Geophys. Res.* **1994**, *99*, 14415–14428. [CrossRef]

17. Arnold, J.G.; Srinivasan, R.; Muttiah, R.S.; Williams, J.R. Large area hydrologic modeling and assessment. Part I: Model development. *Nutrition* **1998**, *17*, 70. [CrossRef]

18. Lin, P.; Hopper, L.J.; Yang, Z.-L.; Lenz, M.; Zeitler, J.W. Insights into hydrometeorological factors constraining flood prediction skill during the May and October 2015 Texas Hill country flood events. *J. Hydrometeorol.* **2018**, *19*, 1339–1361. [CrossRef]

19. Naabil, E.; Lamptey, B.L.; Arnault, J.; Kunstmann, H.; Olufayo, A. Water resources management using the WRF-Hydro modelling system: Case-study of the Tono dam in West Africa. *J. Hydrol. Reg. Stud.* **2017**, *12*, 196–209. [CrossRef]

20. Somos-Valenzuela, M.A.; Palmer, R.N. Use of WRF-hydro over the Northeast of the US to estimate water budget tendencies in small watersheds. *Water (Switz.)* **2018**, *10*, 1709. [CrossRef]

21. Xue, Z.G.; Gochis, D.J.; Yu, W.; Keim, B.D.; Rohli, R.V.; Zang, Z.; Sampson, K.; Dugger, A.; Sathiaraj, D.; Ge, Q. Modeling hydroclimatic change in Southwest Louisiana rivers. *Water (Switz.)* **2018**, *10*, 596. [CrossRef]

22. Rojas, R.; Velleux, M.; Julien, P.Y.; Johnson, B.E. Grid scale effects on watershed soil erosion models. *J. Hydrol. Eng.* **2008**, *13*, 793–802. [CrossRef]

23. Rojas Sánchez, R. GIS-Based Upland Erosion Modeling, Geovisualization and Grid Size Effects on Erosion Simulations with Casc2D-Sed. Ph.D. Thesis, Colorado State University, Fort Collins, CO, USA, 2002.

24. Engelund, F.; Hansen, E. *A Monograph on Sediment Transport in Alluvial Streams*; Teknisk Forlag: Copenhagen, Denmark, 1967.

25. Miller, D.A.; White, R.A. A Conterminous United States multilayer soil characteristics dataset for regional climate and hydrology modeling. *Earth Interact.* **1998**, *2*, 2. [CrossRef]

26. McKay, L.; Bondelid, T.; Dewald, T.; Johnston, J.; Moore, R.; Rea, A. NHDPlus Version 2: User Guide. 2012, p. 182. Available online: ftp://ftp.horizon-systems.com/NHDplus/NHDPlusV21/Documentation/NHDPlusV2_User_Guide.pdf (accessed on 26 June 2020).

27. Niu, G.Y.; Yang, Z.L.; Mitchell, K.E.; Chen, F.; Ek, M.B.; Barlage, M.; Kumar, A.; Manning, K.; Niyogi, D.; Rosero, E.; et al. The community Noah land surface model with multiparameterization options (Noah-MP): 1. Model description and evaluation with local-scale measurements. *J. Geophys. Res. Atmos.* **2011**, *116*, 1–19. [CrossRef]

28. Sertel, E.; Robock, A.; Ormeci, C. Impacts of land cover data quality on regional climate simulations. *Int. J. Climatol.* **2010**, *30*, 1942–1953. [CrossRef]

29. Xia, Y.; Mitchell, K.; Ek, M.; Sheffield, J.; Cosgrove, B.; Wood, E.; Luo, L.; Alonge, C.; Wei, H.; Meng, J.; et al. Continental-scale water and energy flux analysis and validation for the North American Land Data Assimilation System project phase 2 (NLDAS-2): 1. Intercomparison and application of model products. *J. Geophys. Res. Atmos.* **2012**, *117*, D03110. [CrossRef]

30. Tolson, B.A.; Shoemaker, C.A. Dynamically dimensioned search algorithm for computationally efficient watershed model calibration. *Water Resour. Res.* **2007**, *43*, 1–16. [CrossRef]

31. Arnault, J.; Wagner, S.; Rummler, T.; Fersch, B.; Bliefernicht, J.; Andresen, S.; Kunstmann, H. Role of runoff-infiltration partitioning and resolved overland flow on land-atmosphere feedbacks: A case study with the WRF-hydro coupled modeling system for West Africa. *J. Hydrometeorol.* **2016**, *17*, 1489–1516. [CrossRef]

32. Yucel, I.; Onen, A.; Yilmaz, K.K.; Gochis, D.J. Calibration and evaluation of a flood forecasting system: Utility of numerical weather prediction model, data assimilation and satellite-based rainfall. *J. Hydrol.* **2015**, *523*, 49–66. [CrossRef]

33. Ehret, U.; Zehe, E. Series distance—An intuitive metric to quantify hydrograph similarity in terms of occurrence, amplitude and timing of hydrological events. *Hydrol. Earth Syst. Sci.* **2011**, *15*, 877–896. [CrossRef]

34. Julien, P.Y. *Erosion and Sedimentation*, 2nd ed.; Cambridge University Press: Cambridge, UK, 2010.

35. Elliott, A.H.; Oehler, F.; Schmidt, J.; Ekanayake, J.C. Sediment modelling with fine temporal and spatial resolution for a hilly catchment. *Hydrol. Process.* **2012**, *26*, 3645–3660. [CrossRef]

36. Fufa, S.D.; Strauss, P.; Schneider, W. Comparison of erodibility of some Hararghe soils using rainfall simulation. *Commun. Soil Sci. Plant Anal.* **2002**, *33*, 333–348. [CrossRef]

37. Vergara, H.; Hong, Y.; Gourley, J.J.; Anagnostou, E.N.; Maggioni, V.; Stampoulis, D.; Kirstetter, P.-E. Effects of resolution of satellite-based rainfall estimates on hydrologic modeling skill at different scales. *J. Hydrometeorol.* **2013**, *15*, 593–613. [CrossRef]

38. Setegn, S.G.; Srinivasan, R.; Melesse, A.M.; Dargahi, B. SWAT model application and prediction uncertainty analysis in the Lake Tana Basin, Ethiopia. *Hydrol. Process.* **2009**, *24*, 357–367. [CrossRef]

39. Rafiei Emam, A.; Kappas, M.; Fassnacht, S.; Linh, N.H.K. Uncertainty analysis of hydrological modeling in a tropical area using different algorithms. *Front. Earth Sci.* **2018**, *12*, 661–671. [CrossRef]

40. Wu, H.; Chen, B. Evaluating uncertainty estimates in distributed hydrological modeling for the Wenjing River watershed in China by GLUE, SUFI-2, and ParaSol methods. *Ecol. Eng.* **2015**, *76*, 110–121. [CrossRef]

41. Yang, J.; Reichert, P.; Abbaspour, K.C.; Xia, J.; Yang, H. Comparing uncertainty analysis techniques for a SWAT application to the Chaohe Basin in China. *J. Hydrol.* **2008**, *358*, 1–23. [CrossRef]

42. Kim, J.; Ivanov, V.Y.; Katopodes, N.D. Modeling erosion and sedimentation coupled with hydrological and overland flow processes at the watershed scale. *Water Resour. Res.* **2013**, *49*, 5134–5154. [CrossRef]

43. Moriasi, D.N.; Arnold, J.G.; Van Liew, M.W.; Bingner, R.L.; Harmel, R.D.; Veith, T.L. Model evaluation guidelines for systematic quantification of accuracy in watershed simulations. *Trans. ASABE* **2007**, *50*, 885–900. [CrossRef]

44. Jia, Y.; Tang, L.; Xu, M.; Yang, X. Landscape pattern indices for evaluating urban spatial morphology—A case study of Chinese cities. *Ecol. Indic.* **2019**, *99*, 27–37. [CrossRef]

45. Ahmadi Mirghaed, F.; Souri, B.; Mohammadzadeh, M.; Salmanmahiny, A.; Mirkarimi, S.H. Evaluation of the relationship between soil erosion and landscape metrics across Gorgan Watershed in northern Iran. *Environ. Monit. Assess.* **2018**, *190*, 643. [CrossRef]

46. Ouyang, W.; Skidmore, A.K.; Hao, F.; Wang, T. Soil erosion dynamics response to landscape pattern. *Sci. Total Environ.* **2010**, *408*, 1358–1366. [CrossRef] [PubMed]

47. Zhou, Z.X.; Li, J. The correlation analysis on the landscape pattern index and hydrological processes in the Yanhe watershed, China. *J. Hydrol.* **2015**, *524*, 417–426. [CrossRef]

48. Routschek, A.; Schmidt, J.; Kreienkamp, F. Impact of climate change on soil erosion—A high-resolution projection on catchment scale until 2100 in Saxony/Germany. *Catena* **2014**, *121*, 99–109. [CrossRef]

49. Mullan, D.; Favis-Mortlock, D.; Fealy, R. Addressing key limitations associated with modelling soil erosion under the impacts of future climate change. *Agric. For. Meteorol.* **2012**, *156*, 18–30. [CrossRef]

50. Kuhnle, R.A.; Bingner, R.L.; Foster, G.R.; Grissinger, E.H. Effect of land use changes on sediment transport. *Water Resour. Res.* **1996**, *32*, 3189–3196. [CrossRef]

Article

Assessing the Impact of Man–Made Ponds on Soil Erosion and Sediment Transport in Limnological Basins

Mario J. Al Sayah [1,2,3], **Rachid Nedjai** [3], **Konstantinos Kaffas** [4], **Chadi Abdallah** [1,*] **and Michel Khouri** [2]

[1] National Council for Scientific Research, Remote Sensing Center, Beirut 11-8281, Lebanon; mario.alsayah@gmail.com

[2] Centre de Recherches en Sciences et Ingénierie, Lebanese University Faculty of Engineering II, Roumieh 1205, Lebanon; mkhuri@ul.edu.lb

[3] Centre d'Études et de Développement des Territoires et de l'Environnement, Université d'Orléans, 45100 Orléans, France; rachid.nedjai@univ-orleans.fr

[4] Faculty of Science and Technology, Free University of Bozen, 39100 Bozen, Italy; Konstantinos.Kaffas@unibz.it or kostaskaffas@gmail.com

* Correspondence: chadi@cnrs.edu.lb; Tel.: +961-3-534-436 or +961-4-409846; Fax: +961-4-409845

Received: 17 October 2019; Accepted: 27 November 2019; Published: 29 November 2019

Abstract: The impact of ponds on basins has recently started to receive its well-deserved scientific attention. In this study, pond-induced impacts on soil erosion and sediment transport were investigated at the scale of the French Claise basin. In order to determine erosion and sediment transport patterns of the Claise, the Coordination of Information on the Environment (CORINE) erosion and Soil and Water Assessment Tool (SWAT) models were used. The impact of ponds on the studied processes was revealed by means of land cover change scenarios, using ponded versus pondless inputs. Results show that under current conditions (pond presence), 12.48% of the basin corresponds to no-erosion risk zones (attributed to the dense pond network), while 65.66% corresponds to low-erosion risk, 21.68% to moderate-erosion risk, and only 0.18% to high-erosion risk zones. The SWAT model revealed that ponded sub-basins correspond to low sediment yields areas, in contrast to the pondless sub-basins, which yield appreciably higher erosion rates. Under the alternative pondless scenario, erosion risks shifted to 1.12%, 0.52%, 76.8%, and 21.56% for no, low, moderate, and high-erosion risks, respectively, while the sediment transport pattern completely shifted to higher sediment yield zones. This approach solidifies ponds as powerful human-induced modifications to hydro/sedimentary processes.

Keywords: soil erosion; sediment transport; SWAT; CORINE erosion model; ponds; Brenne; limnology; land cover change

---

1. Introduction

Understanding the impact of land occupation (land use/cover) on basin processes, such as rainfall-runoff and soil erosion, is an integral part of land and water management-oriented decisions [1,2]. The processes of soil erosion and sediment transport take part in van Rijn's (1993) [3] sedimentary cycle, and often are the main causes of soil loss in basins [4]. Although these processes are of natural origin [5], the interaction between climate, soil, topography, land use, and land cover significantly influences erosion rates and sediment loads [6,7]. Soil loss due to these processes is a frequent problem that hydrologists, land planners, and basin managers will need to contend with [8]. Accordingly, soil erosion quantification in erosion-prone areas, with the highest accuracy possible, provides a complete knowledge of soil loss hotspots and allows prioritized treatment measures for supporting land processes in the concerned basin [9]. Subsequently, the reliable assessment and representation of sediment yields—which depend on the cascading effect of soil erosion—allows an in-depth understanding of the soil erosion-sediment link at the basin scale [9].

For representing both processes, several models have been developed and used extensively to replace the conventional assessment methods, i.e., the Water Erosion Prediction Project (WEPP) [10,11], the Universal Soil Loss Equation (USLE) [12], the Modified Universal Soil Loss Equation (MUSLE) [13], the Revised Universal Soil Loss Equation (RUSLE) [14,15] and the Soil and Water Assessment Tool (SWAT) [16]. SWAT is one of the most widely used basin models. It has been applied extensively in modeling the impact of land occupation changes, under different scenarios and different contexts [17,18]. The widespread use of SWAT can be justified by its sensitivity and flexibility towards the land occupation input [19], its adaptability to different contexts—even to those with data scarcity [20]—its simple data requirements and ease of computation [21], as well as the straightforward calibration through its stand-alone SWATCUP interface [22].

Despite abundant research regarding the impact of land occupation on soil loss, few studies focus on the particular case of small water bodies and their effect as a land occupation class [23]. Small water bodies, like ponds and wetlands, are considered as the most amplified form of human-induced modifications to the hydro-sedimentological system of basins [24]. Ponds represent a total of over 90% of global standing water bodies, 30% of global standing waters by surface area [25], and form the most widespread aquatic habitat dominating the continental standing waters in Europe [26]. Despite their well-documented significance [27], abundant numbers, and increasing proliferation [28], ponds have not received considerable scientific attention compared to rivers and lakes [26]. It is worth mentioning that research regarding ponds in Europe has tripled in the last decade [29], where results showed that ponds contribute significantly to several basin related processes [30]. Examples of these processes are sediment interception [31], removal of pollutants for river protection [32], nutrient recycling [33], greenhouse gas emission [34], regulation of hydrological flows [35], biogeochemistry [30], and climate [36]. In addition to their environmental role, ponds have a well-known value for housing and sustaining biodiversity, supporting livelihoods, local economies, and taking part in the socio-cultural heritage of the settings in which they are located [37].

Under the hydro-sedimentological scope, particularly, ponds have shown to retain as much as 90% of sediments transported in basins [38]. Consequently, ponds have been heavily blamed for rupturing the ecological and sedimentary continuum of the basins to which they belong [39]. The disruptive effect of ponds is due to the increase of residence time of waters, resulting in a decline in the temporal variation of the main discharge [40]. Accordingly, the deceleration of overland flow allows suspended particles to settle under the effect of their weight, causing a reduction in the amount of sediments entrained by water, making ponds sediment sinks [40]. However, this effect strongly depends on their position in the basin, their depth, volume, slope [41], as well as the surrounding land occupation [42]. Winfield Fairchild and Velinsky (2006) [43] showed that ponds located upstream of rivers—considering their sediment retention capacity—are capable of creating a state of imbalance in the geochemical and hydro-sedimentary status of the underlying rivers. Consequently, the Directive Cadre sur l'Eau (DCE) [44] stresses the need to assess the impact of hydromorphological elements that are capable of influencing hydrologic pathways, river morphology, width, and continuity.

Beyond the contribution of isolated ponds, connected networks of ponds were found to contribute to basin processes at higher rates than lakes or even rivers [45,46]. Particularly in France, ponds are mainly concentrated in three regions: the Sologne region, Brenne (Central France), and Dombes (Eastern France). In response to DCE recommendations, this study aims to assess the impact of man–made ponds on soil erosion and sediment transport, at the scale of the Indre portion of the Claise basin. This part corresponds to the Brenne Natural Regional Park that houses 4500 waterbodies (ponds, marshes, and small water surfaces), 2179 of which are located in Claise, being part of an interconnected network. To evaluate erosion risks in the Claise, the Coordination of Information on the Environment CORINE (1992) Erosion Risk model will be used, since it presents a simplification of the reliable USLE model [1] and given that no-erosion field data, for the Claise, were available for the study. SWAT is employed to assess the impact of ponds on the Claise's hydro-sedimentary regime. This choice is owed to SWAT's ability to simulate the physical processes that occur in ponds, which in turn allows an accurate representation of pond containing basins [38]. This is achieved through the SWAT Pond (.pnd) input file that makes SWAT one of the few hydrological models having an input for ponds [47]. Related studies have assessed the behavior of ponds using SWAT [38,48] and highlighted the efficiency of SWAT on this part.

The impact of the Claise's ponds on erosion and sediment transport will be assessed by testing alternative scenarios, where the land occupation input for both models will be simulated with and without ponds. By this approach, a quantification of the pond impact can be obtained. The presented work serves as a decision-oriented tool for basins similar to the Claise, where pond proliferation has been halted until a proper understanding of their effect is established. In addition, analysis of soil erosion risks and sediment transport is useful for conservation measures that aim towards prolonging the useful life of these small water bodies or for ceasing their proliferation.

2. Materials and Methods

2.1. Study Area

The Brenne portion of Claise basin (Figure 1) covers an area of 707 km^2 and is one of the three entities housing the 4500 ponds of the Brenne Natural Regional Park, which are mostly grouped in the form of interconnected chains [36]. Within the Claise, 2179 of the Brenne waterbodies, along with the Five Bonds channel (Blizon), are the main contributors [49] to the 87.6 km long Claise River having an average discharge of 4.5 m^3/s [50]. The basin is mostly dominated by a degraded oceanic climate with a mean temperature of 11 °C and average annual precipitation of 700 mm [51]. As to its topographical profile, the basin is considered to be a flat plain, with 99% of its surface falling into the 0%–5% slope class, while its altitude varies between 76 m and 181 m. The six poorly-permeable soil classes of the basin are mostly dominated by Luvisols [36,52]. At a combined state, the Claise's challenging pedology, flat topographical setting, and quasi–impermeable lithology have resulted in the stagnation of incoming water leading to the formation of natural ponds [53]. According to Bennarrous (2009) [37], however, these ponds are not only a product of natural processes but also an anthropogenic adaptation to a poorly drained domain and a source of economic livelihood (aquaculture) in an environment of limited productivity. As a result of intensive pond proliferation throughout time, the basin has acquired a particular hydrographic network characterized by an abundance of different kinds of water bodies. Despite its richness, nonetheless, the hydrographic network of the basin is randomly organized and presents severe fragmentation [54]. In contrast to the evolving pond proliferation, the land occupation setting of the Claise has been relatively unchanged for the last 19 years, mainly displaying a dominance of an interlocked mosaic of forests, grasslands, and agricultural areas [54].

2.2. The SWAT Model: Basins and Impoundments

SWAT is one of the few hydrological models that have an input for ponds [47], which makes it an ideal tool for this study. SWAT is a semi-physical deterministic distributed and continuous hydrologic model that functions at a daily time step with options for sub-hourly routing, as well [55]. It is a quite complex model with significant input data demands. However, the basic components can be readily obtained and are relatively simple [56]. Essential inputs consist of weather data, digital elevation model (DEM), soil, and land occupation maps [57]. Based on these inputs, SWAT divides the basin into sub-basins, which further divides into smaller hydrological response units (HRUs) [58]. HRUs are defined as units with homogeneous land occupation, topography, and pedological properties [59]. These are generated in order to lump somewhat similar areas scattered through the basin into a single unit, simplifying the model's run by avoiding unpractical simulations while accounting for the diversity of different factors in the basin [60]. Within SWAT, ponds are defined as waterbodies, being an integral part of a sub-basin's hydrological network; they are capable of intercepting surface runoff [60], thereby modifying the hydro-sedimentary behavior of the basin. Since this work targets the erosion/sedimentary behavior of the SWAT model, only their related equations will be presented. Further details can be found in Neitsch et al. (2011) [47].

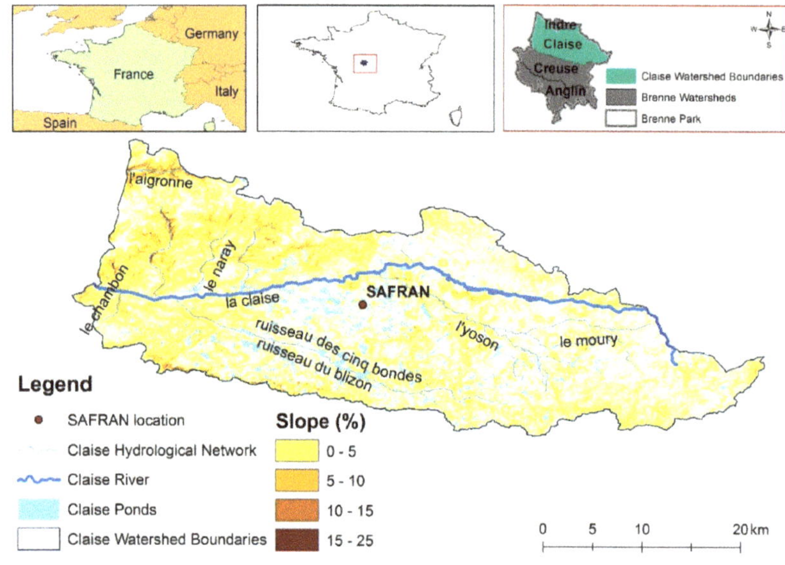

Figure 1. Study area.

Soil erosion is calculated based on the Modified Universal Soil Loss Equation (MUSLE) using the following formula [55]:

$$sed = 11.8 \times (Q_{surf} \cdot q_{peak} \cdot area_{hru})^{0.56} \times K_{USLE} \times C_{USLE} \times P_{USLE} \times LS_{USLE} \times CFRG \tag{1}$$

where sed is the HRU sediment yield (t); Q_{surf} is the surface runoff volume (mm); q_{peak} is the peak runoff rate (m^3/s); $area_{hru}$ is the area of the HRU (ha); K_{USLE} is the Universal Soil Loss Equation (USLE) soil erodibility factor; C_{USLE} is the USLE cover and management factor; P_{USLE} is the USLE support practice factor; LS_{USLE} is the USLE topographic factor; and $CFRG$ is the coarse fragment factor.

It should be noted that MUSLE virtually calculates the sediment yields due to soil erosion that reach the main streams of the sub-basins in a unit of time. This is the reason why sediment delivery ratios are not required, contrarily to USLE and RUSLE [61].

The mass balance equation of sediment in ponds is described by the following formula [47]:

$$sed_{wb} = sed_{wbi} + sed_{flowin} - sed_{stl} - sed_{flowout} \tag{2}$$

where sed_{wb} is the amount of sediment at the end of the day; sed_{wbi} is the amount of sediment at the day's beginning; sed_{flowin} is the amount of sediment provided from inflows; sed_{stl} is the amount of settled sediments; and $sed_{flowout}$ is the amount of sediment transported as outflow. All components are expressed in metric tons.

2.3. The CORINE Erosion Model

The CORINE erosion model is a simplification of the USLE model. In the CORINE model, erosion risks are classified on a scale of 0–3, with 0 corresponding to the no-erosion class, 1 to low erosion risks, 2 to moderate erosion risks, and 3 to high-erosion risks [62]. For the estimation of erosion using the

CORINE model, parameters such as soil erodibility, climate erosivity, topography (slope), and LU/LC (vegetation cover) are required [61]. Each parameter, in turn, consists of several sub-factors, and is classified according to the CORINE model into respective indices (Figure 2). Once established, the soil erodibility index is combined with climate erosivity and slope, to yield the potential soil erosion risk map. Subsequently, the potential soil erosion risk map indices (0–3) are crossed with those of the vegetation cover to yield the actual soil erosion map.

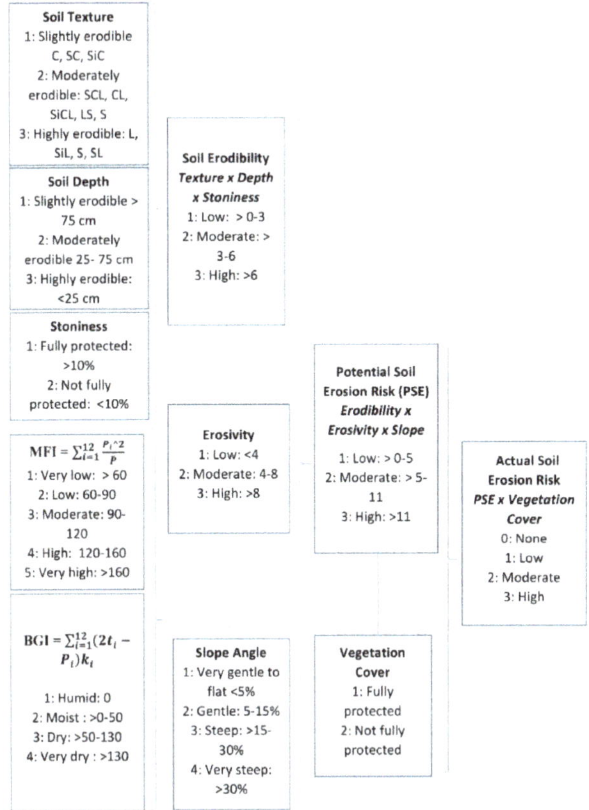

Figure 2. CORINE model framework, MFI: Modified Fournier Index [63]; P_i is the total precipitation in month i and $\overline{P}$ is the mean annual total precipitation; BGI: Bagnouls–Gaussen Index [64]; t_i the mean temperature for the month i; P_i the total precipitation in month i and k_i the proportion of the month i in which $2t_i - P_i > 0$ [62].

2.4. Input Data and Adaptation to the SWAT and CORINE Models

The input data used in this study, along with a short description, are summarized in Table 1.

Table 1. CORINE erosion model and SWAT input data.

Input Data	Source	Date [†]	Description
Topography—DEM	Institut Géographique National (IGN)—France	2010	25 m resolution
Soil map	Harmonized World Soil Database (HWSD) [52]	2008	30 arc-second raster database
Weather data	Weatherlink Pro 2 weather stations, source: R. Nedjai (Dynétangs)	2016–2018	Daily time step
	Système d'Analyse Fournissant des Renseignements Adaptés à la Nivologie (SAFRAN) model (Durand et al., 1993)	1970–2018	
Land Use/Land Cover map	Digitized from aerial photography and cross-checked against ancillary maps	2018	Ortho-rectified aerial photography-Resolution: 0.5 m verified against ancillary maps: CORINE 2012 land use/land cover maps

[†] The correspondence of the Claise basin to a natural park renders any human induced modifications on soils minor. Further, the Claise's land occupation pattern is relatively stable. Therefore, climate is the only variable factor in the study area. Hence, the temporal difference of the utilized datasets does not cause temporal induced biases.

ArcSWAT version 2012.10_3.19 was applied in an ArcGIS 10.3 (Environmental Systems Research Institute, Inc. (Esri), 380 New York Street, Redlands, California, USA) environment to perform simulations at a daily time step for the period 1970–2018. The first seven years of the simulation, from 1970 to 1976, were used for the model's warm up. The calibration and validation phases of the model were carried out by means of comparison between simulated discharges and measured discharge data from the station L6202040–La Claise au Grand–Pressigny (Pont de Fer), obtained from Eau France Banque HYDRO–Ministère de l'Ecologie, de l'Energie, du Développement Durable et de l'Aménagement du Territoire (MEEDDAT)/Direction Générale de la Prévention des Risques (DGPR)/Service des Risques Naturels et Hydrauliques (SRNH). Due to the unavailability of sediment records for this study, only a hydrologic calibration was performed. However, surface runoff and stream flow—regardless of the approach used to model the sedimentary cycle—are the driving forces of runoff and streambed erosion, as well as of overland and stream sediment transport [65]. Hence, one might say that the calibration of the sediment part of the study might be implicit, but nonetheless robust. According to Hallouz et al. (2018) [66], a calibrated hydrologic model gives a certain degree of reliability to the sedimentary output. However, the availability of sediment data is often a constraint in relative studies, as measurements, either in the form of gross or net erosion or in the form of total sediment discharge, in streams, are often nonexistent. Even in cases where such data is available, it is in the form of sparse discrete measurements and rarely in the form of continuous sediment graphs, where they could be used for a robust calibration. In the case of the Claise, this would be a very challenging task, given the large number of ponds. For this reason, SWAT was calibrated according to the calibration scheme of Jalowska and Yuan (2019) [38], where a complete representation of an impoundment effect (here, ponds) is ensured. Nonetheless, the absence of sediment records poses some short of limitation that needs to be considered by the decision-makers of the basin.

2.4.1. DEM: Topographic Effect and Use in Both Models

In the CORINE model, DEM is used to extract the topographic parameter by means of slope computation. The slope angles of the study area were obtained using the "slope" tool of ArcGIS. Subsequently, the slope raster was classified, with respect to the CORINE's model indices, into (1) very gentle to flat (<5%), (2) gentle (5%–15%), (3) steep (15%–30%), and (4) very steep (>30%).

In SWAT, the DEM is used for the extraction of the topographic parameters and for representing the basin's physical parameters such as slopes, flow direction and accumulation, delineation of the hydrologic network, and basin partitioning [67]. After inputting the DEM into SWAT, the basin was initially divided into 25 sub-basins using SWAT's watershed delineator. However, given the presence of a great number of waterbodies, and their impact on output accuracy, the number of sub-basins was increased to 35. The purpose of increasing the sub-basins' number was to increase the spatial resolution for a more detailed representation of the basins' processes [68]. Since the SWAT model allows only one pond per sub-basin [39] during the sub-basins' delineation, the largest number of ponds was included in each in order to ensure maximal representation of pond-induced processes. Further, all ponds within each sub-basin were lumped and the outlets of the sub-basins were chosen to coincide with those of the ponds in order to maximize their representation and account for their effect. The SWAT delineated hydrologic network was, in turn, verified against a pre-defined stream network in order to ensure and improve its accuracy.

2.4.2. Pedology: Adaptation to the Different Requirements of the Models Used

The pedological composition of the Claise was determined from the Harmonized World Soil Database (HWSD) [52]. For the CORINE model, soil texture was determined using the USDA textural triangle after inputting the respective percentages of sand, silt, and clay for each soil group obtained from the HWSD. Texture was then classified with respect to the CORINE indices. The parameters of depth and stoniness were treated and classified similarly. After obtaining the texture, depth, and stoniness sub-factors, the three layers were input into the "raster calculator" in analogy to Figure 2.

For the SWAT model, the hydrologic soil group of each class was assigned following the United States Department of Agriculture [69] Soil Conservation Service (SCS) soil survey. Bulk densities were computed using the Soil Water Characteristics software following Saxton and Rawl (2006) [70] equations that were shown to have adequate accuracy for bulk density computation by Al Sayah et al. (2019) [71]. Likewise, the available water capacity and saturated hydraulic conductivities were also computed using the same software. Organic carbon content was determined by multiplying the organic matter content derived from the HWSD by 0.58, since organic carbon forms around 58% of the soil's organic matter as a rule of thumb [72]. Sand, silt, and clay percentages, as well as the rock fragment contents were extracted from the HWSD. The USLE_K factor was computed following the formula presented in the SWAT documentation [55], while soil surface albedo was determined using the following formula [73]:

$$Soil\ albedo\ (0.3 - 2.8\ \mu m) = 0.069 \cdot (color\ value) - 0.114 \tag{3}$$

After building the soil database, all parameters were reclassified in SWAT using user-adapted look-up tables for integration into the SWAT database.

2.4.3. Weather Data: Forcing on-Field Weather Data to a Meteorological Model

Weather data was obtained from two sources, on-field weather station data and Météo France's Système d'Analyse Fournissant des Renseignements Adaptés à la Nivologie (SAFRAN) model [74]. The SAFRAN model was used since it contains records for the period 1970–2018 while the weather stations, located next to a pond network, were used to account for the climatic regulating effect reported by Nedjai et al. (2018) [36]. SAFRAN was used in order to provide a large time span for the study. Moreover, the SAFRAN was validated by test-correlation with weather station data using Pearson's correlation coefficient "r" [75]. Prior to correlation, harmonization of both datasets was performed since the weather stations record parameters at the hourly time step, while SAFRAN simulates at the daily time step. Results of the correlation are presented in Table 2.

Table 2. Validation testing of the SAFRAN model against measured on field-data for revealing SAFRAN's validity for use.

Weather Stations Parameters	SAFRAN Parameters	Parameter Label	Correlation, r
Temp Out	T_Q	Average temperature (°C)	0.98
Hi Temp	TSUP_H_Q	Maximal temperature (°C)	0.98
Low Temp	TINF_H_Q	Minimal temperature (°C)	0.97
Out Hum	HU_Q	Relative humidity (%)	0.97
Wind Speed	FF_Q	Wind speed (m/s)	0.83
Rain	PE_Q	Efficient rainfall (mm)	0.64
Solar Rad.	SSI_Q	Incoming solar radiation (J/cm^2)	0.98

For the CORINE model, the rainfall and temperature parameters are used for computation of the MFI and BGI indices following their respective formulas (Figure 2); these indices were then used to calculate erosivity following the workflow of Figure 2. In SWAT, the weather database was built using the WGN maker macro-tool [76] and input into the SWAT database.

2.4.4. Land Use and Land Cover: A Particularly Rich Natural Limnological Setting

The land occupation map of the Claise was obtained by on-screen digitizing of aerial photography at 0.5 m resolution. Such fine scale was used in order to ensure a detailed representation of the study area, particularly to account as accurately as possible for the scale of ponds.

For the CORINE model, the obtained land occupation map was reclassified into fully protected (forest, permanent pasture, and scrublands) and not-fully protected (cultivated or bare land) areas as demonstrated in Table 3.

Table 3. Numerical distribution of the land occupation setting of the Claise basin.

Land Occupation Class	Area (km^2)	Percentage (%)	CORINE Vegetation Cover Index
Clear broad—leaved forest	2.21	0.31	1
Clear mixed forest	0.90	0.13	1
Coniferous forest	26.94	3.81	1
Dense broad—leaved forest	163.58	23.11	1
Dense mixed forest	27.00	3.82	1
Field crops in medium to large terraces	19.14	2.70	2
Fruit trees	0.20	0.03	2
Grassland	207.04	29.26	1
Inland marshes	4.01	0.57	1
Low density urban tissue	3.24	0.46	2
Medium density urban tissue	1.76	0.25	2
Mineral extraction site	0.09	0.01	2
Non–irrigated field crops	151.17	21.36	2
Pond	79.47	11.23	0
River	0.55	0.08	0
Scrubland	2.80	0.40	1
Scrubland with some bigger dispersed trees	15.36	2.17	1
Urban expansion site	0.01	0	2
Urban sprawl on clear wooded lands	0.01	0	2
Urban sprawl on field crops	1.01	0.14	2
Urban sprawl on grassland	1.20	0.17	2

In the case of SWAT, land use and land cover classes were reclassified into SWAT's classes. Particular attention was given to the water (in SWAT terms WATR) classes. As reported by Jalowska and Yuan (2019) [38], the reason for this is that though SWAT allows the creation of HRUs with WATR, water bodies should be modeled either as reservoirs or ponds. Almendinger et al. (2014) and Jalowska and Yuan (2019) [38,48] have shown that an accurate representation of basin processes requires the integration of the SWAT model's impoundments function. Furthermore, Wang et al. (2008) [77] highlighted the importance of considering impoundments, such as ponds, by testing scenarios of impoundment integration versus impoundment disregarding in a basin covered by only 3% of impoundments. Their results confirmed that simulations were considerably affected even with this small cover. In the same manner, Jalowska and Yuan (2019) [38] simulated different scenarios following integration or absence of impoundments like reservoirs rather than just a normal "water" land occupation class. They noted that disregarding impoundments leads to a series of uncertainties starting by an inaccurate SWAT performance, which in turn leads to inefficient calibration efforts, and overall inaccurate model performance.

After computing the slope, erosivity and soil erodibility indices, the potential soil erosion risk map was constructed. By combining the potential soil erosion risk map to the vegetation cover layer in the "raster calculator" tool and crossing each ones indices, the actual soil erosion risk map was obtained.

3. Results

3.1. CORINE Erosion Model Outputs for the Claise Basin

In the following sections, the erosion assessment for the Claise is presented and each component is detailed accordingly.

3.1.1. Soil Erodibility

Soil erodibility parameters are presented in Figure 3. The dominant textural class in the basin was found to be loam (68.4%) followed by loamy sand (28.75%), while the remainder percentages are clay (2.48%) and sand (0.34%). In terms of texture, since most of the study area is covered by loam with respect to [62], the study area predominantly falls into the highly erodible class. In terms of soil depth, 94.6% of the Claise fits to the slightly erodible class (1000 mm > 750 mm), while the remainder 5.4% rests within the moderately erodible class (300 mm, corresponding to the 250–750 mm class). As far as stoniness is concerned, only 6% of the Claise is under the fully protected cover.

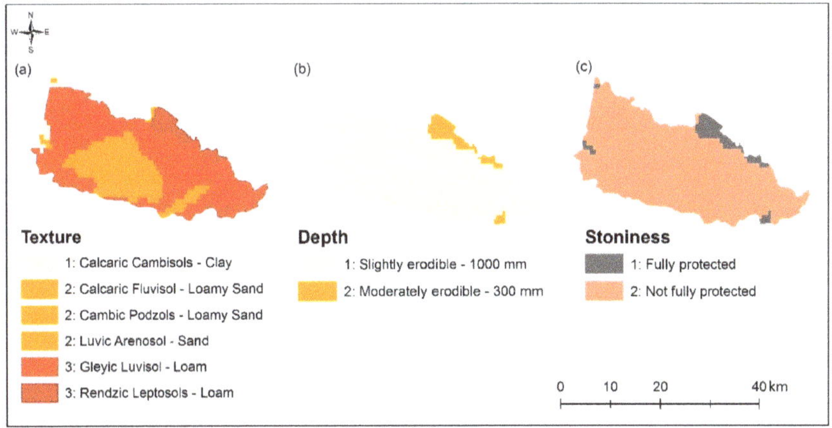

Figure 3. Spatial distribution of soil erodibility parameters: (**a**) texture, (**b**) depth, (**c**) stoniness, and class descriptions of the Claise with their respective indices.

Table 4 provides the numerical description of the soil's erodibility components.

Table 4. Numerical distribution of soil erodibility parameters and corresponding covered areas.

Parameter	CORINE Class	Area (km^2)	Percentage (%)
Soil Texture	1: Slightly erodible (clay)	17.53	2.48
	2: Moderately erodible (loamy sand and sand)	205.88	29.12 (28.77 and 0.35)
	3: Highly erodible (loam)	483.59	68.4
	Total	**707**	**100**
Soil Depth	1: Slightly erodible (>1000 mm)	668.82	94.6
	2: Moderately erodible (250–750 mm)	38.18	5.4
	Total	**707**	**100**
Stoniness	1: Fully protected (>10%)	43.26	6.12
	2: Not fully protected (<10%)	663.73	93.88
	Total	**707**	**100**

By inputting the texture, depth and stoniness parameters into the "raster calculator" tool of ArcGIS, the soil erodibility raster was generated and then reclassified using the "reclassify" tool (Figure 4).

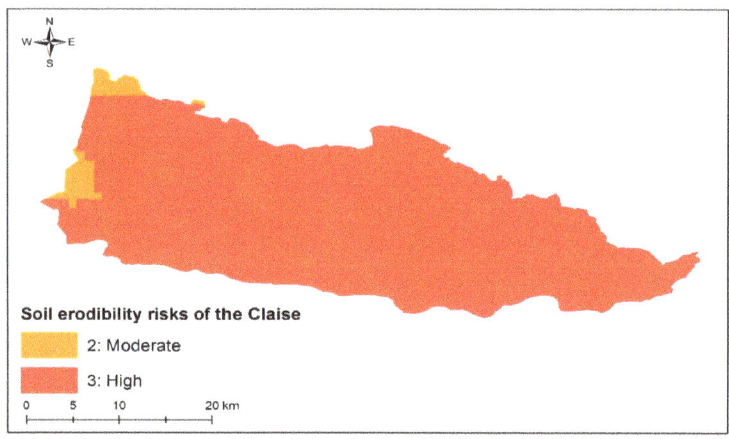

Figure 4. Soil erodibility map of the Claise classified according to the CORINE indices.

As seen in Figure 4, the generated soil erodibility map revealed that about 96.83% (684.37 km^2) of the study area is covered by highly erodible soils, while the remainder 3.17% (22.63 km^2) correspond to moderate-risk areas. This fact is mainly due to the textural distribution of soil and the little stone cover of the Claise basin.

3.1.2. Erosivity in a Degraded Oceanic Climate Setting

Meteorological data of the study area is presented in Table 5. As seen, no dry month exists in the Claise and highest temperatures are recorded during the month of July making the basin subject to continuous precipitation.

Table 5. Average temperature and precipitation for the Claise basin (1970–2018).

Month	January	February	March	April	May	June	July	August	September	October	November	December
Average temp. (°C)	4.7	5.2	7.8	10.4	14.3	17.9	20.1	19.7	16.1	11.8	7.7	4.8
Precipitation (mm)	416	677	1106	1526	1782	2045	2108	1826	1356	818	482	361

MFI was found to be 80, corresponding to the low variability class. This signifies evenly distributed rainfall, thus reducing risks of climate-induced soil erosion [78]. The Bagnouls–Gaussen Aridity Index (BGI) on the other hand, was found to be "0", signifying that the study area corresponds to the humid regime as a result of its oceanic influence. In terms of CORINE erosivity, the Modified Fournier Index (MFI) belongs to class 2 variability (low), while BGI corresponds to class 1 (humid). Accordingly, the erosivity index of the Claise watershed was found to be 2, corresponding to the low erosivity class.

3.1.3. Topography (Slope): A Reduced Effect in a Flat Setting

Using the "slope" tool in ArcGIS, slope angles were extracted from the DEM. Figure 5 displays the slope of the Claise basin and the adapted reclassified raster following CORINE classification.

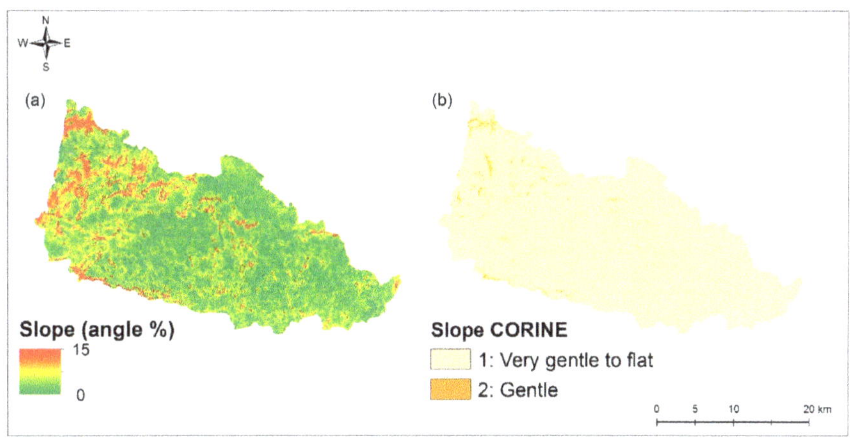

Figure 5. (**a**) Slope angle of the Claise and (**b**) the corresponding CORINE indices description.

From Figure 5, it can be observed that the Claise presents a dominantly flat topography with a maximum slope of 15°. In CORINE terms, 99.3% (702.051 km^2) of the study area corresponds to "very gentle to flat" slope class (<5°) and 0.7% (4.94 km^2) to the "gentle" slope category (5°–15°). The particular setting of low relief and small slope significantly plays a role in the reduction of erosion generated by runoff [79], despite the high soil erodibility risks.

3.1.4. Potential Soil Erosion Map

By overlaying soil erodibility (Figure 4), topography and erosivity, using the "raster calculator" tool, the potential soil erosion risk map was obtained (Figure 6), following the formula presented in Figure 2.

From Figure 6, low-erosion risks present 3% (21.22 km^2), moderate risks present 96.5% (682.25 km^2) and high risks form 0.5% (3.53 km^2). These percentages, however, reflect only the potential soil erosion risks which according to CORINE (1992) [62] do not take into consideration the vegetative cover at this stage. The slope effect and equal repartition of precipitation, as reflected by the MFI values and the absence of arid periods (low BGI), are potentially responsible for moderating the soil erodibility map yielding the dominantly moderate potential soil erosion risks.

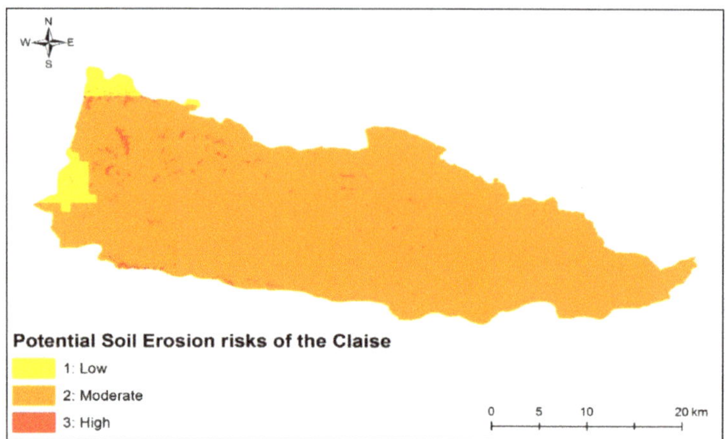

Figure 6. Potential soil erosion (PSE) risk map of the Claise basin.

3.1.5. Vegetation Cover: The Presence/Absence Effect of Ponds

This factor is the focal point of this study since by slightly changing this parameter the outcome changes significantly. At this point, two vegetative cover scenarios are presented: the first presents the actual setting of the Claise accounting for the presence of ponds, while the second simulates a scenario where ponds are removed to assess the difference in erosion outcomes with and without their presence. This last step allowed to quantify the impact of ponds on erosion in the Claise basin. In the second scenario, the land occupation group "ponds" was changed to their surrounding class (grasslands).

The large area of ponds, which makes up around 11% of the Claise (Table 3), displays their potential role as modifiers basin processes. Figure 7 presents the two considered vegetation covers as inputs for crossing with the potential soil erosion risk map to yield two different actual soil erosion risk maps. These maps were then compared to evaluate the effect of the presence and absence of ponds and their role on erosion.

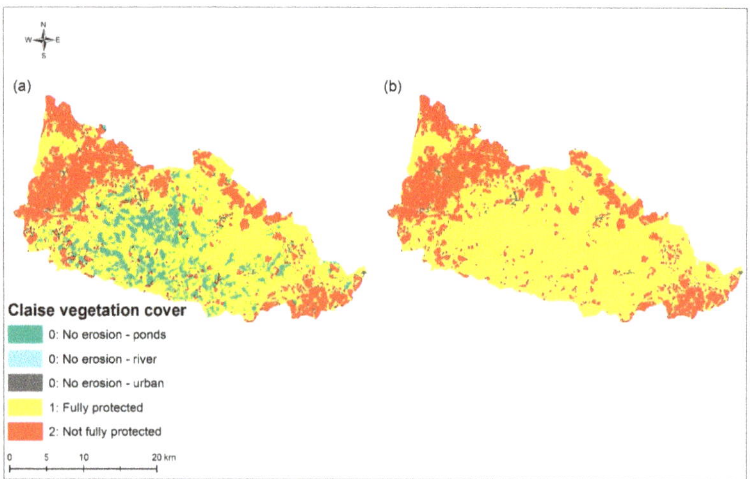

Figure 7. (**a**) Current Claise vegetation cover and (**b**) alternative vegetation cover and corresponding CORINE classification; 1: fully protected, 2: not fully protected.

Class 0 refers to land cover categories that are not considered in CORINE; these categories are urban areas and water bodies, while classes 1 and 2 refer to the fully protected and not fully protected covers. From Figure 7b, 72.5% of the Claise basin corresponds to the fully protected class, while 27.5% of the study area is occupied by not fully protected cover. To assess the impact of pond presence, the second scenario of replacing ponds by their surrounding dominant cover was performed.

3.1.6. Actual Soil Erosion Maps Under Current and Alternative Scenarios

The two actual soil erosion risks maps were produced by multiplying the respective indices of the potential soil erosion risk map and the two vegetation cover scenarios using the "raster calculator" tool. Figure 8 reveals the outcome under both scenarios.

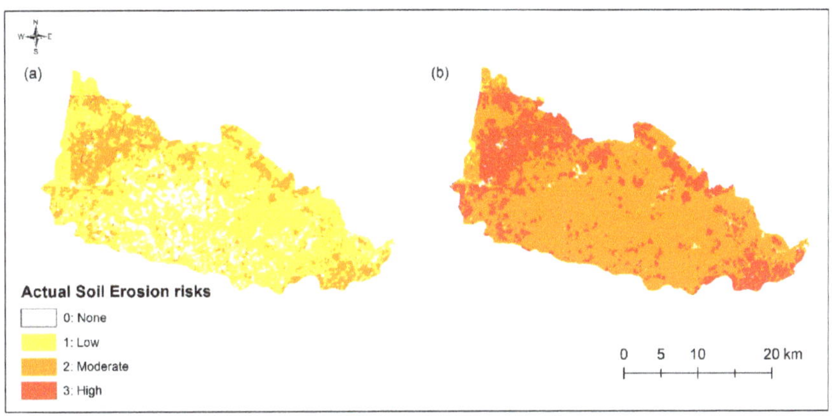

Figure 8. Actual soil erosion risk map of the Claise under (**a**) current vegetation cover and (**b**) the alternative pondless scenario.

The Actual soil erosion map was cross-checked against: Institut National de la Recherche Agronomique (INRA) 2000 erosion maps and the combined GIS sol–INRA–SOeS 2011 maps [80]. These were produced following the Modèle d'Évaluation Spatiale de l'Aléa d'Érosion des Sols (MESALES). The comparison between the INRA maps, and the produced actual soil erosion risk map, is presented in Table 6. By this comparison, it is concluded that there is a good agreement between these maps. In addition, since INRA maps are not completely adequate to be considered at the basin scale [80], the established erosion maps are considered for the no-erosion zones, overcoming, this way, the challenge of coarse representation.

Table 6. Verification of the established erosion map.

Classes	Low	Moderate	High	Validation Points
Low	76	10	0	86
Moderate	15	14	8	37
High	0	0	11	11
Total	91	24	19	134

As can be seen from Table 6, a total of 134 validation points were chosen. These were divided into 86, 37, and 11 low-erosion, moderate-erosion, and high-erosion zones validation points, respectively. A large part of the moderate-erosion class was misinterpreted as the low-erosion class. This discrepancy is due to the fact that the no-erosion zones do not exist in the INRA maps, but are instead classified as low-erosion zones. Therefore, the error margin in the moderate-erosion class from Table 6 is justified by the finer scale representation of the produced maps, compared to the INRA maps. The overall accuracy was determined to be 75%, while the computed Cohen's kappa coefficient [81] was found to be 0.7; this indicates a substantial agreement between the INRA maps, and the produced actual soil erosion risk map. The kappa coefficient was used since it tests inter-rater reliability; i.e., the coefficient represents the extent to which the generated data are correct representations of the measured data. In the case of this study, the generated data is the actual soil erosion risk map, while the measured data consists of the validation points obtained from the INRA maps.

From Figure 8, three main results can be drawn:

(1) The role of vegetation cover in changing erosion risks is solidified. This is particularly reflected by the setting of the Claise basin due to the agricultural and grass cover. By comparing the actual soil erosion risk map with the potential soil erosion risk map and statistics, a shift of erosion risk classes is observed. As mentioned in Section 3.1.4, considering the potential soil erosion risks, and ignoring the vegetation cover, low, moderate and high-risk areas take over 3%, 96.5% and 0.5% of the total basin area, accordingly. When the vegetation cover layer was taken into account the resulting actual soil erosion risk shifted to 65.66%, 21.68% and 0.18%, for low, moderate and high-risk areas. These observations solidify that vegetative cover is the most influential aspect for erosion assessment. In further detail regarding the vegetation cover layer, areas corresponding to agricultural classes are seen to have higher erosion risks than areas with different land cover types. This is in agreement with Verheijen et al. (2009) [82] observations that despite the considerable effect of soil type, topography and climatic conditions, the major influencer of soil erosion is the vegetative cover, especially cultivated areas.

(2) The remainder 12.48% of the actual soil erosion risk map is the no-erosion zone. As seen in Figure 7a, most of the no-erosion zone corresponds to the concentration area of ponds, while the remainder 1.48% represents the Claise River. The ponded area represents 88.23 km^2 of the Claise under no risk of erosion, making these ponds a counter-erosion zone.

(3) At a graphical scale, a complete shift from low to moderate risks, in the greatest part of the basin, is observed. Table 7 presents the statistical difference between the actual soil erosion risks with current vegetation cover and those of the pondless scenario. From Table 7, the effective role of ponds as an erosion counter-measure is revealed.

Table 7. Actual soil erosion risk (ASE) for the Claise basin under current and simulated vegetation cover.

Erosion Risks	ASE with Current Vegetation Cover—Area (km^2)	ASE with Simulated Vegetation Cover (Absence of Ponds)—Area (km^2)	ASE with Current Vegetation Cover—Percentage (%)	ASE with Simulated Vegetation Cover (Absence of Ponds)—Percentage (%)
None	88.23	7.92	12.48	1.12
Low	464.21	3.6	65.66	0.52
Moderate	153.27	543	21.68	76.8
High	1.272	152.4	0.18	21.56

Additionally, the impact of ponds on erosion at the scale of the basin is revealed. Not only did the no-erosion and low-erosion classes decrease by 11.36% and 65.14%, in the absence of ponds scenario, but also the moderate and high-erosion risks increased by 55.12% and 21.38%, respectively. These changes are due to several reasons:

(1) The most evident reason is that ponds effectively and directly nullify splash erosion in the areas they occupy.

(2) Their widespread, yet dense, positioning throughout the basin, counteracts runoff erosion in a twofold way: first, by intercepting eroded soils by overland flow, retaining this way, the transported material and preventing them from reaching the streams, and second, by slowing surface runoff and thus, abating its erosive force. Despite the fact that the low-slope topography does not particularly favor runoff erosion, let alone high velocities of overland flow, this obviously has some effect, especially in cases of intense rainfall events.

(3) Their dense aggregation in the basin attributes them the role of cascade check dams, containing sediments.

(4) Their large density and chain sequence where the retention effect is greatly amplified (factor of 2179 ponds) [38].

(5) The highly erodible setting of the basin resulting from a challenging pedology.

3.2. Sediment Transport in a Limnologically Rich Setting

Sediment transport in the Claise was simulated using the SWAT model. The model was calibrated using the SUFI-2 algorithm of SWATCUP following Jalowska and Yuan (2019) [38] proposed sequential calibration for settings characterized by the presence of water impoundments. As mentioned previously, due to the unavailability of sediment measurements for this study, only a hydrologic calibration was performed with the most related sensitive parameters (Table 8). Results of the hydrologic calibration yielded an R^2 of 0.7 during calibration and 0.67 during validation.

Table 8. Sensitivity analysis for calibration of the SWAT model.

Parameter Name	t-Stat	p-Value	Fitted Value
R__WET_NVOL.pnd	−37.52	0.00	0.823
V__GW_DELAY.gw	−31.39	0.00	52.325
R__PND_SED.pnd	1.62	0.11	318.100
V__GWQMN.gw	1.59	0.11	1.461
R__SPCON.bsn	−1.35	0.18	0.005
R__CN2.mgt	1.29	0.20	0.564
R__PND_NSED.pnd	1.23	0.22	2420.81
R__USLE_K.sol	−0.97	0.33	0.209
V__ALPHA_BF.gw	−0.90	0.37	0.529
R__NDTARG.pnd	0.44	0.66	49.890
R__PND_FR.pnd	0.15	0.88	0.644
R__USLE_P.mgt	−0.11	0.91	0.188
R__IGRO.mgt	−0.04	0.97	0.080

As a result of an exhaustive hydrologic calibration, the average sediment yield of the Claise basin for the studied period under current and alternative conditions is presented in Figure 9.

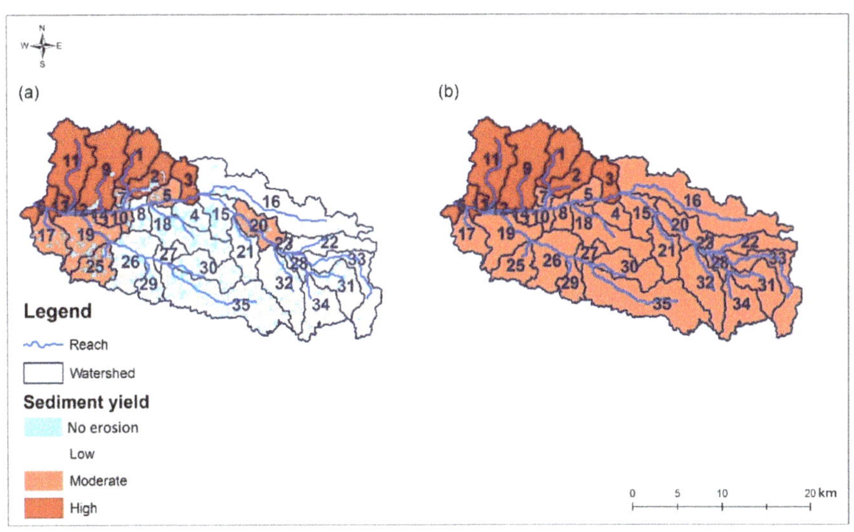

Figure 9. Sediment yields of the Claise basin under (**a**) current conditions and (**b**) the pondless scenario.

As can be seen from Figure 9, low sediment yields correspond to ponded sub–basins in contrast to pondless sub-basins which present lower soil loss rates. However, this gradient varies according to the setting of the ponds in these sub–basins. Despite that sub-basins 1 and 9 contain a small number of ponds, these correspond to areas of high sediment yield as a result of their dominant agricultural cover and their location in the basin's steepest areas. This shows that the presence of ponds alone is not sufficient to alter the sediment yields. In fact, the setting of ponds under the form of a collective dense network is the main modifier of sediment transport patterns. In order to further highlight the impact of ponds, alternative scenario testing was performed where a pondless land occupation setting was re-inputted to the calibrated SWAT model in order to determine subsequent sediment transport changes. Results of sediment yields, highlighting the impact of ponds, are presented in Table 9.

Table 9. Shifts of sediment yields as a result of pond presence or absence.

Sediment Yield	Pond Presence (Area km^2; Percentage %)	Pond Absence (Area km^2; Percentage %)
Low	365.07; 63.31	0; 0
Moderate	89.23; 15.47	454.30; 78.79
High	122.27; 21.20	122.27; 21.20

In the low sediment yield sub-basins, where the vast majority of ponds lies, occupying a large portion of the sub-basins they are located in, the land cover change from pond to grasslands is far too great and the effect of ponds presence/absence is quite obvious. Accordingly, the role of ponds as integral features of the landscape processes is solidified. Having determined their effect, an increased exposure for their integration into management plans is recommended, while an understanding of basin scale sediment dynamics offers insights regarding the role of ponds in water resources. Such implications could matter not only for the Claise and similar basins, but also for expectations regarding the landscape role of sediment retention basins, and for the general understanding of sediment transport in rain-dominated contexts. This, in turn, could be considered as a contribution to hydrologic modelling efforts, where small waterbodies, especially artificial ones, are often neglected, or even smoothed out, from digital elevation models.

4. Conclusions

An investigation of the pond-induced effects on soil erosion and sediment transport of limnologically rich basins was presented. Through this task, recommendations of the European framework for the Thematic Strategy on Soil Protection were addressed by revealing the different levels of soil loss, represented by providing an insight to the investigation of erosion-prone regions and sediment yield zones of different levels. Furthermore, recommendations of the DCE regarding the behavioral understanding of hydromorphological alternating factors (ponds as hydro-sedimentary elements), at the basin scale, were also considered. Despite the Claises' weakly structured pedology, resulting in high erodibility, the Claise was found to have low erosion risks and sediment transport rates due to several reasons, like the evenly distributed rainfall, the relatively flat topography, and most importantly due to the presence of its dense pond network that acts as a natural measure against soil erosion. This was solidified by simulating a scenario where ponds were substituted by grasslands. After replacing ponds by a protective vegetative cover, all erosion risks and sediment yield classes of the Claise significantly varied: no and low-erosion risk zones decreased, while moderate and high-erosion risk zones increased. Regarding sediment transport, the replacement of ponds by grasslands led to a complete disappearance of low sediment yield zones and considerable increases of moderate and high sediment yield zones. Accordingly, the "safe" soil loss status of the Claise can be attributed to the low soil loss rates of the Claise basin to the presence of these ponds. Despite their protection against soil erosion, however, their presence in very large numbers might cause a distortion in the sediment balance of the underlying rivers. Such cases may lead to sediment starvation and force the river to engage in increasing streambed and bank erosion [83]. Therefore, the use of the presented approach may serve as an efficient tool towards the orientation of future decisions regarding the proliferation or cease of ponds depending on their effect.

Author Contributions: Conceptualization: R.N. and M.J.A.S.; data curation: M.J.A.S.; methodology: M.J.A.S. and R.N.; project administration: R.N., C.A., and M.K.; supervision: R.N., K.K., C.A., and M.K.; validation: M.J.A.S.; visualization: M.J.A.S. and K.K.; writing—original draft: M.J.A.S. and K.K.; writing—review and editing: M.J.A.S., K.K., R.N., C.A., and M.K.

Funding: This research is part of a PhD thesis funded by the National Council of Scientific Research—Lebanon (CNRS—L), Agence Universitaire de la Francophonie (AUF), Lebanon, and the Lebanese University. It is also part of the Dynétangs project funded by the French Centre–Val–de–Loire region.

Acknowledgments: The authors would like to express their gratitude for the funding agencies, the editor, and reviewers for leveraging the quality of this work and to the Brenne Natural Park for their help.

Conflicts of Interest: The authors declare no conflict of interest.

References

1. Gyamfi, C.; Ndambuki, J.M.; Salim, R.W. Simulation of sediment yield in a semi–arid river basin under changing land use: An integrated approach of hydrologic modelling and principal component analysis. *Sustainability* **2016**, *8*, 1133. [CrossRef]

2. Dwarakish, G.S.; Ganasri, B.P. Impact of land use change on hydrological systems: A review of current modeling approaches. *Cogent Geosci.* **2015**, *1*, 1115691. [CrossRef]

3. van Rijn, L.C. *Principles of Sediment Transport in Rivers, Estuaries and Coastal Seas*; Aqua Publications: Amsterdam, The Netherlands, 1993; p. 690.

4. Esa, E.; Assen, M.; Legass, A. Implications of land use/cover dynamics on soil erosion potential of agricultural watershed, northwestern highlands of Ethiopia. *Env. Syst. Res.* **2018**, *7*, 21. [CrossRef]

5. Aksoy, H.; Mahe, G.; Meddi, M. Modeling and Practice of Erosion and Sediment Transport under Change. *Water* **2019**, *11*, 1665. [CrossRef]

6. Zhang, S.; Li, Z.; Lin, X.; Zhang, C. Assessment of Climate Change and Associated Vegetation Cover Change on Watershed–Scale Runoff and Sediment Yield. *Water* **2019**, *11*, 1373. [CrossRef]

7. Bussi, G.; Dadson, S.J.; Prudhomme, C.; Whitehead, P.G. Modelling the future impacts of climate and land–use change on suspended sediment transport in the River Thames (UK). *J. Hydrol.* **2016**, *542*, 357–372. [CrossRef]

8. Ffolliott, P.F.; Brooks, K.C.; Neary, D.G.; Tapia, P.R.; Chevesich, P.G. *Soil Erosion and Sediment Production on Watershed Landscapes: Processes and Control, UNESCO Special Technical Publication No. 32*; USDA: Washington, DC, USA, 2013; p. 73.

9. Aga, A.O.; Chane, B.; Melesse, A.M. Soil erosion modelling and risk assessment in data Scarce Rift Valley Lake Regions, Ethiopia. *Water* **2018**, *10*, 1684. [CrossRef]

10. Laflen, J.M.; Lane, L.J.; Foster, G.R. WEPP: A new generation of erosion prediction technology. *J. Soil Water Conserv.* **1991**, *46*, 34–38.

11. Flanagan, D.C.; Gilley, J.E.; Franti, T.G. Water Erosion Prediction Project (WEPP) Development History, Model Capabilities and Future Enhancements. *Am. Soc. Agric. Biol. Eng.* **2007**, *50*, 1603–1612. [CrossRef]

12. Wischmeier, W.H.; Smith, D.D. *Predicting Rainfall Erosion Losses from Cropland East of the Rocky Mountains–Guide for Selection of Practices for Soil and Water Conservation, Agriculture Handbook No. 282*; USDA: Washington, DC, USA, 1965.

13. Williams, J.R. Sediment–Yield Prediction with Universal Soil Loss Equation Using Runoff Energy Factor. Present and Prospective Technology for Predicting Sediment Yields and Sources. In Proceedings of the Sediment Yield Workshop, Oxford, MS, USA, 17–18 January 1975; pp. 244–252.

14. Renard, K.G.; Foster, G.R.; Weesies, G.A.; Mc Cool, D.K.; Yoder, D.C. *Predicting Soil Erosion by Water: A Guide to Conservation Planning with the Revised Universal Soil Loss Equation (RUSLE)*; US Government Printing Office: Washington, DC, USA, 1997; Volume 703.

15. Renard, K.G.; Yoder, D.C.; Lightle, D.T.; Dabney, S.M. . Universal Soil Loss Equation and Revised Universal Soil Loss Equation. In *Handbook of Erosion Modelling*; Morgan, R.P.C., Nearing, M., Eds.; Blackwell Publishing: Hoboken, NJ, USA, 2011; pp. 137–166.

16. Arnold, J.G.; Srinivasan, R.; Muttiah, R.S.; Williams, J.R. Large area hydrologic modeling and assessment part I: Model development. *J. Am. Water. Resour. Assoc.* **1998**, *34*, 73–89. [CrossRef]

17. Samaras, A.G.; Koutitas, C.G. Modeling the impact of climate change on sediment transport and morphology in coupled watershed–coast systems: A case study using an integrated approach. *Int. J. Sediment. Res.* **2015**, *29*, 304–315. [CrossRef]

18. Kaffas, K.; Righetti, M.; Avesani, D.; Spiliotis, M.; Hrissanthou, V. Coupling CFSv2 with ArcSWAT for seasonal hydrological forecasting in a Mediterranean basin. In Proceedings of the 11th World Congress on Water Resources and Environment (EWRA 2019), Madrid, Spain, 25–29 June 2019; pp. 459–460.

19. Napoli, M.; Massetti, L.; Orlandini, S. Hydrological response to land use and climate changes in a rural hilly basin in Italy. *Catena* **2017**, *157*, 1–11. [CrossRef]

20. Alemaw, D.; Ayana, E.K.; Legesse, E.S.; Moges, M.M.; Tilahun, S.A.; Moges, M.A. Estimating reservoir sedimentation using bathymetric differencing and hydrologic modeling in data scarce Koga watershed, Upper Blue Nile, Ethiopia. *J. Agric. Environ. Int. Dev.* **2016**, *110*, 413–427. [CrossRef]

21. Dhami, B.S.; Pandey, A. Comparative Review of Recently Developed Hydrologic Models. *J. Indian Water Resour. Soc.* **2013**, *33*, 34–42.

22. Abbaspour, K.C.; Vaghefi, S.A.; Srinivasan, R. A guideline for successful calibration and uncertainty analysis for soil and water assessment: A review of papers from the 2016 international SWAT conference. *Water* **2017**, *10*, 6. [CrossRef]

23. Biggs, J.; von Fumetti, S.; Kelly–Quinn, M. The importance of small waterbodies for biodiversity and ecosystem services: Implications for policy makers. *Hydrobiologia* **2016**, *793*, 3–39. [CrossRef]

24. Mujere, N.; Eslamian, S. Climate Change Impacts on Hydrology and Water Resources. In *Handbook of Engineering Hydrology Modeling, Climate Change and Variability*; Eslamian, S., Ed.; CRC–Press: Boca Raton, FL, USA, 2014; pp. 114–125.

25. Downing, J.A.; Prairie, Y.T.; Cole, J.J.; Duarte, C.M.; Tranvik, L.J.; Striegl, R.G.; McDowell, W.H.; Kortelainen, P.; Caraco, N.F.; Melack, J.M.; et al. The global abundance and size distribution of lakes, ponds, and impoundments. *Limnol. Ocean.* **2006**, *51*, 2388–2397. [CrossRef]

26. Miracle, M.R.; Oertli, B.; Céréghino, R.; Hull, A. Preface: Conservation of European ponds–current knowledge and future needs. *Limnetica* **2010**, *29*, 1–8.

27. Céréghino, R.; Biggs, J.; Oertli, B.; Declerck, S. *The Ecology of European Ponds: Defining the Characteristics of a Neglected Freshwater Habitat, Hydrobiologia*; Springer: Berlin/Heidelberg, Germany, 2008; Volume 597, pp. 1–6. [CrossRef]

28. Williams, P.; Whitfield, M.; Biggs, J. *How Can We Make New Ponds Biodiverse? A Case Study Monitored Over 7 Years, Hydrobiologia*; Springer: Berlin/Heidelberg, Germany, 2008; Volume 597, pp. 137–148. [CrossRef]

29. Céréghino, R.; Boix, D.; Cauchie, H.M.; Martens, K.; Oertli, B. *The Ecological Role of Ponds In a Changing World, Hydrobiologia*; Springer: Berlin/Heidelberg, Germany, 2014; Volume 723, pp. 1–6. [CrossRef]

30. Sharpley, A.N.; Bergström, L.; Aronsson, H.; Bechmann, M.; Bolster, C.H.; Börling, K.; Djodjic, F.; Jarvie, H.P.; Schoumans, O.F.; Stamm, C.; et al. Future agriculture with minimized phosphorus losses to waters: Research needs and direction. *Ambio* **2015**, *44*, 163–179. [CrossRef]

31. Oertli, B.; Frossard, P.A. *Mares et Etangs: Ecologie, Conservation, Gestion, Valorisation*, 1st ed.; Presses Polytechniques et Universitaires Romandes: Lausanne, Switzerland, 2013; p. 512.

32. Hill, M.J.; Hassall, C.; Oertli, B.; Fahrig, L.; Robson, B.J.; Biggs, J.; Samways, M.J.; Usio, N.; Takamura, N.; Krishnaswamy, J.; et al. New policy directions for global pond conservation. *Conserv. Lett.* **2018**, *11*, 8. [CrossRef]

33. Indermuehle, N.; Oertli, B.; Biggs, J.; Céréghino, R.; Grillas, P.; Hull, A.; Nicolet, P.; Scher, O. Pond conservation in Europe: The European Pond Conservation Network (EPCN). *Int. Vereinigung Theor. Angew. Limnol. Verh.* **2008**, *30*, 446–448. [CrossRef]

34. Xiao, S.; Yang, H.; Liu, D.; Zhang, C.; Lei, D.; Wang, Y.; Peng, F.; Li, Y.; Wang, C.; Li, X.; et al. Gas transfer velocities of methane and carbon dioxide in a subtropical shallow pond. *Tellus B Chem. Phys. Meteorol.* **2014**, *66*, 23795. [CrossRef]

35. Mujere, N.; Eslamian, S. Impact of Urbanization on Runoff Regime. In *Handbook of Engineering Hydrology Modeling, Climate Change and Variability*; Eslamian, S., Ed.; CRC–Press: Boca Raton, FL, USA, 2014; pp. 114–125.

36. Nedjai, R.; Azaroual, A.; Chlif, K.; Bensaid, A.; Al Sayah, M.; Ysbaa, L. Impact of Ponds on Local Climate: A Remote Sensing and GIS Contribution Application to the Ponds of Brenne (France). *J. Earth. Sci. Clim. Chang.* **2018**, *9*, 10.

37. Bennarrous, R. *La Grande Brenne aux Périodes Préindustrielles (Indre) Contribution à L'histoire des Paysages, des Etangs et des Relations Sociétés/Milieux Dans une Zone Humide Continentale. Approches Historique, Archéologique et Paléo–Environnementale*; Université de Paris I–Panthéon Sorbonne: Paris, France, 2009.

38. Jalowska, A.M.; Yuan, Y. Evaluation of SWAT Impoundment Modeling Methods in Water and Sediment Simulations. *J. Am. Water. Resour. Assoc.* **2019**, *55*, 209–227. [CrossRef]

39. Clifford, C.C.; Heffernan, J.B. Artificial Aquatic Ecosystems. *Water* **2018**, *10*, 1096. [CrossRef]

40. Ebel, J.D.; Lowe, W.H. Constructed Ponds and Small Stream Habitats: Hypothesized Interactions and Methods to Minimize Impacts. *J. Water. Resour. Prot.* **2013**, *5*, 723–731. [CrossRef]

41. Ismail, T.; Othman, M.A.; Fadzil, A.B.; Zainuddin, Z.M. Deposition of Sediments in Detention Pond. *Malaysian. J. Civ. Eng.* **2010**, *22*, 95–118.

42. Declerck, S.; De Bie, T.; Ercken, D.; Hampel, H.; Schrijvers, S.; Van Wichelen, J.; Gillard, V.; Mandiki, R.; Losson, B.; Bauwens, D.; et al. Ecological characteristics of small farmland ponds: Associations with land use practices at multiple spatial scales. *Biol. Conserv.* **2006**, *131*, 523–532. [CrossRef]

43. Fairchild, G.W.; Velinsky, D.J. Effects of Small Ponds on Stream Water Chemistry. *Lake. Reserv. Manag.* **2006**, *22*, 321–330. [CrossRef]

44. Commission Européenne. *La Directive–Cadre Européenne sur l'Eau*; Commission Européenne: Brussels, Belgium, 2000.

45. Biggs, J.; Williams, P.; Whitfield, P.; Nicolet, P.; Weatherby, A. 15 years of pond assessment in Britain: Results and lessons learned from the work of Pond Conservation. *Aquat. Conserv. Mar. Freshw. Ecosyst.* **2005**, *15*, 693–714. [CrossRef]

46. Jeffries, M. *Local–Scale Turnover of Pond Insects: Intrapond Habitat Quality and Inter–Pond Geometry are Both Important, Hydrobiologia*; Springer: Berlin/Heidelberg, Germany, 2005; Volume 543, pp. 207–220.

47. Neitsch, S.L.; Arnold, J.G.; Kiniry, J.R.; Williams, J.R. Soil & Water Assessment Tool Theoretical Documentation Version 2009. Texas Water Resources Institute Technical Report No.406: College Station, TX, USA, 2011.

48. Almendinger, J.E.; Murphy, M.S.; Ulrich, J.S. Use of the Soil and Water Assessment Tool to Scale Sediment Delivery from Field to Watershed in an Agricultural Landscape with Topographic Depressions. *J. Environ. Qual.* **2014**, *43*, 9–17. [CrossRef] [PubMed]

49. Guichané, R. L'aménagement hydraulique de la Claise tourangelle et de ses affluents du Moyen–Âge à nos jours/Mills on the Claise and its tributaries in Indre–et–Loire from the Middle Ages to modern times. *Rev. Archeol. Centre France* **1993**, *32*, 109–152. [CrossRef]

50. SANDRE. *Fiche Cours D'eau la Claise (L6—0200)*; SANDRE: Paris, France, 2012.

51. Joly, D.; Brossard, T.; Cardot, H.; Cavailhes, J.; Hilal, M.; Wavresky, P. Les types de climats en France, une construction spatiale–Types of climates on continental France, a spatial construction. *Cybergéo Eur. J. Geogr.* **2010**, *501*, 1–23.

52. Fischer, G.; Nachtergaele, F.; Prieler, S.; van Velthuizen, H.T.; Verelst, L.; Wiberg, D. *Global Agro–Ecological Zones Assessment for Agriculture (GAEZ 2008)*; IIASA: Laxenburg, Austria; FAO: Rome, Italy, 2008.

53. Dauphin, P.; Mansons, J.; Pellé, B.; Airault, V.; Trotignon, J.; Boyer, P.; Chatton, T.; Issa, N. Document D'objectifs des Sites Natura 2000 FR2410003 "Brenne" et FR2400534 "Grande Brenne.". Available online: http://natura2000.mnhn.fr/uploads/doc/PRODBIOTOP/1847_DOCOB_BRENNE.pdf (accessed on 29 November 2019).

54. Bouscasse, H.; Defrance, P.; Amand, B.; Grandmougi, B.; Strosser, P.; Beley, Y. Amélioration des connaissances sur les fonctions et usages des zones humides: Évaluation économique sur des sites tests le cas des étangs de la Grande Brenne. Available online: http://oaidoc.eau-loire-bretagne.fr/exl-doc/DOC00021327_s3.pdf (accessed on 29 November 2019).

55. Arnold, J.G.; Kiniry, J.R.; Srinivasan, R.; Williams, J.R.; Haney, E.B.; Neitsch, S.L. *Soil & Water Assessment Tool Input/Output Documentation Version 2012*; Texas Water Resources Institute: College Station, TX, USA, 2012; p. 541.

56. Kaffas, K.; Hrissanthou, V.; Sevastas, S. Modeling hydromorphological processes in a mountainous basin using a composite mathematical model and ArcSWAT. *Catena* **2018**, *162*, 108–129. [CrossRef]

57. Gassman, P.W.; Reyes, M.R.; Green, C.H.; Arnold, J.G. The Soil and Water Assessment Tool: Historical Development, Applications, and Future Research Directions. *Am. Soc. Agric. Biol. Eng.* **2007**, *50*, 1211–1250. [CrossRef]

58. Epelde, A.M.; Cerro, I.; Sánchez–Pérez, J.M.; Sauvage, S.; Srinivasan, R.; Antigüedad, I. Application of the SWAT model to assess the impact of changes in agricultural management practices on water quality. *Hydrol. Sci. J.* **2014**, *60*, 825–843. [CrossRef]

59. Srinivasan, R.; Zhang, X.; Arnold, J. SWAT ungauged: Hydrological budget and crop yield predictions in the Upper Mississippi river basin. *Trans. ASABE* **2010**, *53*, 1533–1546. [CrossRef]

60. Neitsch, S.L.; Arnold, J.G.; Kiniry, J.R.; Rinivasan, R.S.; Williams, J.R. *Soil and Water Assessment Tool User's Manual–Version 2000*; Texas Water Resources Institute: College Station, TX, USA, 2002.

61. Kaffas, K.; Hrissanthou, V. Annual sediment yield prediction by means of three soil erosion models at the basin scale. In Proceedings of the 10th World Congress of EWRA "Panta Rhei", Athens, Greece, 5–9 July 2017; pp. 1346–1352.

62. CORINE. *CORINE Soil Erosion Risk and Important Land Resources-in the Southern Regions of the European Community*; European Community-European Environment Agency: Copenhagen, Denmark, 1992; p. 541.

63. Fournier, F. *Climat et Erosion: La Relation Entre l'Erosion du sol par l'Eau et Les Precipitations Atmospheriques*; Presses Universitaires De France: Paris, France, 1960; p. 201.

64. Gaussen, H. *Bioclimatic Map of Mediterranean Zone*; UNESCO and FAO: Paris, France, 1963.

65. Arnold, J.G.; Moriasi, D.N.; Gassman, P.W.; Abbaspour, K.C.; White, M.J.; Srinivasan, R.; Santhi, C.; Harmel, R.D.; van Griensven, A.; van Liew, M.W.; et al. SWAT: Model use, Calibration, and Validation. *Am. Soc. Agric. Biol. Eng.* **2012**, *55*, 1491–1508.

66. Hallouz, F.; Meddi, M.; Mahé, G.; Alirahmani, S.; Keddar, A. Modeling of discharge and sediment transport through the SWAT model in the basin of Harraza (Northwest of Algeria). *Water. Sci.* **2018**, *32*, 79–88. [CrossRef]

67. Buakhao, W.; Kangrang, A. DEM Resolution Impact on the Estimation of the Physical Characteristics of Watersheds by Using SWAT. *Adv. Civ. Eng.* **2016**, *2016*, 9. [CrossRef]

68. Polanco, E.I.; Fleifle, A.; Ludwig, R.; Disse, M. Improving SWAT model performance in the upper Blue Nile Basin using meteorological data integration and subcatchment discretization. *Hydrol. Earth. Syst Sci.* **2017**, *21*, 4907–4926. [CrossRef]

69. USDA. *USDA Soil Conservation Service. National Engineering Handbook, Section 4: Hydrology*; NEH Notice 4–102; USDA: Washington, DC, USA, 1972.

70. Saxton, K.; Rawls, W.J. Soil Water Characteristic Estimates by Texture and Organic Matter for Hydrologic Solutions. *Soil. Sci. Soc. Am. J.* **2006**, *70*, 1569–1578. [CrossRef]

71. Al Sayah, M.J.; Abdallah, C.; Khouri, M.; Nedjai, R.; Darwich, T. Application of the LDN concept for quantification of the impact of land use and land cover changes on Mediterranean watersheds–Al Awali basin–Lebanon as a case study. *Catena* **2019**, *176*, 264–278. [CrossRef]

72. Périé, C.; Ouimet, R. Organic carbon, organic matter and bulk density relationships in boreal forest soils. *Can. J. Soil. Sci.* **2008**, *88*, 315–325. [CrossRef]

73. Post, D.F.; Fimbres, A.; Matthias, A.D.; Sano, E.E.; Accioly, L.; Batchily, A.K.; Ferreira, L.G. Predicting Soil Albedo from Soil Color and Spectral Reflectance Data. *Soil. Sci. Soc. Am. J.* **2000**, *64*, 1027–1034. [CrossRef]

74. Durand, Y.; Brun, E.; Mérindol, L.; Guyomarc'h, G.; Lesaffre, B.; Martin, E. A meteorological estimation of relevant parameters for snow models. *Ann. Glaciol.* **1993**, *18*, 65–71. [CrossRef]

75. Pearson, K. Notes on regression and inheritance in the case of two parents. *Proc. R. Soc. Lond.* **1985**, *58*, 240–242.

76. SWAT. SWAT Weather Data and Tools [Internet]. Available online: https://swat.tamu.edu/software/ (accessed on 8 March 2019).

77. Wang, X.; Yang, W.; Melesse, A.M. Using Hydrologic Equivalent Wetland Concept within SWAT to Estimate Streamflow in Watersheds with Numerous Wetlands. *Trans. Am. Soc. Agric. Eng.* **2008**, *51*, 55–72. [CrossRef]

78. Ahmed, S.I.; Rudra, R.P.; Gharabaghi, B.; Mackenzie, K.; Dickinson, W.T. Within–Storm Rainfall Distribution Effect on Soil Erosion Rate. *ISRN Soil. Sci.* **2012**, *2012*, 1–7. [CrossRef]

79. Akale, A.T.; Dagnew, D.C.; Belete, M.A.; Tilahun, S.A.; Mekruia, W.; Steenhuis, T.S. Impact of Soil Depth and Topography on the Effectiveness of Conservation Practices on Discharge and Soil Loss in the Ethiopian Highlands. *Land* **2017**, *6*, 78. [CrossRef]

80. Gis Sol; INRA; SOeS. *Aléa D'érosion des Sols en 2000*; Institut national de la recherche agronomique: Orléans, France, 2000.
81. Cohen, J. A coefficient of agreement for nominal scales. *Educ. Psychol. Meas.* **1960**, *20*, 37–46. [CrossRef]
82. Verheijen, F.G.A.; Jones, R.J.A.; Rickson, R.J.; Smith, C.J. Tolerable versus actual soil erosion rates in Europe. *Earth Sci. Rev.* **2009**, *94*, 23–38. [CrossRef]
83. Kondolf, G.M. Hungry water: Effects of dams and gravel mining on river channels. *Environ. Manag.* **1997**, *21*, 533–551. [CrossRef]

Article

Assessing the Impact of Terraces and Vegetation on Runoff and Sediment Routing Using the Time-Area Method in the Chinese Loess Plateau

Juan Bai [1,2]**, Shengtian Yang** [2,*]**, Yichi Zhang** [1,2]**, Xiaoyan Liu** [3] **and Yabing Guan** [1,2]

[1] State Key Laboratory of Remote Sensing Science, Faculty of Geographical Science, Beijing Normal University, Beijing 100875, China; baijuanaction@163.com (J.B.); yichizhang@mail.bnu.edu.cn (Y.Z.); guanyabingbj@163.com (Y.G.)

[2] College of Water Sciences, Beijing Normal University, Beijing 100875, China

[3] Yellow River Conservancy Commission, Ministry of Water Resources, Zhengzhou 450003, China; liuxiaoyansci@126.com

* Correspondence: yshengtian@gmail.com; Tel.: +86-10-5880-5034

Received: 12 February 2019; Accepted: 13 April 2019; Published: 18 April 2019

Abstract: Terracing and vegetation are an effective practice for soil and water conservation on sloped terrain. They can significantly reduce the sediment yield from the surface area, as well as intercept the sediment yield from upstream. However, most hydrological models mainly simulate the effect of the terraces and vegetation on water and sediment reduction from themselves, without considering their roles in the routing process, and thus likely underestimate their runoff and sediment reduction effect. This study added the impact of terraces and vegetation practice on water and sediment routing using the time-area method. The outflow in each travel time zone was revised in each time step by extracting the watershed of the terrace units and the vegetation units and calculating the water or sediment stored by the terraces or held by the vegetation. The revised time-area method was integrated into the Land change Model-Modified Universal Soil Loss Equation (LCM-MUSLE) model. Pianguanhe Basin, in the Chinese Loess Plateau, was chosen as the study area and eight storms in the 1980s and 2010s were selected to calibrate and verify the original LCM-MUSLE model and its revised version. The results showed that the original model was not applicable in more recent years, since the surface was changed significantly as a result of revegetation and slope terracing, while the accuracy improved significantly when using the revised version. For the three events in the 2010s, the average runoff reduction rate in routing process was 51.02% for vegetation, 26.65% for terraces, and 71.86% for both terraces and vegetation. The average sediment reduction rate in routing process was 32.22% for vegetation, 24.52% for terraces, and 53.85% for both terraces and vegetation. This study provides a generalized method to quantitatively assess the impact of terraces and vegetation practice on runoff and sediment reduction at the catchment scale.

Keywords: terrace; vegetation; time-area method; MUSLE; soil and water loss; the Loess Plateau

---

1. Introduction

Soil erosion and water loss is a serious environmental problem in the Chinese Loess Plateau [1–3]. It can cause soil deterioration and loss of sustainable productivity in croplands [2,4,5]. In addition, sediment yields and chemical loadings associated with soil erosion can cause severe degradation of surface water quality [6–8]. To control soil and water losses, terrace engineering was implemented since the 1980s and the Grain for Green (GFG) project was launched in 1999 [9–12]. Therefore, it is necessary to quantitatively analyze the effect of terraces and vegetation on runoff and sediment in the Loess Plateau.

Slope terracing and vegetation planting are common practices for soil and water conservation on sloped terrain susceptible to water erosion, and have been proven to be effective at retaining water and soil [6,13–20]. Terraces and vegetation can reduce the sediment yield from themselves significantly. Terraces reduce the peak runoff rate by reducing the slope gradient and slope length of the hillside [21,22]. Owing to the topographic slope and the embankment, terraced fields have a certain storage capacity similar to a reservoir, which can intercept surface runoff and sediment and promote runoff infiltration, evaporation, and sediment deposition [23–25]. The vegetation canopy can intercept rainfall directly and influence the rainfall kinetic energy and erosion rates. The stems, roots, and litter layer of vegetation can reduce runoff discharge by promoting infiltration, increasing surface roughness, and slowing down the overland flow and peak runoff. Vegetation also can reduce soil erosion by reducing surface flow volume and increasing sediment trapping through reducing flow velocity [21,26,27]. In addition to this, both terraces and vegetation can intercept the sediment yield from upstream and achieve sediment reduction in the valley by reducing runoff flowing from the slope into the valley [16,24,28,29].

Numerous studies have provided many insights into how terraces and vegetation control water erosion at local scales using observational experiments [20,27]. Terraces could be classified into level terraces, slope terraces, slope-separated terraces, and zig terraces according to their structure [23]. In the Loess Plateau of China, the main terrace type is level terraces. Yao [30] found a terraced field could reduce soil erosion by 92–100% compared with sloped farmland, while Wu [31] found the average benefit of level terraces on soil and water conservation were 86.7% and 87.7%, respectively. Huo and Zhu [32] combined soil water moisture and 137Cs content analysis and found the average soil and water conservation benefit of level terraces was 53%_ENREF_18, while Pan and Shangguan [27] reported that grassplots had 14–25% less runoff and 81–95% less sediment yield compared to a bare soil plot. Meng et al. [33] conducted a series of laboratory flume simulation experiments and the results showed that vegetation could reduce the mean velocity by 31–65%_ENREF_36. As the effectiveness of terraces and vegetation is limited by many factors, such as climate, soil properties, topography, land use, vegetation type, and spatial patterns, the diversity and natural variability of previously conducted erosion studies limit their potential extrapolation to the catchment scale [23,34–36]. Thus, how to assess the effects of terraces and vegetation on water erosion control at the catchment scale remains a crucial issue.

Estimation of rainstorm-generated sediment yield by means of a hydrological model is an important way to quantitatively evaluate the effect of soil and water conservation measures, such as by using the Soil and Water Assessment Tool (SWAT) model [37–39], the Agricultural Non-Point Source pollution (AGNPs) model [40], and the Agricultural Policy/ Environmental eXtender (APEX) model [41]. In these models, accounting for the impact of terraces and vegetation on runoff and sediment yields has focused on reduction from themselves through adjusting the key input variables, such as the Soil Conservation Service Curve Number (SCS-CN), slope gradient, slope length, Universal Soil Loss Equation (USLE) support practice factor (P-factor), and cover and management factor (C-factor) [11,21,22,25,29,42,43], without considering the roles of water and sediment reduction in the routing process. Then, the runoff and sediment reduction effect of terraces and vegetation is likely to have been underestimated. Therefore, it is necessary to take into account the water and sediment reduction effect of terraces and vegetation in the model's routing process.

Explicitly simulating the interaction between conservation practice and watershed response is difficult. The time-area method, with its inherently distributed concept, explains which parts of a watershed contribute to runoff during a specific period [44]. It provides a simple and useful tool to understand runoff mechanisms and is widely used at the catchment scale [44]. It has also been indicated that the routing of sediment through time-area segments in a catchment produces better results than the conventional routing through a network of cells [45,46]. Her and Heatwole [47] revised the time-area method and considered the upstream contribution for routing sediment, such that the new method provided detailed spatial representation_ENREF_15. These studies mainly focused on the effect of surface heterogeneity on routing time or flow velocity [48,49]. Topographic features are delineated through a Digital Elevation Model (DEM), slope gradient, flow direction, and so on. However, the lack of representation of a specific terrace and vegetation process makes it difficult to quantitatively distinguish vegetation and terraced fields and their integrated ability for water and sediment reduction.

Thus, the purpose of this study was to incorporate the effect of terraces and vegetation on runoff and sediment routing in the time-area method and assess their impact in water and sediment reduction. The outflow in each travel time zone was revised in each time step by extracting the watershed of the terrace units and the vegetation units, and calculating the water or sediment stored by the terraces or retained by the vegetation based on their properties.

2. Materials and Methods

2.1. Study Area

In this study, we selected the Pianguanhe River in the key soil erosion region of the Loess Plateau as our case study. The Pianguanhe River is located in the hilly and gully region of the Loess Plateau (111°26′ E–112°17′ E, 39°15′ N–39°42′ N) (Figure 1). It is a tributary of the Yellow River with a watershed area of 2089 km², with altitude ranging from 984 to 2162 m. The study area features a semi-arid continental climate, with an average annual rainfall of 429 mm. The uneven seasonal distribution of precipitation results in more than 80.9% of the annual precipitation occurring from May to September [50]. The average annual runoff and sediment discharge is 39.48 million m³ and 12.58 million t, respectively. The sediment modulus is 65.23 t/ha/yr [50]. The overall loss of soil and water is a serious environmental issue.

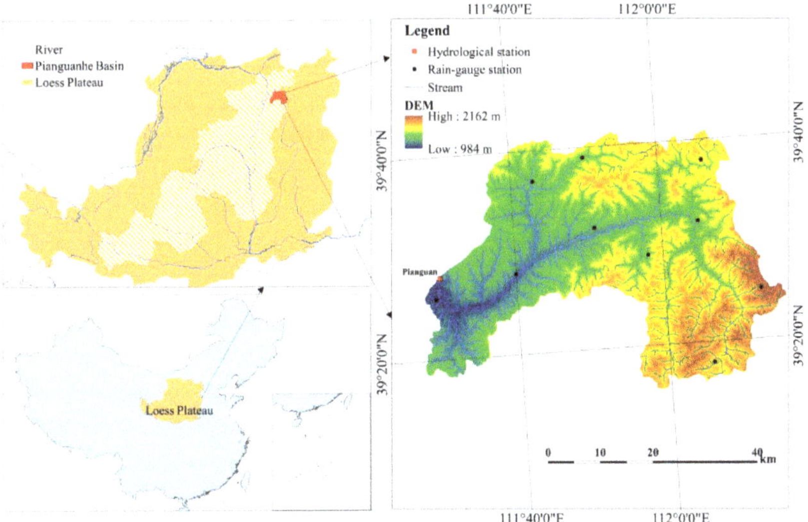

Figure 1. Location of the study area.

2.2. Methods

The serious soil and water loss in the Loess Plateau has always occurred because of the surface runoff caused by heavy rains. In this paper, runoff generation was simulated based on the Land Catchment Model (LCM), which is a flood event forecasting model that was developed through more than 300 artificial rainfall experiments in the Loess Plateau [51,52]. The model has been modified into a distributed model [53–55]. The Modified Universal Soil Loss Equation (MUSLE) was chosen to estimate the sediment yield of individual heavy rainfall events [56], and numerous studies had proven its applicability to estimate the sediment yield in the Loess Plateau [57–59]. The main functions of LCM-MUSLE model are listed in Tables 1 and 2. The reader is referred to Luo et al. [59] for a detailed description of the LCM-MUSLE model, which is not described here. Runoff and sediment generation from sub-catchments was routed to the outlet by considering overland flow and stream channels. Time-area method was used for overland routing [49] and Muskingum method was used for channel routing [60,61].

Table 1. Functions and parameters related to runoff generation in the Land Catchment Model-Modified Universal Soil Loss Equation (LCM-MUSLE) model.

No.	Module Name	Equations Reference	Equations Reference
1	Canopy interception	$S_v = V_f \times S_{\max} \times [1 - e^{-\eta \frac{P}{S_{\max}}}]$ $S_{\max} = 0.935 + 0.498 \times LAI - 0.00575 \times LAI^2$ $\eta = 0.046 \times LAI$ $P' = P - S_v$	[62]
2	Surface runoff	$Q_d = P' - f = P' - R \times P'^r$	[51]
3	Interflow	$Q_l = L_a(W_s/W_{sm})f$	[51]
4	Base flow	$Q_b = K_b(GW_s + REC)$ $REC = R_c(W_s/W_{sm})(f - Q_l)$	[51]

Annotation: S_v: interception (mm); V_f: vegetation coverage (%); S_{max}: canopy storage capacity (mm); η: a correction coefficient; P: precipitation (mm); LAI: leaf area index; Q_d: surface runoff (mm); P': effective rainfall (mm); f: infiltration (mm); R and r: both infiltration coefficient; L_a: interflow coefficient; W_s: unsaturated soil moisture storage (mm); W_{sm}: maximum soil moisture storage capacity of the soil layer (mm); Q_l: interflow (mm); Q_b: base flow (mm); K_b: base flow coefficient; GW_s: groundwater storage (mm); REC: groundwater recharge (mm); R_c: groundwater recharge coefficient.

Table 2. Functions and parameters related to sediment yield in the Land Catchment Model-Modified Universal Soil Loss Equation (LCM-MUSLE) model.

No.	Module Name	Equations Reference	Equations Reference
1	Sediment yield	$Sed = 11.8 \times (Rs \times q_{peak} \times A_{pixel})^{0.56} \times K \times C \times P \times LS \times CFRG$	[63]
2	Runoff factor	$q_{peak} = \dfrac{\alpha_{tc} \times Rs \times A_{pixel}}{3.6 \times t_{conc}}$	[37]
3	Soil erodibility factor	$K = \left\{0.2 + 0.3 \exp\left[-0.0256 S_d(1 - \frac{S_i}{100})\right]\right\}\left(\frac{S_i}{C_i+S_i}\right)^{0.3}\left[1 - \frac{0.25C}{C + \exp(3.72 - 2.95C)}\right]$ $\left[1 - \frac{0.7(1 - S_d)}{1 - S_d + \exp(-5.51 + 22.9(1 - S_d))}\right]$	[63]
4	Topographic factor	$L = (L_{slp}/22.1)^{0.44}$ $S = \begin{cases} 10.8\ \sin\theta + 0.03,\ \theta < 5° \\ 16.8\ \sin\theta - 0.5, 5° \leq \theta < 10° \\ 21.9\ \sin\theta - 0.96,\ \theta > 10° \end{cases}$	[64,65]
5	Cover and management factor	$C = \begin{cases} 1,\ c_v = 0 \\ 0.6508 - 0.3436 \lg c_v,\ 0 < c_v \leq 78.3\% \\ 0,\ c_v > 78.3\% \end{cases}$	[66]
6	Coarse fragment factor	$CFRG = \exp(-0.053 \times rock)$	[63]

Annotation: Sed: sediment yield (t); Rs: surface runoff (mm); q_{peak}: peak runoff rate (m³/s); A_{pixel}: area of the grids (ha); K: soil erodibility factor (0.013·t·m²·h/(m³·t·cm)); C: cover and management factor; P: support practice factor; LS: topographic factor; $CFRG$: coarse fragment factor; α_{tc}: fraction of rainfall that occurs during the time of concentration (for event modeling the value of α_{tc} is 1); t_{conc}: the time of concentration for the grid; S_d: percent sand content (%); S_i: percent silt content (%); C_i: percent clay content (%); C: percent organic carbon content of the layer (%); L_{slp}: slope length (m); θ: gradient of the slope (°); c_v: vegetation coverage (%); $rock$: percent rock in the top soil layer (%).

With the purpose of simulating the influence of terrace and vegetation units on water and sediment reduction at confluences, as the critical hydrologic process, both terrace and vegetation modules were added to the time-area method. In the time-area method, the catchment is divided into a number of travel time zones via isochrones [67]. By extracting the watershed of terrace units and vegetation units, and calculating the water stored by the terraces or intercepted by vegetation, the runoff yield and the outflow in each travel time zone are revised. This represents the spatiotemporally varied flow in the routing simulation. Sediment reduction was achieved in a similar way.

The integrated structure of the model is shown in Figure 2. Specific equations and methods are discussed in the following sections. The integrated model can represent landscape heterogeneity in detail, if a higher spatial resolution DEM is used.

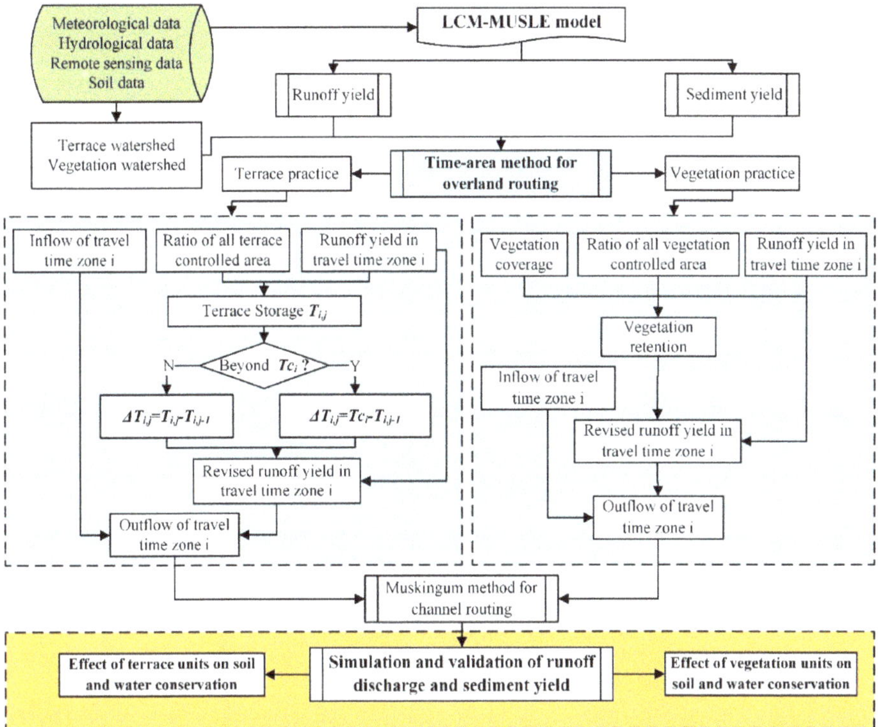

Figure 2. Framework of revised integrated model. $T_{i,j}$ is the amount of stored water or sediment in all terraces in travel time zone i at present time step j (m^3 or t); $T_{i,j-1}$ is the amount of stored water or sediment in all terraces in i at previous time $j - 1$ (m^3 or t); and Tc_i is water storage capacity of all terraces in i (m^3).

2.2.1. Time-Area Method for Overland Routing

Isochrones are defined as the contours of equal travel time to the catchment outlet, and the travel time zone is the area between two adjacent isochrones [67]. The isochrones of the whole basin are calculated, and then these are adjusted within each sub-catchment, as shown in Figure 3. It is assumed that the area of each time zone is ΔA_1, ΔA_2, ⋯, ΔA_n, and their corresponding travel time is τ_1, τ_2, ⋯, τ_n, respectively. Then, we set $\Delta \tau = \tau_i - \tau_{i-1}$ (i = 1, 2, ⋯, n). It is assumed that m is the time of runoff generation in hours, ΔR_r (r = 1, 2, … , m) is the runoff depth in m, and the runoff discharge of the outlet at time step j is:

For $n \geq m$

$$Q_j = \frac{1}{\Delta \tau} \sum_{r=1}^{k} \Delta R_r \Delta A_{j-r+1}, \ (j = 1, 2, 3, \ldots, n+m-1) \tag{1}$$

If $j < m$, $k = j$; if $j \geq m$, $k = m$, and if $j > n$, $\Delta A = 0$.
For $n < m$:

$$Q_j = \frac{1}{\Delta \tau} \sum_{r=1}^{k} \Delta A_r \Delta R_{j-r+1}, \ (j = 1, 2, 3, \ldots, n+m-1) \tag{2}$$

If $j < n$, $k = j$; if $j \geq n$, $k = n$, and if $j > m$, $\Delta R = 0$.
For travel time zone i, its runoff discharge at time step j can be calculated as follows:

$$Q_{i,j} = Q_{i+1,j-1} + \Delta R_j \Delta A_i \tag{3}$$

where $Q_{i,j}$ is the outflow from travel time zone i at time step j (m^3); and $Q_{i+1,j-1}$ is the outflow from time zone $i + 1$ at time step $j - 1$ (m^3).

Unlike runoff routing, sediment transport simulation gave consideration to the sediment particle size and the hydraulic characteristic of the basin. It was calculated using Equation (4), as suggested by Williams [68]:

$$RY = \sum_{i=1}^{n} Y_i e^{-BT_i \sqrt{D50_i}} \tag{4}$$

where RY is the sediment yield for the entire basin (t); Y_i is the sediment yield for the sub-catchment i before routing to channel (t); B is the routing coefficient; T_i is the travel time from the sub-catchment i to the basin outlet (h); and $D50_i$ is median particle diameter of the sediment for sub-catchment i (mm).

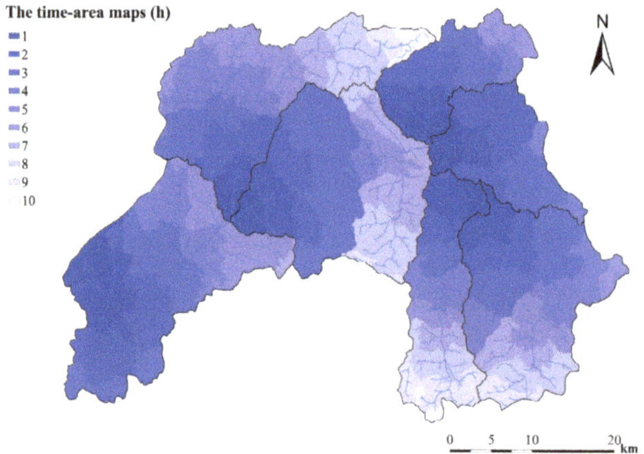

Figure 3. Isochrones of Pianguanhe's sub-catchments.

2.2.2. Extraction of the Terrace Units and Vegetation Units Watershed

The impact of terrace and vegetation units on water and sediment flow depends upon their size and position in the catchment. The flow direction of the natural hillslope is shown in Figure 4a. When there is a terrace in the hillslope, the connectivity of its original flow pathways is broken, and the whole slope could be divided into three segments: upstream section, terrace section, and downstream section (Figure 4b). Here, we define the upstream section as the watershed of terrace (or runoff contributing area). The area impacted by the terrace units in terms of water and sediment reduction is the total of the terrace and its watershed.

The terrace watershed was extracted by searching upward from the terrace cell based on deterministic-8 (D8) flow direction method [69]. For each cell, once there was a route that water could follow to reach the terrace cells, the cell was considered to belong to the terrace watershed. The extraction of vegetation watershed was the same as that for the terrace watershed.

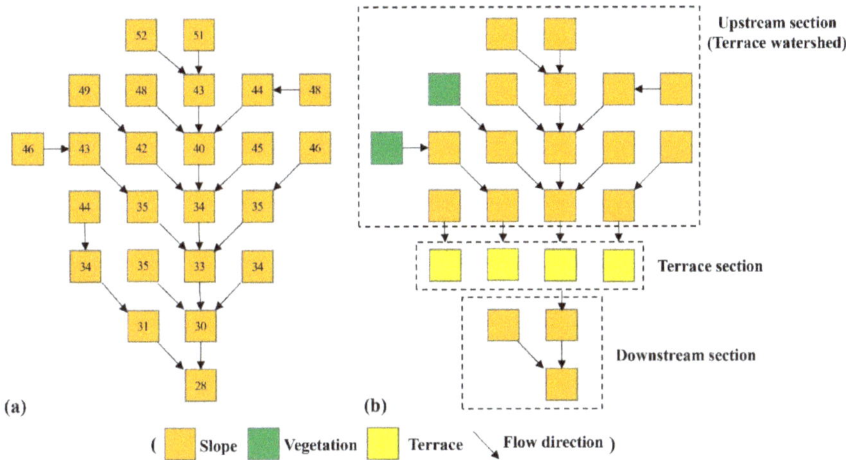

Figure 4. Schematic diagram of flow direction of (**a**) natural hillslope, and (**b**) the hillslope with vegetation and terrace practice. Numbers in the grid are elevations. Orange represents natural hillslope unit. Green represents vegetation unit. Yellow represents terrace unit. Arrows represent flow direction.

2.2.3. Consideration of Terrace Units in the Time-Area Method

The revised time-area method simplifies a level terrace as a dynamic water tank, and its storage capacity is the product of the terrace area and the embankment height (Figure 5a,b). In this study, we made the assumption that the stream flow in a terrace should only be considered as the overflow, regardless of the drainage discharge. Namely, when the water trapped by the terrace exceeds the terrace storage capacity during heavy rainfall events, the surplus water will overflow (Figure 5c,d).

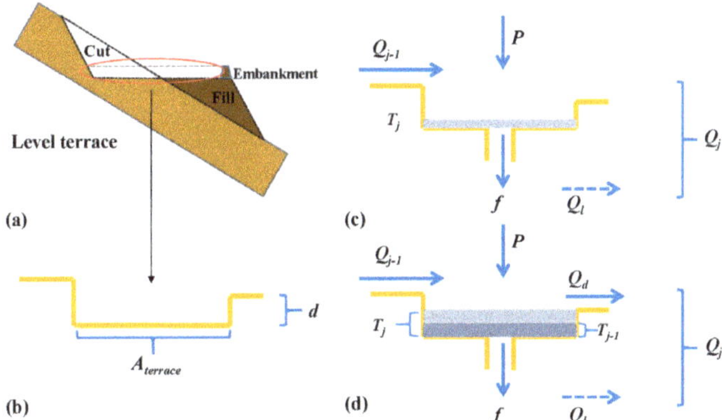

Figure 5. Schematic diagrams of the hydrological processes and flow distribution of the level terrace unit. (**a**) Section profile of a level terrace. (**b**) The generalization of the level terrace in (**a**). The flow distribution assumes that the soil profile of the terrace is deep enough for subsurface flow generation. The influence of the flow in different scenarios is listed from (**c**,**d**). $A_{terrace}$ is the area of terrace; d is embankment height of terrace; P is rainfall; f is the infiltration; Q_l is the interflow; Q_{j-1} is the inflow; Q_d is the overflow; T_j is the water stored in the terrace at the present time step j; T_{j-1} is the water stored in terrace at previous time $j-1$; and Q_j is the outflow.

Referring to Equation (3), when there are terrace units occurring in time zone i, the outflow $Q_{i,j}$ from time zone i at time step j is revised, as follows:

$$Q_{i,j} = Q_{i+1,j-1} + \Delta R_j \Delta A_i - \Delta T_{i,j} \tag{5}$$

where $Q_{i+1,j-1}$ is the outflow from time zone $i + 1$ at time step $j - 1$ (m^3); $\Delta T_{i,j}$ is the increased water ponded in terraces that are located in time zone i at time step j (m^3). When $\Delta T_{i,j}$ equals 0, the storage capacity of the terraces is filling up and they can no longer have a retention function.

Before filling up, the current terrace storage volume is the sum of its previous storage volume and the current terrace watershed inflows. Due to the possible distribution of several terraces in a time zone and due to the fact that each terrace may correspond to multiple cells, the statistics of the water interception of each terrace requires a large amount of calculations. In order to simplify the complex problem, we estimated the terrace watershed inflows as a percentage of the runoff generation in the time zone, and the percentage equals the area ratio of all terraces control area to the time zone. Then $\Delta T_{i,j}$ is calculated as follows:

$$\Delta T_{i,j} = T_{i,j} - T_{i,j-1} \tag{6}$$

$$T_{i,j} = \begin{cases} Tc_i & \text{(Filling up)} \\ T_{i,j-1} + \Delta R_j \Delta A_i \times Tu_i & \text{(Unfilling up)} \end{cases} \tag{7}$$

$$Tu_i = \Delta A_{Tcontrol,i} / \Delta A_i \tag{8}$$

$$Tc_i = \Delta A_{terrace,i} \times d_i \tag{9}$$

where $T_{i,j}$ is the amount of stored water in all terraces in time zone i at present time step j (m^3); $T_{i,j-1}$ is the amount of stored water in all terraces in time zone i at previous time step $j - 1$ (m^3); Tc_i is the maximum water storage of all terraces in the time zone i (m^3); Tu_i is the ratio of all terrace control area in the time zone i; $\Delta A_{Tcontrol,i}$ is the area of all terrace control area in time zone i (m^2); $\Delta A_{terrace,i}$ is the total area of terraces in time zone i (m^2); and d_i is embankment height of terrace (m). As runoff is the main carrier of sediment, the increased sediment stored by terraces in time zone i at time step j equals the sediment yield in time zone i at time step j multiplied by the runoff trapped rate of terraces in time zone i at time step j (namely, the ratio of $\Delta T_{i,j}$ to $\Delta R_j \Delta A_i$).

Taking Equation (6) and Equation (7) into Equation (5), we can get the outflow $Q_{i,j}$ as follows:

$$Q_{i,j} = \begin{cases} Q_{i+1,j-1} + \Delta R_j \Delta A_i - Tc_i + T_{i,j-1} & \text{(Filling up)} \\ Q_{i+1,j-1} + \Delta R_j \Delta A_i \times (1 - Tu_i) & \text{(Unfilling up)} \end{cases} \tag{10}$$

2.2.4. Consideration of Vegetation Units in the Time-area Method

Unlike terraces, which can act as a water tank, vegetation does not have a direct storage volume, but it can resist the confluence process of runoff and sediment. Vegetation coverage and canopy density are the main indicators influencing vegetation's impact on water and soil conservation [70–72]. In this study, we chose vegetation cover that was easily derived to incorporate the vegetation module into the time-area method.

Similar to Equation (5), when there are vegetation units occurring in time zone i, the outflow $Q_{i,j}$ from time zone i at time step j is as follows:

$$Q_{i,j} = Q_{i+1,j-1} + \Delta R_j \Delta A_i - \Delta Veg_{i,j} \tag{11}$$

where $\Delta Veg_{i,j}$ is the water or sediment trapped by vegetation units in time zone i at time step j (m^3). $\Delta Veg_{i,j}$ is calculated as follows:

$$\Delta Veg_{i,j} = (Q_{i+1,j-1} + \Delta R_j \Delta A_i) \times Vu_i \times \overline{f(Veg)} \tag{12}$$

$$Vu_i = \Delta A_{Vcontrol,i} / \Delta A_i \tag{13}$$

where Vu_i is the ratio of all the vegetation control area in the time zone i; $\Delta A_{Vcontrol,i}$ is the total area of vegetation in time zone i (m^2); Veg is vegetation coverage (%); $f(Veg)$ is function of Veg, it refers to runoff or sediment retention rate of the forestland or grassland with certain vegetation coverage, and $\overline{f(Veg)}$ is the average value of runoff or sediment retention rates in the time zone i.

Most flume test researches in the Loess Plateau lack the complete information about the runoff and sediment retention rate of grassland and forestland with different cover [73–76]. Xiong et al. [77] systematically deconstructed the experimental data from different slope runoff plots in the Loess Plateau, and summarized benefit indices of runoff and sediment reduction by forestland and grassland of different qualities in years with different runoff and sediment levels [77], as shown in Table 3. This paper refers to the study of Xiong et al. [77]. For the convenience of distributed calculation, we fitted the data in Table 3 to get the runoff and sediment reduction functions of forestland and grassland under different conditions, as shown in Tables 4 and 5, x means vegetation coverage (%) and y means runoff and sediment reduction rates.

Table 3. The reduction percentage in runoff and sediment generation of forestland and grassland of different quality.

	Vegetation Coverage (%)	Dry Year		Normal Year		Wet Year	
		Runoff Reduction (%)	Sediment Reduction (%)	Runoff Reduction (%)	Sediment Reduction (%)	Runoff Reduction (%)	Sediment Reduction (%)
Forest	70	100	100	100	98	76.5	57.7
	60	100	100	96.5	92.9	72.2	51
	50	99	99	90.1	86.9	64.2	46.2
	40	94	96	73.2	69.8	48.8	33.3
	30	80	89	52	48.2	28.4	19.2
	20	55	73	26.7	20.2	11.1	6.4
Grass	70	100	100	96.3	94.4	64.8	50
	60	100	100	92.6	89.9	59.3	45.1
	50	98	99	83.7	82.5	51.2	40
	40	86	95	67.8	66.5	37.7	30
	30	72	85	42.7	41.8	22.1	16.9
	20	45	69	19.5	18.6	8.2	5.9

Table 4. The reduction function of runoff by forestland and grassland.

Land	Dry Year	Normal Year	Wet Year
Forest	$y = 0.3619\ln(x) - 0.468$	$y = 0.6145\ln(x) - 1.557$	$y = 0.5551\ln(x) - 1.565$
Grass	$y = 0.4537\ln(x) - 0.854$	$y = 0.6498\ln(x) - 1.748$	$y = 0.4733\ln(x) - 1.357$

Table 5. The reduction function of sediment by forestland and grassland.

Land	Dry Year	Normal Year	Wet Year
Forest	$y = 0.2136\ln(x) + 0.133$	$y = 0.6429\ln(x) - 1.701$	$y = 0.4239\ln(x) - 1.222$
Grass	$y = 0.2527\ln(x) - 0.028$	$y = 0.6384\ln(x) - 1.721$	$y = 0.3669\ln(x) - 1.053$

2.2.5. Model Performance Evaluation Criteria

Nash–Sutcliffe efficiency (NSE) was used to evaluate the performance of the simulation:

$$\text{Nash} = 1 - \sum_{i=1}^{N}(O_i - E_i)^2 \Big/ \sum_{i=1}^{N}(O_i - \overline{O})^2 \tag{14}$$

where O_i is observed data (runoff discharge or sediment discharge); E_i is simulated data (runoff discharge or sediment discharge); $\overline{O}$ is average observed data; N is number of values. NSE varies from negative infinity to 1, where NSE closer to 1 indicates a better simulation.

A total of eight isolated storms with observed runoff and sediment yield were selected to calibrate and verify the model. There were five in the 1980s, and three in the 2010s, and each storm was encoded with its start time as Nos. year/day/hour. Among them, Nos. 1981/203/17, 1983/215/22 and 1983/235/16 were used for calibration, while Nos. 1988/199/13, 1989/203/19, 2006/195/5, 2006/224/8 and 2010/263/20 were used for validation.

The runoff and sediment simulation was implemented in four cases: O1—original simulation without considering terrace and vegetation practice; R1—revised with vegetation module; R2—revised with terrace module; R3—revised with vegetation and terrace modules. For the five events in the 1980s, as the terrace data was unavailable and terraces just accounted for a small proportion of the area, only O1 and R1 were simulated. For three events in the 2010s, all four cases were simulated. The terraces in the study area are all level terraces with good quality [78], and we made an assumption that all terraces had an embankment height of 20 cm. For vegetation, the retention functions of runoff and sediment were chosen according to the rainfall amount of each event and the rainfall condition of the two days before each event. Here, for Nos. 1983/215/22, 1983/235/16, 1988/199/13, 2006/224/8 and 2010/263/20 functions of normal period were chosen in Tables 4 and 5, while for Nos. 1981/203/17, 1989/203/19 and 2006/195/5 functions of wet period were chosen.

2.3. Data Source

Input data for the model included hourly precipitation, DEM, land use, vegetation coverage, soil data, and particle size of the sediment. In addition, observed runoff and sediment discharge of hydrological stations were used in the model's calibration and validation.

(1) Hourly precipitation was collected from 10 rain-gauge stations (Figure 1). The gauge data were interpolated by the inverse distance weighted (IDW) method to acquire the spatial data.

(2) Advanced Spaceborne Thermal Emission and Reflection Radiometer Global Digital Elevation Model (ASTER GDEM) was selected to extract topographical information, river network, sub-basins and isochrones (http://www.gscloud.cn/).

(3) Land use and vegetation coverage were derived from Landsat images (http://www.gscloud.cn/). Land use was classified into 11 types, including cropland (slope < 6°), cropland (6° < slope < 25°), cropland (slope ≥ 25°), forest, shrub, open woodland, immature forest land and orchard, grassland, water area, developed land, and other land. Vegetation coverage was derived based on its relation with normalized difference vegetation index (NDVI) [79], while the vegetation coverage of cropland was not included in routing process. Land use and vegetation coverage data from 1978 and 2010 were used to represent data for 1980s and 2010s, respectively.

(4) Terrace data in 2012 was acquired from the Yellow River Conservancy Commission (YRCC), which was interpreted from ZY-3 images with a spatial resolution of 2.5 m and an accuracy of 94% [78].

(5) The soil types, sand content, silt content, clay content, organic carbon, and gravel content were derived from the Harmonized World Soil Database (HWSD), and the soil saturated moisture of each soil type were determined from the software Soil-Plant-Atmosphere-Water Field & Pond Hydrology (SPAW).

(6) The measured discharge, sediment concentration and median of sediment particle size (D_{50}) at the Pianguanhe hydrological station were acquired from the YRCC.

All data were reproduced at 30 × 30 m spatial resolution and projected to Albers, using the World Geodetic System-84 (WGS84) datum. Data were processed by ENVI5.1 (Harris Corporation, Melbourne, FL, USA), GisNet [80], ArcGIS10.4.1 (Environmental Systems Research Institute, Redlands, CA, USA), or programs written in IDL (Interactive Data Language).

3. Results

3.1. Watershed of Terrace Unit and Vegetation Unit

The terrace ratio is the proportion of terrace area to water and soil loss area [78]. In 2012, the total terrace area of Pianguanhe was 189.32 km^2, accounting for 9.87% of the whole basin and giving a terrace ratio of 10.18%. The terrace ratio of Pianguanhe was 1.80% in 1979 [50]. In view of data availability, we chose terrace data from 2012 (to represent the 2010s) and extracted its terrace watershed. The terrace watershed was 169.90 km^2 in 2012. The terrace control area is the sum of terrace area and terrace watershed and accounted for 1.89 times the area of the terrace, as shown in Figure 6a.

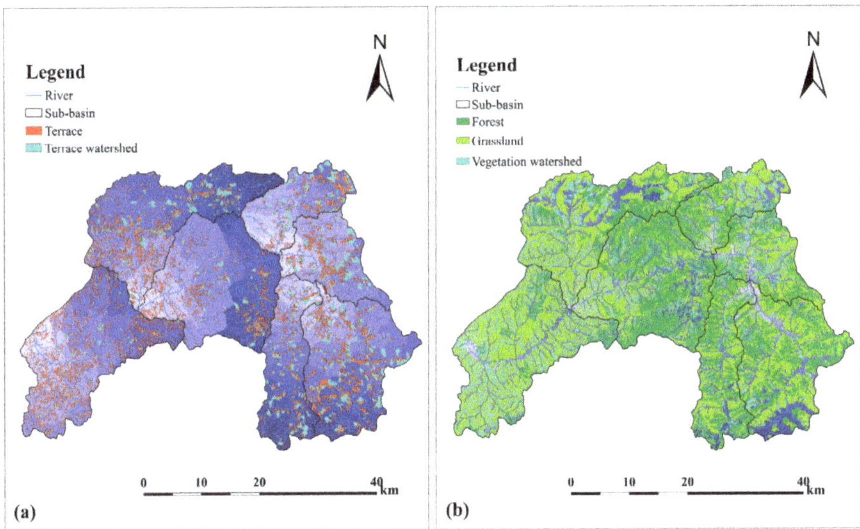

Figure 6. Extraction of terrace watershed (**a**) and vegetation watershed (**b**) of Pianguanhe in 2010s.

The area of forestland and grassland in Pianguanhe was 1328.70 km^2 in 1980s and 1349.17 km^2 in the 2010s, giving an increase of 20.47 km^2. The average vegetation coverage was 22.27% in 1980s and 62.74% in the 2010s, giving an increase of 40.47%. Thus, the vegetation quality demonstrated a significant increase compared to the area. We extracted the vegetation watersheds in the 1980s and 2010s. Figure 6b shows the vegetation watershed of Pianguanhe in the 2010s. Vegetation control area was the sum of vegetation units and vegetation watershed. It was 1675.77 km^2 in the 1980s and 1693.96 km^2 in the 2010s, indicating that it did not change very much.

3.2. Validation of Runoff Discharge

Figure 7a presents the event-based comparison between the measured runoff discharge and the runoff discharge predicted under different cases. The simulated runoff discharge in most events were comparable to their measured values. For the five events in the 1980s, the peak values of R1 were lower than those of O1. For the three events in the 2010s, the peak values rank from lowest to highest: R3 < R1 < R2 < O1 (Figure 7a). This shows that the revised simulation with vegetation can significantly reduce the runoff peak compared to the simulation with terrace, and that the simulation considering both terrace and vegetation was closest to the estimated value.

Comparison of the simulated runoff with the measured runoff is shown in Figure 7b–e. Figure 7b shows the validation of O1, in which the five events in the 1980s achieved an average NSE of 0.46, while the three events in the 2010s achieved an average NSE of −15.29. The data points of the 1980s are distributed close to the 1:1 line, while the data points of the 2010s are distributed farther away from the

1:1 line. It shows that the original model did not achieve good performance for recent years, since the surface had undergone significant change. Figure 7c shows that the average NSE of the 1980s and 2010s both increased in R1, with the average NSE of the 2010s showing a greater increase (value of −1.32), and the data points of the 2010s distributed closer to the 1:1 line than in the O1 validation shown in Figure 7b. Figure 7d shows the R2 validation, and indicates that the average NSE of the 2010s has also increased compared with that of O1, but that the extent of the increase is less than for R1. Figure 7e shows the R3 validation and indicates that the average NSE of the 2010s is 0.39, which for the first time is positive, and that the data points of the 2010s are closely distributed around the 1:1 line. Overall, the simulation accuracy of the 2010s is highest in R3, followed by R1. These figures show that the revised model was better at simulating the runoff in recent years and can reflect the effect of significant surface change (i.e., slope terracing and revegetation) on runoff.

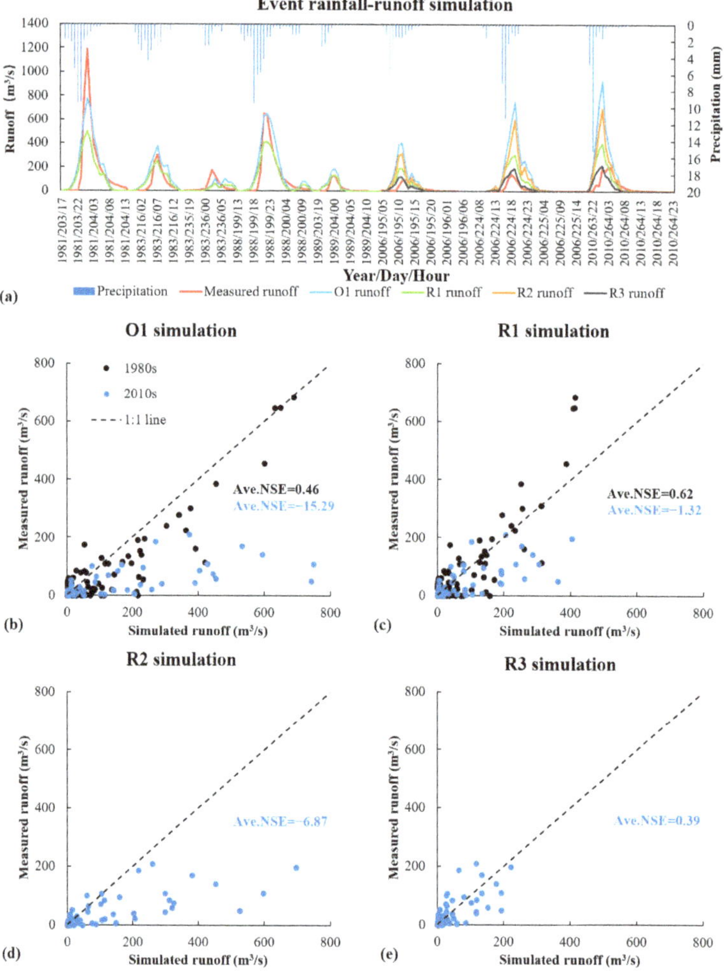

Figure 7. Comparison of observed and simulated runoff for eight events in the Pianguanhe Basin. (**a**) Event rainfall-runoff simulation; (**b**) Runoff in O1 simulation; (**c**) Runoff in R1 simulation; (**d**) Runoff in R2 simulation; (**e**) Runoff in R3 simulation.

3.3. Validation of Sediment Discharge

The hourly measured sediment concentration (kg/m^3) along with the measured runoff discharge were converted to get sediment discharge (t/h). Figure 8a shows the event-based comparison between the measured and predicted sediment discharge under different cases. The simulations of sediment discharge were in good agreement with their measured values in most events. The difference between the sediment discharge of O1 and measured values is smaller than the runoff discharge between O1 and measured values in recent years; it was because the sediment reduction was partly achieved in the sediment yield process through the C and P factors in MUSLE. For the five events in the 1980s, the peak values of R1 were lower than those of O1. For the three events in the 2010s, the peak values rank from lowest to highest were: R3 < R1 < R2 < O1 (Figure 8a). This shows that the revised simulation with vegetation can significantly reduce the sediment peak compared to the simulation with terrace, and that the simulation considering both terrace and vegetation was closest to the estimated value.

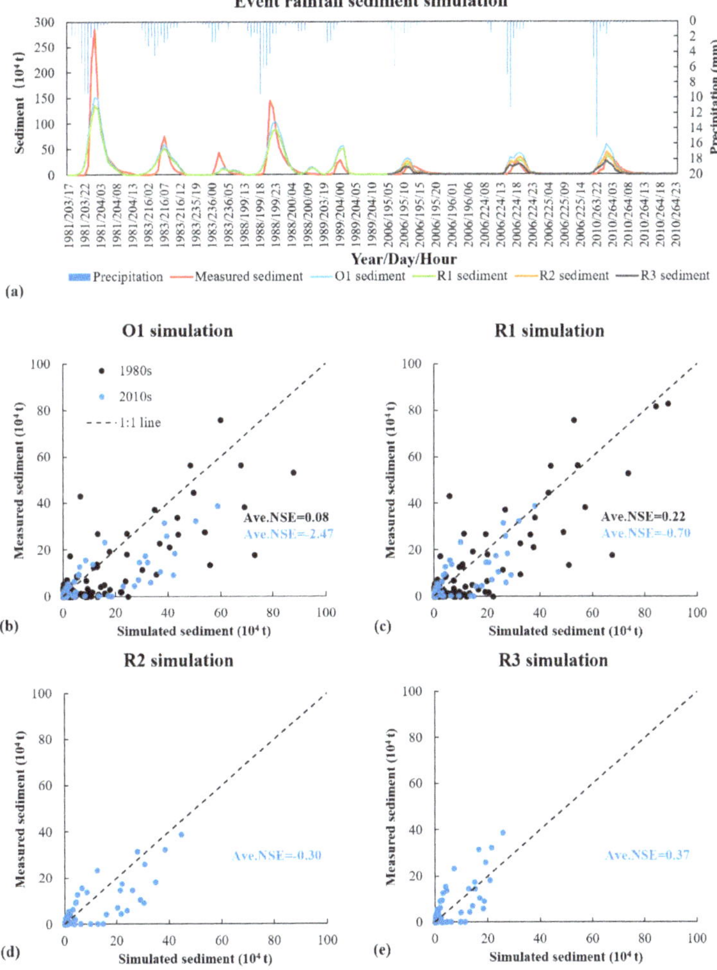

Figure 8. Comparison of observed and simulated sediment discharge for eight events in the Pianguanhe Basin. (**a**) Event rainfall sediment simulation; (**b**) Sediment in O1 simulation; (**c**) Sediment in R1 simulation; (**d**) Sediment in R2 simulation; (**e**) Sediment in R3 simulation.

NSE was also used to evaluate the performance of the simulated sediment discharge, as shown in Figure 8b–e. Figure 8b shows the O1 validation, in which the five events in the 1980s achieved an average NSE of 0.08, while the three events in the 2010s achieved an average NSE of −2.47. It shows that the original model did not achieve good performance in the recent years. Figure 8c shows the R1 validation, in which the average NSE of the 1980s and 2010s both increased. The data points of the 1980s and 2010s were distributed closer to the 1:1 line than in Figure 8b. Figure 8d shows the R2 validation, in which the average NSE of the 2010s have also increased, and greater than that of R1, which indicates that terrace can significantly reduce sediment. Figure 8e shows the R3 validation, in which the average NSE of the 2010s is 0.37, and the data points of the 2010s are distributed closely around the 1:1 line. Overall, the accuracy of the simulation in the 2010s is highest in R3, followed by R2. These figures show that the revised model performs better in recent years, and can reflect the effect of significant surface change (i.e., slope terracing and revegetation) on sediment yield.

4. Discussion

4.1. Effect of Terraces and Vegetation on Runoff Reduction

To quantify the reduction effects of vegetation and terracing during runoff and sediment routing process, the absolute reduction in unit area (AR) and the reduction rate (RR) were calculated. In the case of runoff, for example, we first summed the hourly runoff discharge to obtain the total runoff of each rainfall, and counted the total simulated runoff of each event per case. We then calculated the AR of R1 and R2, and the RR of R1, R2, and R3, as shown in Table 6. The AR and RR were calculated as follows:

$$AR_i = \frac{y - y_i}{A_i}, \ (i = 1, 2) \tag{15}$$

$$RR_i = \frac{y - y_i}{y} \times 100\%, \ (i = 1, 2, 3) \tag{16}$$

where y is total simulated runoff of O1 (m^3); and y_i is total simulated runoff of R1, R2 or R3 (m^3), A_i is total area of R1 or R2 (km^2).

Table 6. The efficiency of runoff reduction by vegetation and terraces in Pianguanhe Basin.

No. of Storm (year/day/hour)	Class	Practice						
		Vegetation		Terrace		Vegetation and Terrace		
		RR_1 (%)	AR_1 (m^3/km^2)	RR_2 (%)	AR_2 (m^3/km^2)	RR_3 (%)	$RR_1 + RR_2$ (%)	Difference (%)
1981/203/17		35.95	4096.15	-	-	-	-	-
1983/215/22	Calibration	30.42	1642.02	-	-	-	-	-
1983/235/16		39.60	642.21	-	-	-	-	-
1988/199/13		35.77	3959.85	-	-	-	-	-
1989/203/19		26.18	494.47	-	-	-	-	-
2006/195/5	Validation	48.22	3318.46	26.13	9122.72	69.02	74.35	−5.33
2006/224/8		53.17	5034.97	26.26	18,060.69	73.26	79.43	−6.17
2010/263/20		51.66	5223.21	27.57	20,244.11	73.30	79.23	−5.93

Annotation: RR_1 and AR_1 mean the RR and AR of vegetation, respectively; RR_2 and AR_2 mean the RR and AR of terrace, respectively; RR_3 means the RR of both vegetation and terrace; Difference denotes RR_3 minus the sum of RR_1 and RR_2.

As shown in Table 6, the average RR of R1 on runoff was 33.58% in the 1980s. For the three events in the 2010s, the average RR of R1 was 51.02%, the average RR of R2 was 26.65%, and the average RR of R3 was 71.86%. The average AR of R1 was 3051.42 m^3/km^2 per event, and it was 15809.18 m^3/km^2 for R2. The results show that the runoff reduction rate of vegetation was significantly higher than that of terraces, as the area of vegetation is seven times larger than that of terraces. However, terraces could reduce more runoff per unit area. Influenced by the revegetation and increase in vegetation coverage, the average RR of R1 in recent years increased by 17.44% over that in the 1980s. In general, the effectiveness of runoff reduction was highest under R3. Besides, the sum of RR_1 and RR_2 is inconsistent with RR_3. The reason for the latter might be that when the terraces had played the role of water reduction, the water reducing capability of vegetation in the same time zone would be decreased. The assumption that all terraces in the Pianguanhe Basin are of good quality and have the same embankment height may be an oversimplification that overestimated the runoff reduction efficiency of the terraces.

4.2. Effect of Terraces and Vegetation on Sediment Reduction

We counted the total simulated sediment yield of each event per case, and also calculated the AR of vegetation and terraces, and the RR of R1, R2, and R3, as shown in Table 7. The average RR of R1 for sediment was 13.31% in the 1980s. For the three events in the 2010s, the average RR of R1 was 32.22%, the average RR of R2 was 24.52%, and the average RR of R3 was 53.85%. The average RR of R1 in recent years increased by 18.91% over that in the 1980s. The average AR of R1 was 449.95 t/km^2, and 2850.17 t/km^2 for R2. The results also show that the sediment reduction rate of vegetation was higher than that of terraces, but that terraces could reduce more sediment per unit area. In general, the effectiveness of sediment reduction of R3 was highest. Besides, the sum of RR_1 and RR_2 was inconsistent with RR_3. As mentioned previously, the reason might be that when terraces had played the role of sediment reduction, the sediment reducing efficiency of vegetation in the same time zone would be decreased. The assumption that all terraces in the Pianguanhe Basin are of good quality and have the same embankment height may have overestimated the sediment reduction efficiency of terraces.

Table 7. The efficiency of sediment reduction by vegetation and terraces in the Pianguanhe Basin.

No. of Storm (year/day/hour)	Class	Practice						
		Vegetation		Terrace		Vegetation and Terrace		
		RR_1 (%)	AR_1 (t/km^2)	RR_2 (%)	AR_2 (t/km^2)	RR_3 (%)	$RR_1 + RR_2$ (%)	Difference (%)
1981/203/17		13.66	796.05	-	-	-	-	-
1983/215/22		11.52	299.66	-	-	-	-	-
1983/235/16	Calibration	17.31	96.91	-	-	-	-	-
1988/199/13		15.37	748.46	-	-	-	-	-
1989/203/19		8.71	129.34	-	-	-	-	-
2006/224/8		33.22	341.94	23.62	1765.63	54.02	56.84	−2.83
2006/195/5	Validation	30.36	528.25	24.08	3041.80	51.75	54.44	−2.69
2010/263/20		33.07	659.02	25.87	3743.10	55.80	58.94	−3.14

Annotation: RR_1 and AR_1 mean the RR and AR of vegetation, respectively; RR_2 and AR_2 mean the RR and AR of terrace, respectively; RR_3 means the RR of both vegetation and terrace; Difference denotes RR_3 minus the sum of RR_1 and RR_2.

4.3. Reliability Analysis of the Water and Sediment Reduction Efficiency of Terraces and Vegetation

Liu et al. [78,81] introduced a runoff coefficient and sediment yield coefficient to discuss the flood and sediment yield for different vegetation conditions at the catchment scale in the Loess Plateau. To analyze the reliability of the simulation results, we also calculated the runoff coefficient and sediment yield coefficient of each event in the Pianguanhe Basin, and compared with the results of Liu et al. [78], as shown in Figure 9a,b. The runoff coefficient is the runoff yield per unit of precipitation per unit area, while the sediment yield coefficient is the sediment yield per unit of precipitation per unit area. In Liu et al. [78], the runoff coefficient was calculated based on annual runoff and precipitation data, and the latter only considered rainfall greater than 25 mm. In this study, the runoff coefficient was calculated based solely on storm event discharge and rainfall data. Thus, there is a disparity between the calculation results of these two studies, but yet the results are comparable to those of Liu et al. [78]. The runoff coefficient decreased along with the increase of the vegetation coverage. The vegetation coverage of the 1980s and 2010s was 16.17% and 46.29%, respectively, and the average runoff coefficient decreased from 0.12 m^3/(mm·km^2) in the 1980s to 0.07 m^3/(mm·km^2) in recent years.

Similarly, the sediment yield coefficients calculated from the event data in this study are different from those of Liu's results based on calculations using annual data, and yet the results are comparable. The sediment yield coefficients also decreased along with the increase of vegetation coverage. The average sediment yield coefficient decreased from 80.68 t/(mm·km^2) in the 1980s to 21.93 t/(mm·km^2) in recent years.

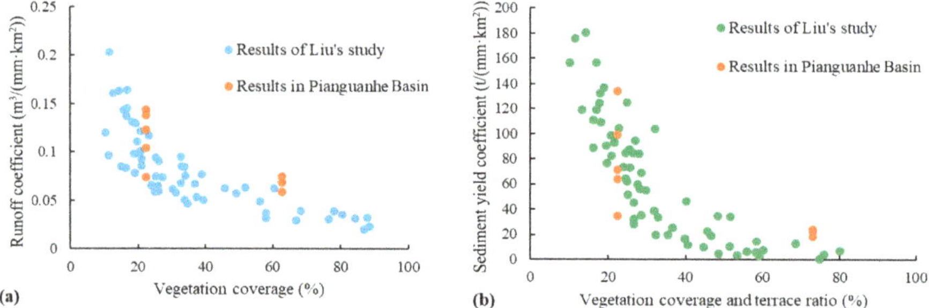

Figure 9. Comparison of runoff generation coefficient (**a**) and sediment yield coefficient (**b**) in this study and in Liu's study.

4.4. Limitations and Potential Improvements

While this study assessed the impact of terraces and vegetation practice on runoff and sediment routing process using the revised time-area method, it still has several assumptions and limitations in the method that need to be clarified and studied further. First, embankment height is a key indicator to evaluate the quality and storage volume of terraces. The assumption that all terraces in the Pianguanhe Basin are of good quality and set the embankment height with a fixed value may overestimate the runoff and sediment reduction efficiency of the terraces. This paper referred to the study of Xiong et al. [77] as the limitation of the flume test data. While the runoff and sediment retention function of vegetation is not independent of slope steepness or vegetation structure, these factors have not been considered yet. In addition, the effect of engineering measures such as check dam and human factors such as mining and road construction should also be considered to further improve simulation efficiency.

5. Conclusions

In this study, we added the impact of terrace and vegetation practice on runoff and sediment routing in the time-area method. The revised time-area method was integrated into the LCM-MUSLE model which is suitable to estimate the water and sediment yield in the Loess Plateau. Eight isolated storm events in the 1980s and 2010s in Pianguanhe Basin were selected to calibrate and verify the original LCM-MUSLE model and its revised version. It is shown that the original model was not applicable in the more recent years, since the surface had changed significantly as a result of revegetation and terrace engineering. The revised model considered the impact of vegetation and terracing on runoff and sediment routing and its accuracy had been improved significantly.

The effect of the level terraces and vegetation was parameterized effectively according to their location, size, embankment height, and vegetation coverage. These parameters could be easily obtained and were used to represent the landscape heterogeneity at the catchment scale. Besides, the revised time-area method was loosely coupled with the LCM-MUSLE model. Therefore, the method could be readily applied in other regions and integrated into other hydrological models and erosion models. Consequently, this study provides a generalized method to quantitatively assess the impact of terrace and vegetation practice on runoff and sediment reduction at the catchment scale, which has great significance in runoff change analysis and implementation of soil and water conservation.

Author Contributions: J.B. conceived and designed the experiments; J.B. and Y.Z performed the experiments; Y.G. contributed analysis tools; J.B., S.Y. and X.L. analyzed the data; J.B. and Y.Z. prepared figures; J.B. wrote the main manuscript text.

Funding: The authors are also thankful for the financial support from the National Key Project for R&D (Nos. 2016YFC0402403, 2016YFC0402409), the National Natural Science Foundation of China (Grant Nos. U1603241, 51779099), and the Fundamental Research Funds for Central Public Welfare Research Institutes (HKY-JBYW-2017-10).

Acknowledgments: We acknowledge all of the editors and reviewers for their valuable advice.

Conflicts of Interest: The authors declare no conflict of interest. The founding sponsors had no role in the design of the study; in the collection, analyses, or interpretation of data; in the writing of the manuscript; and in the decision to publish the results.

References

1. Schnitzer, S.; Seitz, F.; Eicker, A.; Güntner, A.; Wattenbach, M.; Menzel, A. Estimation of soil loss by water erosion in the Chinese Loess Plateau using Universal Soil Loss Equation and GRACE. *Geophys. J. Int.* **2013**, *193*, 1283–1290. [CrossRef]

2. Zhou, J.; Fu, B.; Gao, G.; Lü, Y.; Liu, Y.; Lü, N.; Wang, S. Effects of precipitation and restoration vegetation on soil erosion in a semi-arid environment in the Loess Plateau, China. *Catena* **2016**, *137*, 1–11. [CrossRef]

3. Gao, H.; Li, Z.; Jia, L.; Li, P.; Xu, G.; Ren, Z.; Pang, G.; Zhao, B. Capacity of soil loss control in the Loess Plateau based on soil erosion control degree. *J. Geogr. Sci.* **2016**, *26*, 457–472. [CrossRef]

4. Shi, H.; Shao, M. Soil and water loss from the Loess Plateau in China. *J. Arid Environ.* **2000**, *45*, 9–20. [CrossRef]

5. Lal, R. Soil erosion and the global carbon budget. *Environ. Int.* **2003**, *29*, 437–450. [CrossRef]

6. Yang, Q.; Zhao, Z.; Chow, T.L.; Rees, H.W.; Bourque, C.P.A.; Meng, F.-R. Using GIS and a digital elevation model to assess the effectiveness of variable grade flow diversion terraces in reducing soil erosion in northwestern New Brunswick, Canada. *Hydrol. Process.* **2009**, *23*, 3271–3280. [CrossRef]

7. Ouyang, W.; Hao, F.; Skidmore, A.K.; Toxopeus, A.G. Soil erosion and sediment yield and their relationships with vegetation cover in upper stream of the Yellow River. *Sci. Total Environ.* **2010**, *409*, 396–403. [CrossRef]

8. Preiti, G.; Romeo, M.; Bacchi, M.; Monti, M. Soil loss measure from Mediterranean arable cropping systems: Effects of rotation and tillage system on C-factor. *Soil Tillage Res.* **2017**, *170*, 85–93. [CrossRef]

9. Chen, L.; Wang, J.; Wei, W.; Fu, B.; Wu, D. Effects of landscape restoration on soil water storage and water use in the Loess Plateau Region, China. *For. Ecol. Manag.* **2010**, *259*, 1291–1298. [CrossRef]

10. Deng, L.; Shangguan, Z.P.; Sweeney, S. "Grain for Green" driven land use change and carbon sequestration on the Loess Plateau, China. *Sci. Rep.* **2014**, *4*, 7039. [CrossRef] [PubMed]

11. Fu, B.; Liu, Y.; Lü, Y.; He, C.; Zeng, Y.; Wu, B. Assessing the soil erosion control service of ecosystems change in the Loess Plateau of China. *Ecol. Complex.* **2011**, *8*, 284–293. [CrossRef]

12. Wang, Z.-J.; Jiao, J.-Y.; Rayburg, S.; Wang, Q.-L.; Su, Y. Soil erosion resistance of "Grain for Green" vegetation types under extreme rainfall conditions on the Loess Plateau, China. *Catena* **2016**, *141*, 109–116. [CrossRef]

13. Marques, M.J.; Bienes, R.; Jimenez, L.; Perez-Rodriguez, R. Effect of vegetal cover on runoff and soil erosion under light intensity events. Rainfall simulation over USLE plots. *Sci. Total Environ.* **2007**, *378*, 161–165. [CrossRef]

14. Huang, D.; Han, J.G.; Wu, J.Y.; Wang, K.; Wu, W.L.; Teng, W.J.; Sardo, V. Grass hedges for the protection of sloping lands from runoff and soil loss: An example from Northern China. *Soil Tillage Res.* **2010**, *110*, 251–256. [CrossRef]

15. Wu, J.Y.; Huang, D.; Teng, W.J.; Sardo, V.I. Grass hedges to reduce overland flow and soil erosion. *Agron. Sustain. Dev.* **2010**, *30*, 481–485. [CrossRef]

16. Tarolli, P.; Preti, F.; Romano, N. Terraced landscapes: From an old best practice to a potential hazard for soil degradation due to land abandonment. *Anthropocene* **2014**, *6*, 10–25. [CrossRef]

17. Cao, L.; Zhang, Y.; Lu, H.; Yuan, J.; Zhu, Y.; Liang, Y. Grass hedge effects on controlling soil loss from concentrated flow: A case study in the red soil region of China. *Soil Tillage Res.* **2015**, *148*, 97–105. [CrossRef]

18. Lambrechts, T.; François, S.; Lutts, S.; Muñoz-Carpena, R.; Bielders, C.L. Impact of plant growth and morphology and of sediment concentration on sediment retention efficiency of vegetative filter strips: Flume experiments and VFSMOD modeling. *J. Hydrol.* **2014**, *511*, 800–810. [CrossRef]

19. Khatavkar, P.; Mays, L.W. Optimization Models for the Design of Vegetative Filter Strips for Stormwater Runoff and Sediment Control. *Water Resour. Manag.* **2017**, *31*, 2545–2560. [CrossRef]

20. Zhao, C.; Gao, J.E.; Huang, Y.; Wang, G.; Zhang, M. Effects of Vegetation Stems on Hydraulics of Overland Flow Under Varying Water Discharges. *Land Degrad. Dev.* **2016**, *27*, 748–757. [CrossRef]

21. Arabi, M.; Frankenberger, J.R.; Engel, B.A.; Arnold, J.G. Representation of agricultural conservation practices with SWAT. *Hydrol. Process.* **2008**, *22*, 3042–3055. [CrossRef]

22. Yang, Q.; Meng, F.-R.; Zhao, Z.; Chow, T.L.; Benoy, G.; Rees, H.W.; Bourque, C.P.A. Assessing the impacts of flow diversion terraces on stream water and sediment yields at a watershed level using SWAT model. *Agric. Ecosyst. Environ.* **2009**, *132*, 23–31. [CrossRef]

23. Chen, D.; Wei, W.; Chen, L. Effects of terracing practices on water erosion control in China: A meta-analysis. *Earth-Sci. Rev.* **2017**, *173*, 109–121. [CrossRef]

24. Arnáez, J.; Lana-Renault, N.; Lasanta, T.; Ruiz-Flaño, P.; Castroviejo, J. Effects of farming terraces on hydrological and geomorphological processes. A review. *Catena* **2015**, *128*, 122–134. [CrossRef]

25. Ben Khelifa, W.; Hermassi, T.; Strohmeier, S.; Zucca, C.; Ziadat, F.; Boufaroua, M.; Habaieb, H. Parameterization of the effect of bench terraces on runoff and sediment yield by SWAT modeling in a small semi-arid watershed in Northern Tunisia. *Land Degrad. Dev.* **2017**, *28*, 1568–1578. [CrossRef]

26. Berendse, F.; van Ruijven, J.; Jongejans, E.; Keesstra, S. Loss of Plant Species Diversity Reduces Soil Erosion Resistance. *Ecosystems* **2015**, *18*, 881–888. [CrossRef]

27. Pan, C.; Shangguan, Z. Runoff hydraulic characteristics and sediment generation in sloped grassplots under simulated rainfall conditions. *J. Hydrol.* **2006**, *331*, 178–185. [CrossRef]

28. Rogger, M.; Agnoletti, M.; Alaoui, A.; Bathurst, J.C.; Bodner, G.; Borga, M.; Chaplot, V.; Gallart, F.; Glatzel, G.; Hall, J.; et al. Land use change impacts on floods at the catchment scale: Challenges and opportunities for future research. *Water Resour. Res.* **2017**, *53*, 5209–5219. [CrossRef]

29. Liu, S.L.; Dong, Y.H.; Li, D.; Liu, Q.; Wang, J.; Zhang, X.L. Effects of different terrace protection measures in a sloping land consolidation project targeting soil erosion at the slope scale. *Ecol. Eng.* **2013**, *53*, 46–53. [CrossRef]

30. Yao, Y. Analysis of Effects of Bench Terraced Field on Reducing Soil Erosion. *Soil Water Conserv. China* **1992**, *12*, 40–41. (In Chinese)

31. Wu, F. Study on the Benefits of Level Terrace on Soil and Water Conservation. *Sci. Soil Water Conserv.* **2004**, *2*, 34–37. (In Chinese)

32. Huo, Y.P.; Zhu, B.B. Analysis on the Benefits of Level Terrace on Soil and Water Conservation in Loess Hilly Areas. *Res. Soil Water Conserv.* **2013**, *20*, 24–28. (In Chinese)

33. Meng, C.; Zhang, H.; Yang, P. Effects of Simulated Vegetation Types and Spatial Patterns on Hydrodynamics of Overland Flow. *J. Soil Water Conserv.* **2017**, *31*, 50–56. (In Chinese)

34. Kovář, P.; Bačinová, H.; Loula, J.; Fedorova, D. Use of terraces to mitigate the impacts of overland flow and erosion on a catchment. *Plant Soil Environ.* **2016**, *62*, 171–177. [CrossRef]

35. Zhao, H.; Fang, X.; Ding, H.; Josef, S.; Xiong, L.; Na, J.; Tang, G. Extraction of Terraces on the Loess Plateau from High-Resolution DEMs and Imagery Utilizing Object-Based Image Analysis. *ISPRS Int. J. Geo-Inf.* **2017**, *6*, 157. [CrossRef]

36. Wei, W.; Chen, D.; Wang, L.; Daryanto, S.; Chen, L.; Yu, Y.; Lu, Y.; Sun, G.; Feng, T. Global synthesis of the classifications, distributions, benefits and issues of terracing. *Earth-Sci. Rev.* **2016**, *159*, 388–403. [CrossRef]

37. Neitsch, S.L.; Arnold, J.G.; Kiniry, J.R.; Srinivasan, R.; Williams, J.R. *Soil and Water Assessment Tool Input/Output File Documentation: Version 2009*; Texas Water Resources Institute Technical Report 365; Texas Water Resources Institute: College Station, TX, USA, 2011.

38. Chiu, Y.-J.; Lee, H.-Y.; Wang, T.-L.; Yu, J.; Lin, Y.-T.; Yuan, Y. Modeling Sediment Yields and Stream Stability Due to Sediment-Related Disaster in Shihmen Reservoir Watershed in Taiwan. *Water* **2019**, *11*, 332. [CrossRef]

39. Chen, Y.-C.; Wu, Y.-H.; Shen, C.-W.; Chiu, Y.-J. Dynamic Modeling of Sediment Budget in Shihmen Reservoir Watershed in Taiwan. *Water* **2018**, *10*, 1808. [CrossRef]

40. Panuska, J.C.; Moore, I.D.; Kramer, L.A. Terrain Analysis: Integration into Agricultural Nonpoint Source (AGNPS) Pollution Model. *J. Soil Water Conserv.* **1991**, *46*, 59–64.

41. Gassman, P.W.; Williams, J.R.; Wang, X.; Saleh, A.; Osei, E.; Hauck, L.M.; Izaurralde, R.C.; Flowers, J.D. Invited Review Article: The Agricultural Policy/Environmental eXtender (APEX) Model: An Emerging Tool for Landscape and Watershed Environmental Analyses. *Trans. ASABE* **2010**, *53*, 711–740. [CrossRef]

42. Shao, H.; Baffaut, C.; Gao, J.E.; Nelson, N.O.; Janssen, K.A.; Pierzynski, G.M.; Barnes, P.L. Development and application of algorithms for simulating terraces within SWAT. *Trans. ASABE* **2013**, *56*, 1715–1730.

43. Mishra, S.K.; Singh, V.P. Soil Conservation Service Curve Number (SCS-CN) Methodology. *Water Sci. Technol. Libr.* **2007**, *22*, 355–362.

44. Her, Y.; Heatwole, C. Two-dimensional continuous simulation of spatiotemporally varied hydrological processes using the time-area method. *Hydrol. Process.* **2016**, *30*, 751–770. [CrossRef]

45. Kothyari, U.C.; Tiwari, A.K.; Singh, R. Estimation of temporal variation of sediment yield from small catchments through the kinematic method. *J. Hydrol.* **1997**, *203*, 39–57. [CrossRef]

46. Darvishan, A.V.K.; Sadeghi, S.H.R.; Gholami, L. Efficacy of Time-Area Method in simulating temporal variation of sediment yield in Chehelgazi Watershed, Iran. *Ann. Wars. Univ. Life Sci. SGGW Land Reclam.* **2010**, *42*, 51–60. [CrossRef]

47. Her, Y.; Heatwole, C. HYSTAR Sediment Model: Distributed Two-Dimensional Simulation of Watershed Erosion and Sediment Transport Using Time-Area Routing. *AWRA J. Am. Water Resour. Assoc.* **2016**, *52*, 376–396. [CrossRef]

48. Melesse, A.M.; Graham, W.D. Storm runoff prediction based on a spatially distributed travel time method utilizing remote sensing and GIS. *AWRA J. Am. Water Resour. Assoc.* **2004**, *40*, 863–879. [CrossRef]

49. Saghafian, B.; Julien, P.Y.; Rajaie, H. Runoff hydrograph simulation based on time variable isochrone technique. *J. Hydrol.* **2002**, *261*, 193–203. [CrossRef]

50. Wang, F.; Mu, X.; Zhang, X.; Li, R. Effect of soil and water conservation on runoff and sediment in Pianguan River. *Sci. Soil Water Conserv.* **2005**, *3*, 10–14. (In Chinese)

51. Liu, C.; Hong, B.; Zeng, M.; Cheng, Y. Experimental study on the relationship between storm and runoff in the Loess Plateau of China. *Chin. Sci. Bull.* **1965**, *10*, 158–161. (In Chinese)

52. Li, J.; Liu, C.M.; Wang, Z.G.; Liang, K. Two universal runoff yield models: SCS vs. LCM. *J. Geogr. Sci.* **2015**, *25*, 311–318. [CrossRef]

53. Liu, C.; Wang, Z.; Zheng, H.; Zhang, L.; Wu, X. Development of Hydro-Informatic Modelling System and its application. *Sci. China Ser. E Technol. Sci.* **2008**, *51*, 456–466. [CrossRef]

54. Liu, C.; Yang, S.; Wen, Z.; Wang, X.; Wang, Y.; Li, Q.; Sheng, H. Development of ecohydrological assessment tool and its application. *Sci. China Ser. E Technol. Sci.* **2009**, *52*, 1947–1957. [CrossRef]

55. Zhang, Y.; Liu, C.; Yang, S.; Liu, X.; Cai, M.; Dong, G.; Luo, Y. Comparison of LCM hydrological models with lumped, semi-distributed and distributed building structures in typical watershed of Yellow River Basin. *Acta Geogr. Sin.* **2014**, *69*, 90–99. (In Chinese)

56. Williams, J.R. Sediment-yield prediction with Universal Equation using runoff energy factor. In *Present and Prospective Technology for Predicting Sediment Yield and Sources*; United States Department of Agriculture: Washington, DC, USA, 1975; pp. 244–252.

57. Ye, A.; Xia, J.; Qiao, Y.; Wang, G. A distributed soil erosion model on the small watershed. *J. Basic Sci. Eng.* **2008**, *16*, 328–340. (In Chinese)

58. Wang, S.P.; Zhang, Z.Q.; Tang, Y.; Guo, J.T. Evaluation of spatial distribution of soil erosion and sediment yield for a small watershed of the Loess Plateau by coupling MIKE-SHE with MUSLE. *Trans. Chin. Soc. Agric. Eng.* **2010**, *26*, 92–98. (In Chinese)

59. Luo, Y.; Yang, S.; Liu, X.; Liu, C.; Zhang, Y.; Zhou, Q.; Zhou, X.; Dong, G. Suitability of revision to MUSLE for estimating sediment yield in the Loess Plateau of China. *Stoch. Environ. Res. Risk Assess.* **2015**, *30*, 379–394. [CrossRef]

60. McCarthy, G.T. The unit hydrograph and flood routing. In Proceedings of the Conference of North Atlantic Division, US Army Corps of Engineers, New London, CT, USA, 24 June 1938; pp. 608–609.

61. Overton, D.E. Muskingum flood routing of upland streamflow. *J. Hydrol.* **1966**, *4*, 185–200. [CrossRef]

62. Aston, A.R. Rainfall interception by 8 small trees. *J. Hydrol.* **1979**, *42*, 383–396. [CrossRef]

63. Williams, J.R.; Singh, V.P. Chapter 25: The EPIC model. In *Computer Models of Watershed Hydrology*; Water Resources Publications: Highlands Ranch, CO, USA, 1995; pp. 909–1000.

64. Liu, B.Y.; Nearing, M.A.; Risse, L.M. Slope Gradient Effects on Soil Loss for Steep Slopes. *Trans. ASAE* **1994**, *37*, 1835–1840. [CrossRef]

65. Liu, B.Y.; Nearing, M.A.; Shi, P.J.; Jia, Z.W. Slope length effects on soil loss for steep slopes. *Soil Sci. Soc. Am. J.* **2000**, *64*, 1759–1763. [CrossRef]

66. Cai, C.F.; Ding, S.W.; Shi, Z.H.; Huang, L.; Zhang, G.Y. Study of Applying USLE and Geographical Information System IDRISI to Predict Soil Erosion in Small Watershed. *J. Soil Water Conserv.* **2000**, *14*, 19–24. (In Chinese)
67. Sabzevari, T. Runoff prediction in ungauged catchments using the gamma dimensionless time-area method. *Arab. J. Geosc.* **2017**, *10*. [CrossRef]
68. Williams, J.R. Sediment routing for agricultural watersheds. *AWRA J. Am. Water Resour. Assoc.* **1975**, *11*, 965–974. [CrossRef]
69. Jenson, S.K. Extracting topographic structure from digital elevation data for geographic information system analysis. *Pe Rs* **1988**, *54*, 1593–1600.
70. Wang, G.Q.; Zhang, C.C.; Liu, J.H.; Wei, J.H.; Xue, H.; Tie-Jian, L.I. Analyses on the variation of vegetation coverage and water/sediment reduction in the rich and coarse sediment area of the Yellow River Basin. *J. Sediment Res.* **2006**, *2*, 10–16. (In Chinese)
71. Cadaret, E.M.; McGwire, K.C.; Nouwakpo, S.K.; Weltz, M.A.; Saito, L. Vegetation canopy cover effects on sediment erosion processes in the Upper Colorado River Basin Mancos Shale formation, Price, Utah, USA. *Catena* **2016**, *147*, 334–344. [CrossRef]
72. Boer, M.; Puigdefabregas, J. Effects of spatially structured vegetation patterns on hillslope erosion in a semiarid Mediterranean environment: A simulation study. *Earth Surf. Processes Landf.* **2005**, *30*, 149–167. [CrossRef]
73. Hua, D.; Wen, Z. Study on Runoff and Sediment Under Simulated Rainstorm Condition of Different Stages of Vegetation Restoration in Loess Hilly Region, China. *J. Soil Water Conserv.* **2015**, *29*, 27–31. (In Chinese)
74. Guo, Y.H.; Zhao, T.N.; Sun, B.P.; Ding, G.D.; Cheng, C.; Hu, F.B. Study on the Dynamic Characteristics of Overland Flow and Resistance to Overland Flow of Grass Slope. *Res. Soil Water Conserv.* **2006**, *13*, 264–267. (In Chinese)
75. Li, M.; Yao, W.; Yang, J.; Chen, J.; Ding, W.; Li, L.; Yang, C. Experimental Study on the Effect of Grass Cover on the Overland Flow Pattern in the Hillslope-gully Side Erosion System. *J. Basic Sci. Eng.* **2009**, *17*, 513–523. (In Chinese)
76. Yu, G.Q.; Li, Z.B.; Pei, L.; Li, P. Difference of Runoff-Erosion-Sediment Yield Under Different Vegetation Type. *J. Soil Water Conserv.* **2012**, *26*, 1–5. (In Chinese)
77. Xiong, Y.; Wang, H.; Bai, Z. Preliminary Study on Benefit Indexes of Runoff and Sediment Reduction by Terraced Field, Forest Land and Grass Land. *Soil Water Conserv. China* **1996**, *8*, 10–14. (In Chinese)
78. Liu, X.Y.; Yang, S.T.; Wang, F.G.; Xing-Zhao, H.E.; Hong-Bin, M.A.; Luo, Y. Analysis on sediment yield reduced by current terrace and shrubs-herbs-arbor vegetation in the Loess Plateau. *J. Hydraul. Eng.* **2014**, *45*, 1293–1300. (In Chinese)
79. Carlson, T.N.; Ripley, D.A. On the relation between NDVI, fractional vegetation cover, and leaf area index. *Remote Sens. Environ.* **1997**, *62*, 241–252. [CrossRef]
80. Mao, Y.; Ye, A.; Xu, J.; Ma, F.; Deng, X.; Miao, C.; Gong, W.; Di, Z. An advanced distributed automated extraction of drainage network model on high-resolution DEM. *Hydrol. Earth Syst. Sci. Discuss.* **2014**, *11*, 7441–7467. [CrossRef]
81. Liu, X.Y.; Yang, S.T.; Xiaoyu, L.I.; Zhou, X.; Luo, Y.; Dang, S.Z. The current vegetation restoration effect and its influence mechanism on the sediment and runoff yield in severe erosion area of Yellow River Basin. *Sci. Sin.* **2015**, *45*, 1052. (In Chinese)

Article

Micro-Watershed Management for Erosion Control Using Soil and Water Conservation Structures and SWAT Modeling

Ghulam Nabi [1], Fiaz Hussain [2,3], Ray-Shyan Wu [2,*], Vinay Nangia [4] and Riffat Bibi [5]

[1] Centre of Excellence in Water Resources Engineering, University of Engineering and Technology, Lahore 54890, Pakistan; gnabi60@yahoo.com
[2] Department of Civil Engineering, National Central University, Chung Li 32001, Taiwan; engr.fiaz@uaar.edu.pk
[3] Department of Agricultural Engineering, PMAS-Arid Agriculture University Rawalpindi, Rawalpindi 46000, Pakistan
[4] Soil, Water, and Agronomy, International Centre for Agriculture Research in the Dry Areas (ICARDA), Rabat 10010, Morocco; v.nangia@cgiar.org
[5] Soil and Water Conservation Research Institute (SAWCRI), Chakwal 48800, Pakistan; riffat_ises@yahoo.com
* Correspondence: raywu@ncu.edu.tw; Tel.: +886-3-4227151 (ext. 34126)

Received: 27 March 2020; Accepted: 14 May 2020; Published: 19 May 2020

Abstract: This study evaluated the effectiveness of soil and water conservation structures for soil erosion control by applying a semi-distributed Soil and Water Assessment Tool (SWAT) model in various small watersheds of the Chakwal and Attock districts of Pothwar, Pakistan. The validated model without soil conservation structures was applied to various ungauged small watershed sites with soil conservation stone structures. The stone bund-type structure intervention was used in the model through the modification of the Universal Soil Loss Equation (USLE) to support the practice factor (P-factor), the curve number, and the average slope length for the sub-basin (SLSUBBSN). The structures had significant effects, and the average sediment yield reduction caused by the soil conservation stone structures at these sites varied from 40% to 90%. The sediment yield and erosion reductions were also compared under conditions involving vegetation cover change. Agricultural land with winter wheat crops had a higher sediment yield than fallow land with crop residue. The fallow land facilitated sediment yield reduction, along with soil conservation structures. The slope classification analysis indicated that 60% of the agricultural area of the Chakwal and Attock districts lie in a slope range of 0–4%, where considerable potential exists for implementing soil conservation measures by installing soil conservation stone structures. The slope analysis measured the suitability of conservation structures in the semi-mountainous Pothwar area in accordance with agriculture practice on land having a slope of less than 5%. The SWAT model provides reliable performance for erosion control and watershed management in soil erosion-prone areas with steep slopes and heavy rainfall. These findings can serve as references for policymakers and planners.

Keywords: SWAT modeling; soil erosion; land management; soil conservation stone structures

---

1. Introduction

Soil is a precious natural resource that covers Earth's land surfaces, and it contributes to basic human needs like food, clean water, and clean air, as well as being a major carrier for biodiversity. There have been antecedents (from 3500 B.C to 17th century) of soil knowledge and its relationship with human practices before soil scientific studies. Soil is an integrated discipline within soil sciences, geography, and land management, and it was developed in parallel with agriculture [1,2]. In the globalized world of the 21st century, soil sustainability depends not only on management choices by farmers, foresters, and land planners but also on political decisions on rules and regulations; it also requires a large effort of awareness raising and the communication of issues related to the degradation of soils and land by scientists, civil society organizations, and policy makers [3]. Estimations have shown that worldwide, 75% of land is degraded due to physical, chemical, and biological processes [4]. Soil erosion has a severe impact on the degradation of quality fertile topsoil. Worldwide, soil erosion losses are the highest in the agro-ecosystems of Asia, Africa, and South America, averaging 30–40 t ha^{-1} year^{-1}, and it is the lowest in the United States, Europe, and Australia, averaging 5–20 t ha^{-1} year^{-1} [5,6]. The multifunctional use of land is needed within the boundaries of the soil–water system to achieve land degradation neutrality, avoid further land degradation, and promote land restoration [7]. Keesstra et al. [7] introduced four concepts (systems thinking, connectivity, nature-based solutions, and regenerative economics) in a more integrated way to accomplish land degradation neutrality in an effort to achieve the soil-related Sustainable Development Goals (SDGs). A robust soil–water system is essential to achieve interlinked SDGs through smart planning based on a socio–economical–ecological systems analysis [3,7].

Agricultural land degradation in rainfed mountainous areas is a major onsite problem (the removal of top soil) that also causes offsite effects, such as downstream sediment deposition in fields, floodplains, and water bodies. The costliest offsite damages occur when the soil particles enter lakes or river systems [8,9]. Annual soil loss in the middle Yellow River basin of China amounts to 3700 t km^{-2}, the largest sediment-carrying river in the world [10]. The world's 13 large rivers carry 5.8 billion tons of sediments to reservoirs every year [11]. The Indus River in Pakistan ranks third in the world, with an annual sediment load of 435 million tons in the Tarbela dam, which has lost about 35% of initial reservoir capacity (11,600 Mm^3) [12]. Water and soil are the most crucial natural resources for agriculture and livestock production. Globally, water resource deterioration caused by soil erosion is a growing concern. An estimated productivity loss of US$13–28 billion annually in drylands can be attributed to soil erosion as well [13].

In Pakistan, dryland farming is practiced on 12 Mha of the Pothwar Plateau, the northern mountains, and the northeastern plains. Soil erosion is a severe problem due to erratic rainfall, varied soil slopes, and land use. A lot of land has been converted into gullies that are difficult to restore. Different studies related to soil erosion severity have been conducted in the Pothwar region. Hussain et al. evaluated the soil erosion parameters and estimated the annual sediment loss in small watersheds of the Dhrabi River using the Soil and Water Assessment Tool (SWAT) model. The annual sediment yield ranged from 2.6 to 31.1 t hm^{-2} for the non-terraced catchments, while it ranged from 0.52 to 10.10 t hm^{-2} for the terraced catchments [14]. Iqbal et al. studied runoff plots in the Dhrabi watershed in Chakwal Pakistan; cultivated slopes produced the highest soil loss (8.96 Mg ha^{-1}) annually compared to both undisturbed gentle and steep slopes at approximately 2.08 and 4.66 Mg ha^{-1} [15].

Nasir et al. applied the Revised Universal Soil Loss Equation (RUSLE) and a Geographic Information System (GIS) at the small mountainous watershed of Rawal Lake near Islamabad. The predicted annual soil loss ranged from 0.1 to 28 t ha^{-1} [11]. Similarly, Ahmad et al. reported annual soil loss rates of 17–41 t ha^{-1} under fallow conditions, as well as an annual rate of 9–26 t ha^{-1} under vegetative cover in the Fateh Jang watershed with a slope of 1–10% [16]. Saleem et al. assessed the annual soil erosion (70–208 t ha^{-1}) of the Pothwar region using the RUSLE model integrated with a GIS [17]. Bashir et al. estimated the soil erosion risk using the Coordination of Information on the Environment (CORINE) model in the Rawal watershed. The annual soil loss ranged between 24 and 28 t ha^{-1}, with a high erosion risk (26%) in areas with steep slopes and low vegetative cover [18].

The highest estimated record of soil erosion was 150–165 t ha^{-1} year^{-1} in the Dhrabi watershed of the Pothwar region [12]. Nabi et al. reported that in the Soan watershed of Pothwar, the soil loss rates in barren and shrub land were 63.41 and 53.41 t ha^{-1} year^{-1}, respectively, whereas those in low and high cropping intensity land were 34.91 and 25.89 t ha^{-1} year^{-1}, respectively [19]. Vegetation cover on sloped ground helps to reduce soil loss; however, during field preparation and cultivation, surface soil becomes pulverized and easily eroded, causing acute topsoil erosion due to the removal of vegetation cover. Therefore, during the cultivation of sloping land, measures should be adopted to stop fertile surface soil loss caused by substantial rainfall–runoff. If such measures are not applied, agricultural land may turn barren in only a few years. Vegetation cover is a key measure for soil protection against water erosion; it reduces the flow velocity of surface runoff by increasing surface roughness, in addition to increasing the infiltration rate of soil [20–22].

Considerable increase in sediment yield at the expense of soil development poses a major threat to soil and water resource development. Though water erosion is a function of many environmental factors, its assessment and mitigation at the watershed level are complex phenomena; this is due to the unpredictable nature of rainfall and topographic heterogeneities, climate, and land use–land cover variability, as well as other watershed features for the specified areas under study. In addition, inappropriate land management practices and human activities increase the dynamics of these factors.

At present, many models with a broad spectrum of concepts—which are classified as spatially lumped, spatially distributed, empirical, regression, semi-distributed eco-hydrological models, and factorial scoring models—are in use for modelling the rainfall–runoff–soil erosion and sediment transport processes at different scales [23]. The empirical models are generally the simplest, limited to the conditions and parameter inputs for which they have been developed. For example, the Universal Soil Loss Equation (USLE) [24], the Agricultural Non-Point Source Pollution Model (AGNPS) [25], and the Sediment Delivery Distributed (SEDD) model [26]. In conceptual models, a watershed is represented by a storage system, such as the SWAT [27], the Large Scale Catchment Model (LASCAM) [28], or the European Modeling and Simulation Symposium (EMSS) [29]. Physics-based models rely on the solution of fundamental physical equations and are used for the quantification of physical processes. Areal Nonpoint Source Watershed Environment Response Simulation (ANSWERS) [30]; Chemicals, Runoff, and Erosion from Agricultural Management Systems (CREAMS) [31]; the Watershed Erosion Simulation Program (WESP) [32]; Système Hydrologique Européen Sediment (SHESED) [33]; and the European Soil Erosion Model (EUROSEM) [34] are some examples of physically based erosion and sediment transport models.

This research was conducted in ungauged micro-watersheds of the Chakwal and Attock districts of the Pothwar region. Soil erosion and water loss are extreme hazards in this area due to cultivated highland slopes where timely soil and water conservation strategies and remedial measures are required for sustainable crop productivity. A large number of loose stone structures have been built by public departments and farmers themselves to reduce the soil erosion and moisture conservation upside of these structures. There are few measurement points for rainfall and runoff, and most of the watersheds are ungauged; both of these issues hamper model calibration and validations. The purpose of the study was to evaluate the effectiveness of soil and water conservation structure for soil erosion control using the SWAT model. The calibrated and validated model related to soil erosion was adopted from Hussain et al. [14], where an experimental setup was used to monitor the soil and water loss from agricultural catchment. The collected data were used to calibrate and validate soil erosion parameters using the SWAT model.

This validated model was further modified for the application of soil and water conservation structures, eventually to be recommended by this study as a strategy to counteract the soil erosion with soil and conservation structures at a broader scale. Several studies related to soil and water conservation intervention were carried out to control soil erosion at the field and sub-watershed scale within the Gumara-Maksegnit watershed in the northern Highlands of Ethiopia [35–37], while Melaku et al. predicted the impact of soil and water conservation structures on runoff and erosion processes using the SWAT model [38]. However, studies on the impacts of soil and water structures on the erosion process at the watershed scale that have used the SWAT model have been limited.

Our study was localized to the micro-watersheds with soil and water conservation structures installed through the cooperative project coordinated by the International Centre for Agricultural Research in the Dry Areas (ICARDA), the Centre of Excellence in Water Resources Engineering, (CEWRE), and the Soil and Water Conservation Research Institute (SAWCRI). To the best of our knowledge, no study has been conducted to evaluate the effectiveness of these structures to control soil erosion in the Pothwar region. The study results may encourage the stakeholders to extend this practice to a larger scale by knowing the quantitative benefits of soil conservation structures. Therefore, the SWAT model was adopted due to the availability of a comprehensive agricultural management database, as well as a reduced time and cost [39–41]. In this context, the objective of this work was to evaluate the effectiveness of soil and water conservation structures for soil erosion control using the SWAT model in the micro-watersheds of the Chakwal and Attock districts.

2. Materials and Methods

2.1. Pothwar Region and Study Watersheds Description

Determining the relationship between rainfall, runoff, and soil erosion was imperative in the Pothwar rainfed region for creating applicable soil and water conservation mechanisms, as well as for enhancing crop productivity. Considering the long-term sustainability and productivity of eroded land, the present study focused on the Chakwal and Attock districts of the Pothwar plateau between 32°30′ and 34° north latitudes and 71°45′ and 73°45′ east longitudes, as shown in Figure 1. The region has an arid-to-semiarid climate with hot summers and cold winters [42]. The plateau land comprises broken gullies, low hill ranges, and a flat to gently undulating topography. The textural classification varies from sandy to silt and clay loam, and the land consists of poor-to-fertile soil derived from sandstone and loess parent material [43].

The rainfall pattern is unpredictable with a high intensity; 60–70% of the total rainfall occurs during the monsoon season (from mid-June to mid-September). The average annual rainfall varies from 250 to 1675 mm, with a decreasing trend from the north to the south. After rainfall, soil crusting decreases the infiltration rate and aeration and increases the soil strength, which reduces plant emergence and exposes the soil surface to erosion [44]. The soil loss rate becomes relatively high with higher intensity rainfall–runoff over greater slope lengths and steepness levels.

Out of the total 1.82 Mha of the Pothwar region, approximately 0.77 Mha (43%) are cultivated, and the remaining is mostly grazing land. Only 4% of the cultivated area has irrigated agriculture, while the remaining area is under rainfed agriculture [45]. Rainfall plays an important role in crop production. The principal crops of the area are wheat, maize, bajra, barley, pulses, groundnut, fruits, and vegetables. Without adequate protection, the effects of rainfall–runoff erosion on this highly erodible soil are severe, causing extensive fertile soil loss [19], endangered soil and water conservation structures, and reservoir depletion through sedimentation. Moreover, this raises doubts regarding the viability of existing and future soil and water conservation schemes.

The high rate of erosion creates a silting problem in the small dams of the Pothwar area. For the sustainable agricultural and socioeconomic development of the region, the government has started various projects for watershed development in the upstream of storage reservoirs, such as the Watershed Management Program by Pakistan's Water and Power Development Authority (WAPDA).

Similarly, soil and water conservation activities have also been carried out in the Pothwar region for erosion control and land development through a series of Barani area development projects. The application of the loose stone structures project of SAWCRI (Soil and Water Conservation Research Institute, Chakwal) with ICARDA (International Center for Agricultural Research in the Dry Areas) for erosion control resulted in the development of some environmentally friendly and cost-effective resource conservation technologies.

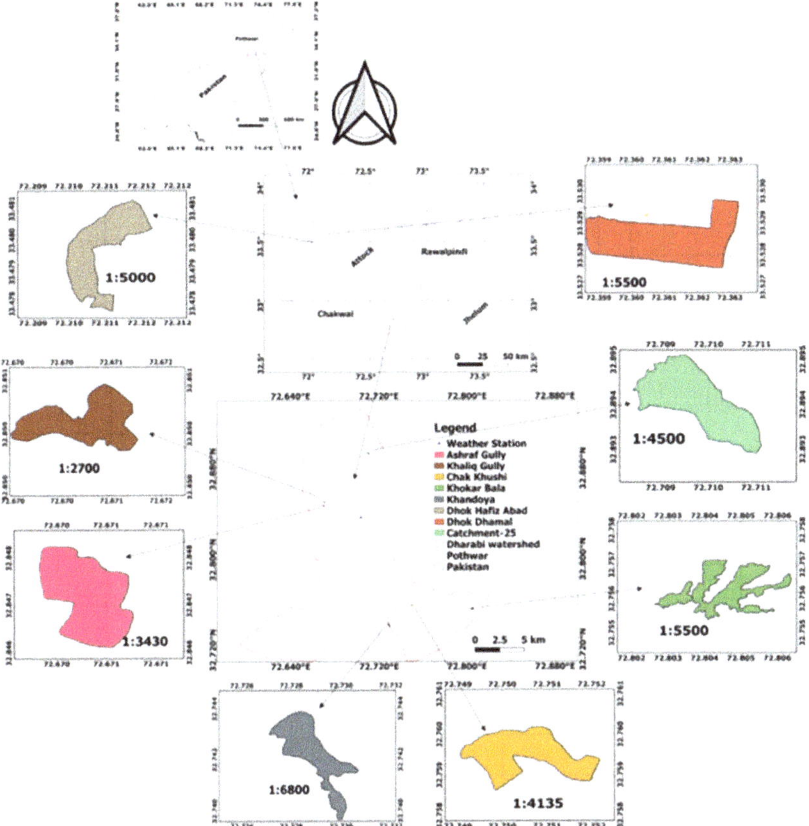

Figure 1. Location map of the Pothwar region, including the Dharabi watershed: Catchment-25; the Kohkar Bala, Chak Khushi, Khandoya, Ashraf Gully, Khaliq Gully, Dhoke Hafiz Abad, and Dhoke Dhamal micro-watersheds.

The soil and water conservation structures were installed in small terraced agriculture fields in the Chakwal and Attock districts by the SAWCRI Chakwal department. Seven small sites were selected to evaluate the effectiveness of these structures on soil erosion control. The description of these sites is given in Table 1, and a location map is shown in Figure 1. The demarcation of the watershed areas was a challenging task and was performed using a Global Positioning System (GPS) survey. During the survey, the point elevation data at different locations were collected within and at the boundary of the watersheds. Using the elevation data in ArcGIS, we performed a topographic analysis and observed that all the watersheds have a land slope of 2–7% where the crops are grown. The location of the conservation structures was also noted for use with the SWAT model setup.

Table 1. Study watershed site descriptions.

Site Name	District	Land Use System	Area (ha)	Mean Land Slope (%)
Kohkar Bala			2.75	7.15
Khandoya		Winter wheat and	5.37	4.35
Khaliq Gully	Chakwal	fallow land	1.25	3.75
Ashraf Gully			2.64	3.52
Chak Khushi			2.33	2.31
Dhoke Dhamal	Attock	Winter wheat and	7.03	3.86
Dhoke Hafiz Abad		fallow land	3.22	4.80
* Catchment-25	Chakwal	Winter wheat	2.0	10.5

* used for SWAT calibration and validation due to the availability of measured flow and sediment data.

2.2. Soil and Water Conservation Structures

In the Pothwar region, the terrace land use system and the wide and deep gullies are used for field crop production. The agriculture fields are usually not flat; however, various field terraces are situated at different elevation levels (Figure 2a). Farmers make earthen embankments (bunds) to retain rainwater and conserve soil moisture. When heavy rainstorms occur, the terrace land use system often fails due to the breaching of the field embankments/bunds. This is mainly caused by the hydraulic shear failure of the soil under saturated conditions. The disturbance of soil organisms can aggravate the impact. Figure 2b shows such terrace failures, which increase the surface runoff and soil erosion, especially in the Pothwar area.

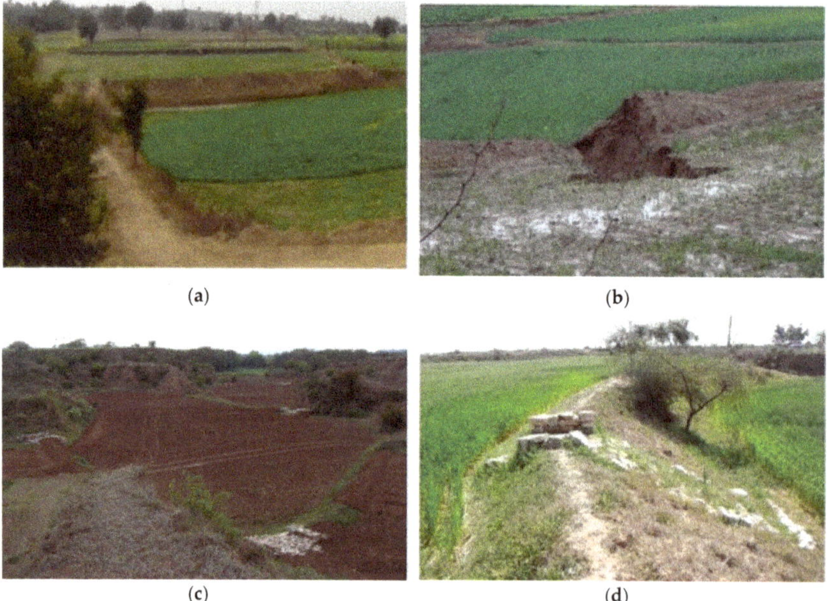

(a) (b)
(c) (d)

Figure 2. Soil and water conservation structures: (**a**) Terraced cultivated lands in Pothwar. (**b**) Breached terrace bund/embankment. (**c**) Loose stone structures system. (**d**) A loose stone structure in the field.

The moving runoff from higher to lower fields takes not only the fertile topsoil but also essential nutrients and organic matter with it, thereby reducing the productive capacity of soils. If the breached bund is not repaired before next rainy season, it leads to the formation of gullies and renders the area out of plough, a great national loss. Crop yields on such eroded lands are poor, and the livelihood of resource-poor farmers is adversely affected. To reduce this problem, the eroded areas need sustainable rehabilitation to ensure food security in the region. With the collaboration work of SAWCRI and ICARDA, loose stone structures were installed in the upper, middle, and lower parts of terraced watersheds, as shown in Figure 2c,d.

The idea is to retain water in a terrace until a certain rainfall amount (without overflowing the terrace) and then to divert the excess rainfall in a non-erosive way. First, this increases the infiltration and improves the amount of plant-available water; secondly, it reduces soil erosion by reducing the amount and kinetic energy of the flowing water. On average, a water height of approximately 4–6 inches can be held back in the fields. The crest of the structures is kept raised 6–9 inches from the soil surface to encourage in situ rainwater conservation. The height of the sidewalls of a structure should be equal to the height of the field bund/embankment where the structure is to be installed. The cross-section view of these structures is shown in Figure 3.

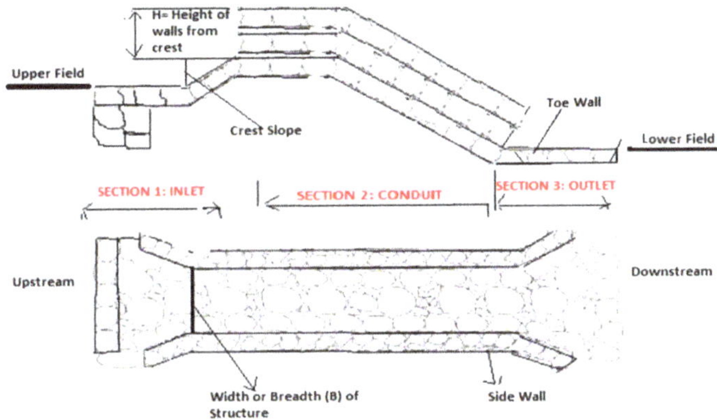

Figure 3. Cross-section of a loose stone structure.

2.3. SWAT Model Description

The SWAT model is a semi-distributed, watershed scale, eco–hydrological model that deals with land–soil–water–plant systems [27]. This model has been tested for a wide variety of watershed and environmental conditions worldwide [46–56]. ArcSWAT jointly developed by USDA Agricultural Research Service (USDA-ARS) and Texas A&M AgriLife Research, was used to spatially link multiple model input data, such as watershed topography (digital elevation model (DEM)), soil, land use, land management, and climatic data. During watershed delineation, the entire watershed was divided into different sub-basins. Then, each sub-basin was discretized into a series of hydrologic response units (HRUs) as the smallest computation unit of a SWAT model, which were characterized by homogeneous soil, land use, and slope combinations. The daily climate input data for defined locations were spatially related to the different sub-basins of the model using a nearest neighbor GIS algorithm. The simulated sediment yield for each HRU was then aggregated and processed to sub-basin level results on a daily time step resolution. The surface runoff computation was performed using modified Soil Conservation Service–Curve Number (SCS–CN) method [57]. Sediment yield levels from each HRU were estimated using the Modified Universal Soil Loss Equation (MUSLE) [58] written as a mass balance equation as follows:

$$SY = 11.8 \left(Q_{surf} \times q_{peak} \times area_{hru} \right)^{0.56} . K_{USLE} . C_{USLE} . P_{USLE} . LS_{USLE} . CFRG ,\qquad (1)$$

where SY is the sediment yield (t), Q_{surf} is the surface runoff (mm ha^{-1}), q_{peak} is the peak discharge (m^3 s^{-1}), and area$_{hru}$ is the area of the hydrological response unit (ha). K_{USLE} (0.013 (t.m^2.hr)/(m^3.t.cm)), C_{USLE}, P_{USLE}, and LS_{USLE} are the USLE parameters. CFRG is the coarse fragment factor.

The sediment transport capacity of the stream channel is a direct function of the channel peak velocity, which is used in the SWAT model, as shown in Equation (2):

$$T_{ch} = \alpha v^b,$$ (2)

where T_{ch} (t m^{-3}) is the transport capacity of a channel, v (m s^{-1}) is the channel peak velocity, and α and b are constant coefficients.

The channel peak velocity was calculated using Manning's formula in a reach segment, as presented in Equation (3):

$$v = \frac{1}{n} R_{ch}{}^{2/3} S_{ch}{}^{1/2},$$ (3)

where n is Manning's roughness coefficient, R_{ch} (m) is hydraulic radius, and S_{ch} (m m^{-1}) is the channel bed slope.

Channel aggradation (Sed_{agg}) and channel degradation (Sed_{deg}) in tons were computed in the channel segment using the criteria presented in Equations (4) and (5):

$$\text{if } sed_i > T_{ch} : Sed_{agg} = (sed_i - T_{ch}) \times V_{ch} \text{ \& } Sed_{deg} = 0,$$ (4)

$$if \ T_{ch} > sed_i : \ Sed_{deg} = (T_{ch} - sed_i) \times V_{ch} \times K_{ch} \times C_{ch} \ \& \ Sed_{agg} = 0,$$ (5)

where sed_i (t m^{-3}) is the initial concentration of sediment, C_{ch} is the channel cover factor, K_{ch} is the channel erodibility factor, and V_{ch} (m^3) is the channel segment water volume.

Sed_{out} (t) is the total sediment transported out of the channel segment, which was computed using Equation (6):

$$Sed_{out} = \left(sed_i + Sed_{deg} - Sed_{agg}\right) \times \frac{V_{out}}{V_{ch}},$$ (6)

where V_{out} (m^3) is the volume of water leaving the channel segment at each time step, sed_i (t) is the sediment inflow concentration at each time step, Sed_{agg} (t) is the channel aggradation, Sed_{deg} (t) is channel degradation at each time step, and V_{ch} (m^3) is volume of channel segment water at each time step.

Soil erosion is a direct function of the slope length and steepness, and it increases due to increases in shear stress. Thus, a major influence of the slope on erosion appears to be exerted through its impact on runoff velocity, and the sediment transport capacity of runoff increases with the increasing flow velocity.

2.4. SWAT Model Input and Setup

The requisite spatial data (DEM, land use, and soil data) and temporal data (rainfall and temperature) were prepared for the SWAT model setup. A physical topographical survey of the watersheds was conducted using a GPS. The DEM of each watershed was generated using point-source elevation data in a geographic information system by applying the inverse distance weighting (IDW) method, as shown in Figure 4. The winter wheat land use classification was used according to cropping practice, and the soil type was sandy loam for all small watersheds based upon the soil textural analysis. The daily precipitation and temperature data were collected from the SAWCRI Chakwal for six years from January 2010 to April 2015.

After the preparation of the requisite data file for model input, ArcSWAT9.3 was used to automatically delineate sub-watersheds and to generate a stream network based on the DEM. An appropriate database of sub-basin parameters and a comprehensive topographic report of the watersheds were generated. The sub-watersheds topographic report was rechecked for area, slope, location of outlet, and soil textural class according to the physical characteristics to make appropriate database changes. SWAT coding conventions were used to reclassify the land use and soil maps into HRUs based on the unique land use, soil class, and slope class in the overlaying section.

The weather station location and lookup tables of daily precipitation and temperature (maximum and minimum) data were loaded to link them with the required files. First, the model was simulated for each watershed with validated parameters adopted from Hussain et al. [14] without the consideration of the conservation structures, and then interventions of the soil and water conservation structures were made by modifying the parameters for surface runoff and sediment yield. The setup of model for each watershed with and without the consideration of the conservation structures is shown as the right and left of Figure 4, respectively.

The locations of each soil and water conservation structure were marked and used for the correct delineations of sub-basins. The demarcated sub-basins indicated the boundary of the agriculture fields, while the structures were the outlet of each field in model setup when the conservation structures were considered. The ideal factors that describe the effect of stone bunds are the USLE support practice factor (P-factor), the curve number, and the average slope length for the sub-basin (SLSUBBSN).

The SLSSUBSN value was modified by editing the HRU (.hru) input table, whereas the P-factor and curve number values were modified by editing the Management (.mgt) input table. Three more parameters were modified, namely the average slope steepness (HRU_SLP) of the HRU input tables and two basin parameters (SPCON and SPEXP) representing the general watershed attributes in the Basin (.bsn) input files. SPCON and SPEXP are linear and exponential channel sediment routing factors, respectively, that affect the movement and separation of sediment fractions in the channel and were used to calculate the maximum amount of sediment re-entrained during channel sediment routing.

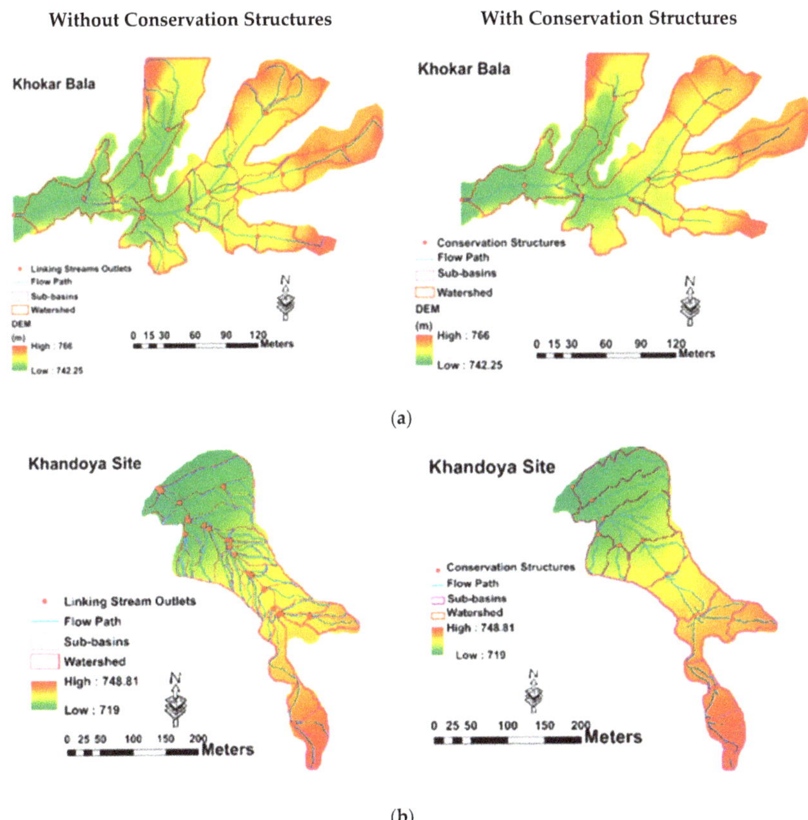

Figure 4. *Cont.*

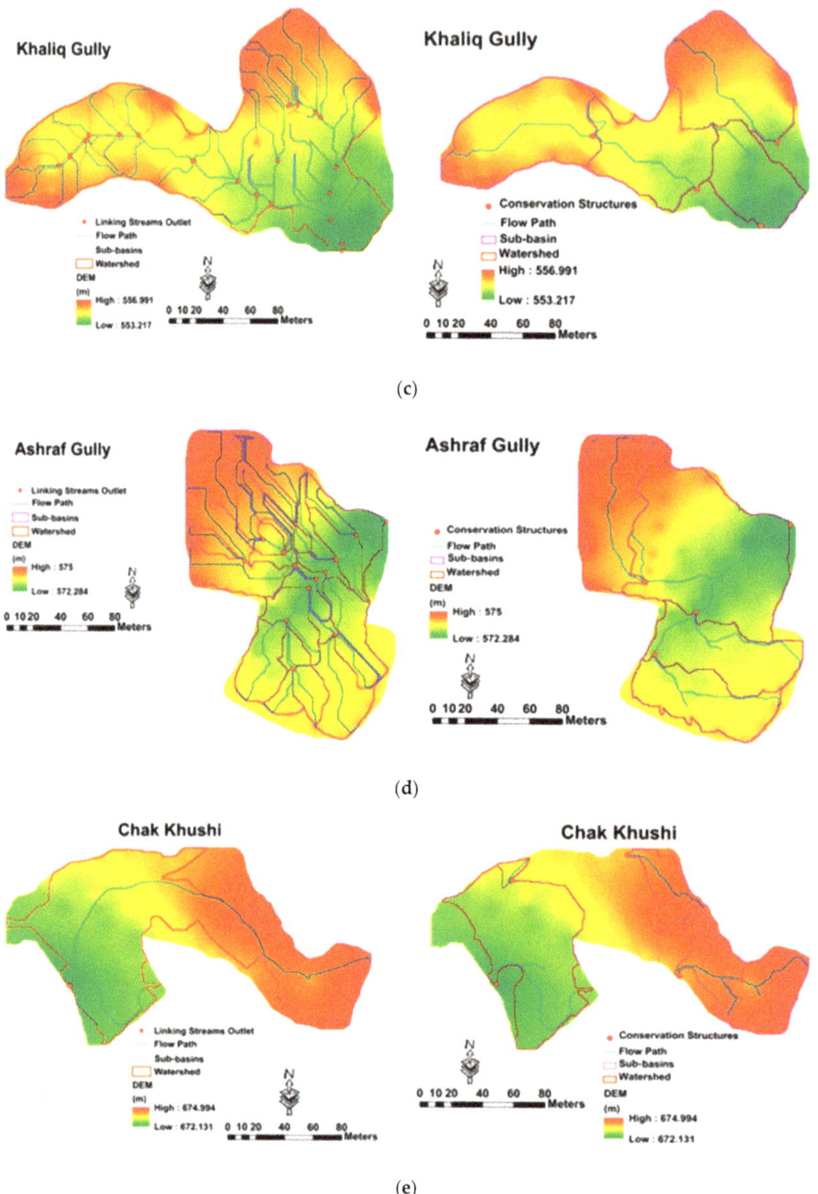

(c)

(d)

(e)

Figure 4. *Cont.*

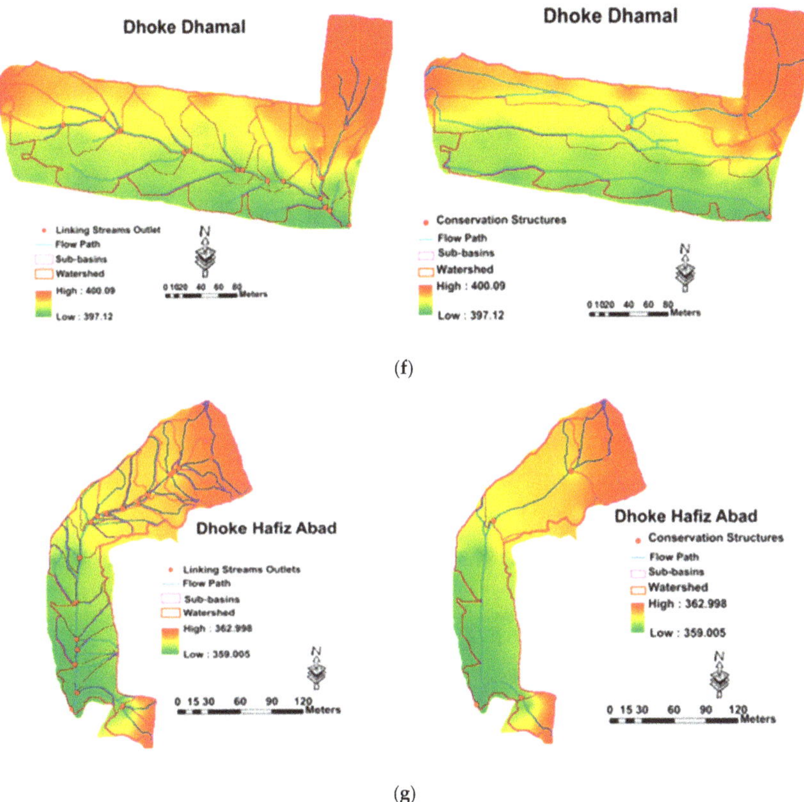

Figure 4. Topographic maps of selected small watersheds in the Chakwal and Attock Districts for model application: (**a**) Khokar Bala watershed, (**b**) Khandoya watershed, (**c**) Khaliq Gully watershed, (**d**) Ashraf Gully watershed, (**e**) Chak Khushi watershed, (**f**) Dhoke Dhamal watershed, and (**g**) Dhoke Hafiz Abad watershed.

2.5. Model Calibration and Validation—Reference to the Previous Study

In this study, the calibrated and validated model was adopted from a previous study [14]. The calibrated parameters were directly used during the simulation of the SWAT model without the consideration of the soil and water conservation structures. Hussain et al. [14] successfully performed the calibration of soil erosion parameters in small watersheds of the Dhrabi River Catchment. In this study, the Catchment-25 parameters were selected, as shown in Table 2 [14]. Catchment-25, having an area of 2.0 ha, is an agricultural watershed consisting of deep gullies, and its average land slope is 10.5%. It has well-defined boundaries and wide gully beds that mimic the full representation of the other study watersheds. The detailed description of Catchment-25 and SWAT model calibration and validation procedure and performance can be seen in the study of Hussain et al. [14].

The SAWCRI collected the surface runoff and sediment yield data at the outlet of Catchment-25. The experimental setup for data collection is shown in Figure 5. The automatic rain gauge and water level recorder were installed for rainfall and runoff depth measurements. The runoff discharge measurement was done using a sharp crested rectangular weir. The settling basin was used for sediment collection. The stilling basin was 3 m wide, 4 m long, and 65 cm deep at the weir and 15 cm deep upstream, in order to trap coarse sediment as bed load, while the suspended load was collected separately in 20 liter plastic buckets covered with a plastic sheet.

The total sediment yield of the catchment for a particular event is the sum of the bed load and suspended sediment. Coarser sediments were trapped in the stilling basin during the runoff event. After each runoff event, the standing water from the stilling basin was drained off through the drainpipe, and the wet sediments were collected and weighed. A composite sample of the wet bed load was obtained after mixing six-to-seven sub-samples collected throughout the stilling basin and oven dried to determine the moisture contents. The moisture contents were deducted from the wet weight to determine the dry weight of the sediment. Finer sediments in the runoff water passing the weir were sampled using vertical sampling tubes with holes. Following the runoff events, the samples present in the container were collected and analyzed. The total suspended sediment loss from the catchment was obtained by multiplying the sediment concentration in the bucket with the runoff volume passing over the weir.

Figure 5. The experimental setup for runoff and sediment yield [14].

2.6. Land Use Scenarios

The scenarios were developed based upon the common cropping practices adopted by the farmers in this area. A common practice for agriculture is the sowing of one or two crops a year. Other than the sowing period, the fields remain uncultivated as fallow land. Based upon this practice, the scenario related to land cover change was adopted—that is, winter wheat to fallow land change. Another other management practice is the use of conservation structures, which are used by the farmers for soil–water conservation and to meet crop water requirements. These structures safely pass the overland flow during the monsoon season and minimize the damages to the terrace ridges and bunds.

The SWAT model was applied based on four scenarios at all watershed sites. The scenarios are described as follows:

Scenario 1 (S1): The model was applied for soil erosion estimation on land without structures under the following conditions: the land use type was determined to be winter wheat; for overland flow, Manning's $n = 0.15$ (for short grass) was used, and for channel flow, Manning's $n = 0.025$ (for natural, earth uniform streams) was used.

Scenario 2 (S2): The model was applied for soil erosion estimation on land with structures under the same conditions as S1.

Scenario 3 (S3): The model was applied for soil erosion estimation on fallow land without structures. Manning's $n = 0.09$ was used for overland flow. The crop residue and channel flow conditions remained the same.

Scenario 4 (S4): The model was applied for soil erosion estimation on land with structures under the same conditions as S3.

3. Results

After the preparation of requisite input file, including different selected parameters, the model was applied to all selected sites for the evaluation of the effectiveness of soil conservation structures. For this purpose, the model was first run without soil conservation structures, and then conversation structures were modeled to see their effectiveness. The model was also applied for the above-mentioned four different scenarios related to field practices being adopted by the farmers in the area.

The modeled period was from 2009 to 2011 for Catchment-25. The runoff and sediment yield data collected during 2009–2010 were used for model calibration, while the 2011 data were used for validation. Some of the appropriate parameters were adjusted (Table 2) until the predicted runoff and sediment yield approximately matched the measured ones at the outlet (Figure 6). To determine the most sensitive parameters for model calibration, the sensitivity analysis was performed in the ArcSWAT interface using five parameters for sediment yield (Table 2): USLE practice factor (P_{USLE}), USLE conservation practice factor (C_{USLE}), USLE soil erodibility factor (K_{USLE}), the linear parameter for calculating the maximum amount of sediment that can be re-entrained during channel sediment routing (SPCON), and the exponent parameter for calculating sediment re-entrained in channel sediment routing (SPEXP). The P_{USLE} factor was found to be the most sensitive parameter during model calibration using sensitivity analysis. Moreover, the obvious correspondence (coefficient of determination (R^2) = 0.80 and Nash–Sutcliffe efficiency (NSE) = 0.70) of the hydrographs of the observed and simulated surface runoff and sediment yield indicated that the SWAT is capable of simulating the hydrological regime of small watersheds in the Pothwar region (Figure 6).

3.1. Model Application without Conservation Structures

After separately setting up the SWAT model for each watershed, the model simulation was performed with the default set of parameters in the default setting. Then, the soil erosion parameters (Table 2) were used for sediment yield simulation in each watershed. The modeled period was from 2010 to 2015. We estimated that all the watersheds generated a maximum sediment yield in 2010, while a minimum sediment yield was simulated in 2012. This indicated that the sediment yield is a direct function of runoff and rainfall intensity. In 2010, Khaliq Gully model estimation was 59.3 t ha^{-1}, while in 2012, it was 2.3 t ha^{-1}. Similarly, the Ashraf Gully, Khokar Bala, Chak Khushi, Dhoke Dhamal, Dhoke Hafiz Abad, and Khandoya watershed models produced annual sediment yields of 25, 37.6, 1.6, 15.3, 32.3, and 45.9 t ha^{-1}, respectively, in 2010 (Table 3).

3.2. Model Application with Conservation Structures

After the model application without conservation structures with calibrated soil erosion parameters, the model was applied to the small watersheds using soil and water conservation structures. The model setting was done in accordance with the location of conservation structures for the correct delineations of sub-basins. The intervention of the soil and water conservation structures was made by modifying the surface runoff and sediment yield parameters, as given in Table 2. The SWAT provides various options to consider soil and water conservation structure impacts [59] including: (i) surface runoff may be modified through the adjustment of the runoff ratio (curve number) and/or the consideration of a micro-pond (pothole) at the related HRU level, which also impacts the soil erosion, and (ii) impacts on the sediment yield levels may be modified via the adjustment of the support P-factor and/or the slope length and steepness factor (LS) of the MUSLE [60]. The ideal factors that describe the effect of stone bunds are the USLE support P-factor, the curve number, and the SLSUBBSN.

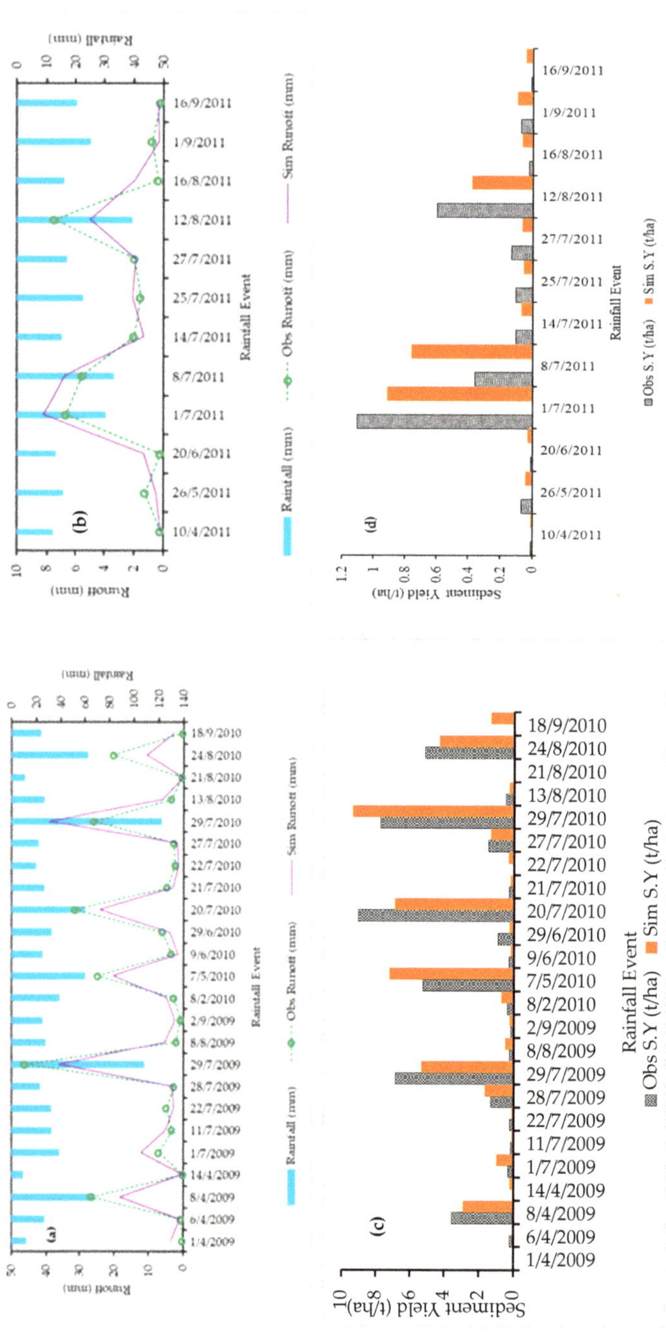

Figure 6. Soil and Water Assessment Tool (SWAT) model calibration (**a,c**) and validation (**b,d**) for surface runoff and sediment yield, respectively of Catchment-25.

Table 2. Soil erosion parameters used during the model's application without conservation structures (pre-condition) also used for Catchment-25 calibration. Post-condition parameters represent the conservation structures. SPEXP: exponential channel sediment routing factor; SPCON: linear channel sediment routing factor; USLE_C: conservation practice factor; USLE_K: soil erodibility factor; SLSUBBSN: the average slope length for the sub-basin; HRU_SLP: average slope steepness of the hydrological response unit (HRU) input tables; CN: curve number.

SWAT Pre-Condition Parameters			SWAT Post-Condition Parameters	
Parameter (Input File)	Default Value	Value Used	Parameter (Input File)	Modified Value
USLE_P (.mgt)	0 to 1	0.65	USLE_P (.mgt)	0.11
SPEXP (.bsn)	1.0 to 2.0	1.0	SPEXP (.bsn)	1.25
SPCON (.bsn)	0.0001 to 0.01	0.0032	SPCON (.bsn)	0.001
USLE_C (crop.dat)	0.001 to 0.5	0.182	SLSUBBSN (.hru)	60 (m)
USLE_K (.sol)	0 to 0.65	0.246	HRU_SLP (.hru)	0.016
			CN2 (.mgt)	65

These small watersheds already have existing soil and water conservation structures for the control of soil erosion. The crests of the structures play a major role in reducing the flow velocity and sediment deposition (erosion reduction) due to ponding upstream of the structures, whereas the downstream sections of the structures prevent channel or gully development. The topography of the region consists of permanent gullies where farmers use these gullies for the cultivation of crops. The farmers manage the gullies in a terraced land use system by making field boundary bunds, as shown in Figure 7 for the example of the Khokar Bala site. During the monsoon season, heavy rainstorms cause the shear failure of terrace edges (field bunds) due to the heavy surface runoff. This problem creates a loss of soil and damage to the crops. To reduce this problem, soil and water conservation structures have been installed to retain water in the terrace up to a certain rainfall amount (without overflowing the terrace) and then to divert the excess rainfall in a non-erosive way. These structures appear as a type of stone bund.

Figure 7. Permanent gullies for the cultivation of crops; an example of Khokar Bala site.

3.3. Soil Erosion Estimation and Effect of Conservation Structures

The sediment yield results were compared under each condition, as shown in Table 3, by modifying the SWAT parameters representing the conservation structures. The six parameters were modified according to the slope characteristics of the small watersheds and field conditions, in addition to being modified according to the terraced and contoured section of the SWAT user's manual [59] and a literature review [61–64]. Soil and water conservation structures, such as stone bunds, act as vital measures in the reduction of flow velocity, surface runoff, soil erosion, and slope length in a watershed system [65]. Suitable parameters that signify the effect and importance of loose stone structures are the SLSUBBSN, land management practice parameter (USLE_P), and the CN2 for rainfall–runoff conversion [61].

The impact of stone bund soil and water conservation structures was simulated through the reduction of the CN2 for surface runoff ratio modification, as well as the adjustment of the P-factor to account for trapped sediments at the stone bunds. Table 3 presents a significant sediment yield reduction achieved by incorporating the parameter values recommended for stone structures. The average annual sediment yield reduction varied from 40% to 98%; the Khokar Bala site showed the maximum reduction. The average five-year sediment yield reduction engendered by structures at various sites varied from 54% to 98%, and these results are relatively comparable to the findings of various studies [61,63,66].

Betrie et al. indicated that 6–69% sediment reductions in the Upper Blue Nile River basin were caused by stone bunds [61]. A field-scale study in the northern part of Ethiopia by Gebremichael et al. indicated a 68% sediment yield reduction was engendered by stone bunds [66]. In addition, Herweg and Ludi conducted a study at plot scale in the Eritrean highlands and Ethiopia, and they reported 72–100% sediment yield reductions engendered by stone bunds [63]. Based on the plot experiments carried out in 2013, stone bund structures were found to reduce surface runoff by approximately 60–80% and sediment yield between 40% and 80% [67]. This is consistent with other plot experimental findings reported by Adimassu et al., where stone bunds were found to reduce the sediment yield by roughly 50% compared to untreated plots [68]. The effect of conservation structures on sediment yield reduction was elucidated by Oweis and Ashraf in the Dhrabi watershed, and it was found that the average soil loss rates in 2009 without and with structures were calculated were 47 and 37.98 t ha^{-1} year^{-1}, respectively, with a 20% reduction. However, the maximum soil loss rates without and with structures were 2716.17 and 1731 t ha^{-1} year^{-1}, respectively, with a 37% reduction [69].

The large variation in sediment reduction with conservation structures was observed due to the watershed topography and the numbers of soil and water conservation structures. For example, the Khokar Bala site showed the maximum 98% reduction because this site has a 90% area at a 0–10% slope (Table 3) and a total of 13 soil and water conservation structures. Based on the field observation findings: (i) The conservation structures require regular maintenance because non-meshing can cause stones to slide, which may lead to the displacement of the whole structure, and (ii) the structures were not designed according to the hydraulic characteristics of the surface flow. Downstream damage of the structures was common due to the non-availability of downstream energy dissipation arrangements.

Table 3. The effect of stone structures on the sediment yield reduction.

Year	Khokar Bala			Khandoya			Khaliq Gully			Ashraf Gully			Chak Khushi			Dhoke Dhamal			Dhoke Hafiz Abad		
	\multicolumn{21}{c}{Sediment Yield (t ha^{-1}) Reduction Due to Stone Structures}
	W.O.S.	W.S.	% Red.	W.O.S.	W.S.	% Red.	W.O.S.	W.S.	% Red.	W.O.S.	W.S.	% Red.	W.O.S.	W.S.	% Red.	W.O.S.	W.S.	% Red.	W.O.S.	W.S.	% Red.
2010	37.6	0.9	97.6	45.9	14.6	68	59.3	30.3	49	25	10.4	58.5	1.6	0.8	49.4	15.3	8.3	45.7	32.3	13.5	58
2011	21.9	0.4	98.1	26.3	7.7	70.5	25.8	15.3	40.6	10.7	2.6	75.8	0.9	0.4	58.8	6.7	2.3	66.3	14.2	4.6	67
2012	3.9	0.1	98.5	5.5	0.5	90	2.3	0	100	0.9	0	100	0 *	0	100	0.6	0	98.2	1.26	0.56	55
2013	28.7	0.7	97.7	21.1	4.1	80.5	32.9	14.6	55.7	14	3.5	75.2	1.1	0.2	78.2	8.9	2.2	75	18.5	9.4	49
2014	13.8	0.2	98.6	18.9	8.7	54	27.6	11.9	57	11.6	2.2	81.1	0.8	0.2	69.7	7.4	1.8	75.4	15.3	5.2	66
2015	21.1	0.3	98.8	32.5	15.6	52	34	25.2	25.9	14.5	3	79	0.9	0.1	92.1	9.4	0.9	90.3	18.6	9.1	51
Ave.	-	-	98.2	-	-	69.2	-	-	54.7	-	-	78.3	-	-	74.7	-	-	75.2	-	-	57.6

W.O.S.: without structures; W.S.: with structures; % Red: percent reduction; 0 * = 0.0001.

3.4. Soil Erosion Estimation under Different Scenarios

The scenarios were developed to estimate the further reduction in soil erosion associated with land use change under soil conservation structures. The scenarios were developed according to the scientific literature of land use and vegetation cover importance to assess soil erosion and farmer's common cropping practices in the study region. Vegetation cover increases the infiltration rate [70], reduces the erosive velocity of surface runoff, and plays a key role in resisting water erosion. A trivial variation in vegetative cover can produce considerable effects in overland flow [71]. Vegetation cover is a key factor in controlling and reducing surface runoff and water erosion on agricultural land [72].

The analysis of the various scenarios (Table 4) revealed that the sediment yield level was higher in S1 and S2 than in S3 and S4. This indicated that the sediment yield level is higher on agricultural land than on fallow land with crop residue. In the comparative analysis of S1 and S2, the average sediment yield decreased to 1.25 t ha^{-1}, whereas in S3 and S4 (fallow land with crop residue), the average sediment yield decreased to 0.85 t ha^{-1}. The results disclosed that land use change facilitates sediment yield reduction, in addition to soil conservation structures.

Table 4. Effect of different scenarios on sediment yield reduction (S represents land use scenario and SY is the sediment yield i.e., the amount of sediment received at the outlet of each watershed in a given period of time).

Watershed Name	S1 (t ha^{-1})	S2 (t ha^{-1})	SY Reduction (t ha^{-1})	S3 (t ha^{-1})	S4 (t ha^{-1})	SY Reduction (t ha^{-1})
Khokar Bala	32.1	29.5	2.6	29.3	28.6	0.7
Khandoya	48.75	47	1.75	42.28	41.18	1.1
Khaliq Gully	25.98	24.75	1.23	17.1	16.5	0.6
Ashraf Gully	10.95	10.15	0.8	7.91	7.04	0.86
Chak Khushi	2.6	2.2	0.4	2.01	1.98	0.03
Dhoke Dhamal	12.6	11.56	1.04	11.9	11.1	0.8
Dhoke Hafiz Abad	24.4	20.8	3.6	18.2	17.3	0.9

Notably, a visual observation of the various structures revealed that the effects of the structures on soil erosion control generally extended to a 4–5 m radius from the center of the structure crests during high flow seasons; the water was accumulated and sediment was deposited upstream of the structures.

3.5. Spatial Analysis of Slope Ranges for Attock and Chakwal Districts

As reported by various researchers, the soil loss is minimal on sloping land with vegetation cover; however, when the available vegetation cover is removed, soil loss becomes more significant as a function of the slope length and slope steepness. The stream power, as a function of the shear stress and flow velocity, is the basic criterion for assessing the erosion of soil particles caused by overland flow. The shear stress and flow velocity are directly proportional to the slope steepness. This means that the steeper the land slope is, the greater the shear stress becomes, consequently increasing the potential for soil erosion.

Additionally, when soil conservation structures are installed in a field, farmers focus on cultivating agricultural crops in the areas above and below such structures. Considering these factors, this section estimated the potential area that would benefit from the installation of structures in Chakwal and Attock. Accordingly, the suitable slopes for stone structures and agricultural practices were analyzed at the district level based on the slope characteristics of selected sites. The areas under various slopes in the small watersheds were calculated and are shown in Table 5.

Table 5. The areas under different slopes in small watersheds of the Chakwal and Attock districts.

Category	Ashraf Gully	Khaliq Gully	Chak Khushi	Dhoke Dhamal	Dhoke Hafiz Abad	Category	Khokar Bala	Khandoya
Slope (%)	Area (%)	Area (%)	Area (%)	Area (%)	Area (%)	Slope (%)	Area (%)	Area (%)
0–2	63	50	97	81	72	0–5	65	70
2–5	30	42	3	17	22	5–10	25	19
5–10	7	8	-	1	6	10–15	10	11

All selected watershed sites were found to have a maximum slope area of less than 5%. This is because the selected sites were used for agricultural production. Farmers have graded the land as suitable for crop production and generating less surface runoff. The agricultural practices are only possible on soil that has a slope of less than 8%; otherwise, land grading must be carried out. The same has been suggested by various authors. A USLE experiment conducted at the SAWCRI office concluded that only a slope of less than 10% is acceptable for agricultural practices under rainfed conditions.

A slope classification analysis was performed to check the areal installation applicability of the soil and water conservation structures on district level, as shown in Table 6. The maximum proportions of the areas in the Attock and Chakwal districts with less than 20% slope were 94% and 94.5%, respectively. The table shows that approximately 60% of the area of the Attock and Chakwal districts lies in a slope range of 0–4%, whereas 30% lies in a slope range of 4–10%. The minimum slope areas were considered according to the findings of Betrie et al., who recommended that stone bunds should be applied in low-slope areas for soil conservation [61]. However, the effectiveness of the structures depends on the local topography and soil and land use–land cover conditions. Considering the topographic conditions, considerable potential exists for implementing soil conservation measures through the installation of stone structures. However, the appropriate maintenance of the structures is crucial for sustaining their effectiveness.

Table 6. Slope classification analysis of the Chakwal and Attock districts.

Slope Category	Chakwal		Attock	
	Area	Area	Area	Area
(%)	(km²)	(%)	(km²)	(%)
0–4	4095	60	3918	61
4.1–10	1913	28	1786	28
10.1–20	547	8	472	7
20.1–40	233	3	165	3
40.1–90	75	1	55	1

4. Discussion

The previous research study conducted by Hussain et al. [14] in the Dhrabi River Catchment indicated that erratic and intensive rainfall during the rainy season generated several peak runoff events, exposing the steep sloped areas to potentially severe soil erosion. For example, in Catchment-25, a total 400 mm rainfall was accumulated from eleven erosive rainstorms in 2009, where a maximum of 108 mm day^{-1} rainstorms generated 46.2 mm runoff and a 6.86 t ha^{-1} sediment yield. Similarly, the total soil loss during the 2010 investigation period was 31.13 t ha^{-1} [14]. In the SWAT, the erosive impact of rainfall is generally estimated in terms of peak runoff generation, so the results obtained during calibration and validation are represented in Figure 6 for surface runoff and sediment yield. The analysis was performed for each total rainfall event and the respective total surface runoff and sediment yield generated by each event. The overall statistical results indicated that the performance of the SWAT was satisfactory and that the simulated values generally matched the corresponding observed values well. However, model adequacy should be further evaluated by how well the model captures high and low rainfall events, specifically regarding the replication of fluctuations in the resulting hydrographs and sediment yields. The graphical results (Figure 6) revealed that the SWAT was able to satisfactorily reproduce most of the low flow and sediment yield events (due to low rainfall events), although some relatively low sediment yields were considerably overpredicted, e.g., sediment yield events on 7 August 2011. In contrast, it was also found that the SWAT typically underestimated or overestimated high flow and sediment yield events in response to high rainfall events. For example, a maximum intensity rainstorm on 29 July 2010 resulted in the overestimation of surface runoff and sediment yields, while another maximum intensity rainstorm on 29 July 2009 resulted in underestimations. These discrepancies may have occurred due to inaccuracies in observed climate, runoff, and sediment data, such as some of the rainfall events, not being measured properly; this, in turn, could have led to underestimations or overestimations of runoff peaks. Another possible reason could be related to short, rapid rainfall events, which could have led lead to an overestimation discrepancy because small catchments have low times of concentration and thus a low capacity to minimize peak runoff. In addition, the CN technique cannot accurately predict runoff for days that experience several storms. The underestimation and/or overestimation of sediment yield was also observed during high intensity rainstorm events, which may have been due to uncertainties in runoff simulation measurements, as well as uncertainties in model parameterization. This may have also been due to the observed data used for model calibration and validation. Relatively short term events with several storms having high intensity may not have been captured well by the sampling of sediment data, including inaccurately high loads being measured during short term events, which led to an overestimation in sediment yield. The literature data findings indicated that the semi-mountainous region of Pothwar is rainfed and soil erosion is a serious issue due to the steep slope and heavy rainfall. The calibration and validation of the SWAT was successfully performed using parameters mentioned in Table 2 for surface runoff and sediment yield. Similar studies, such as the sediment simulation results by Betrie et al. [61] reported good agreement between the model daily sediment predictions and the observed concentrations at the El Diem gauging station (Ethiopia–Sudan border). SWAT studies for smaller watersheds in the northeast and northwest of Ethiopia have tended to show weaker hydrologic results [73,74], which is an indication that it may be difficult to accurately represent the processes and obtain better results for smaller watersheds. Keeping in view the literature studies, the validated SWAT model was applied to small watersheds of the Pothwar area with and without soil and water conservation structures.

The calibrated and validated model parameters were adopted for ungauged small watersheds for the simulation of sediment yield without the consideration of soil and water conservation structures. The soil and water conservation structures were modeled with the modification of appropriate parameters, and then the effectiveness of these structures in terms of reduction of soil erosion was calculated. The results showed that the soil and water conservation structures constructed by the farmers and the SAWCRI department reduced the soil losses in the small watersheds of the Pothwar region. The results showed that the watersheds in the case of without soil and water conservation structures had higher sediment losses than the watersheds with soil and water conservation structures, given similar climatic and land use patterns. The intervention of soil and water conservation structures measures by the mobilization of the community has a significant soil loss reduction to protect their land from the rainfall-driven soil erosion. To the best of our knowledge, no one has reported the effectiveness of soil and water conservation structures for the reduction of soil erosion in the Pothwar region or in Pakistan. This is the first study in this area where the SWAT model has been used for the evaluation of effectiveness of soil and water conservation structures for soil erosion control. For this purpose, appropriate parameters responsible for soil erosion were modified according to the type of soil and water conservation structures, as performed by different researchers in literature such as Betrie et al. [61] in the Upper Blue Nile River basin, Gebremichael et al. [66] in the northern part of Ethiopia, and Melaku et al. [38] in the Gumara Maksegnit watershed in northwest Ethiopia. The SWAT model has been found to be a useful tool for understanding the hydrologic processes and the sediment dynamic in the study area watersheds, and it assessed the impacts of soil and water conservation structures on the erosion process. It was observed that severe erosion led to higher soil losses in some watersheds dominated with gullies, such as the Khokar Bala site. This is substantiated by the photo taken in Figure 7, which shows the development of deep gullies in the upper parts of the watershed that were found to contribute higher soil erosion losses and to generate higher sediment load at the outlet.

The model results indicated that soil and water conservation structures might considerably decrease soil loss by 40–90% in small watersheds of the Pothwar region. Herweg and Ludi conducted a study at the plot scale in the Eritrean highlands of Ethiopia and reported 72–100% sediment yield reductions engendered by stone bunds [63], which was close to the current finding of the soil loss reduction level due to soil and water conservation structures. Similarly, Betrie et al. [61] indicated 6–69% sediment reductions in the Upper Blue Nile River basin caused by stone bunds, and their results were in agreement with our findings. The average annual sediment yield estimated by the SWAT model without the consideration of soil and water conservation structures in all selected sites was in the range of 8.05–30.31 t ha^{-1}. Our findings were in agreement with other studies conducted in the Pothwar region such as those from Hussain et al. [14], who estimated the annual sediment loss (ranged from 2.6 to 31.1 t hm^{-2}) in small watersheds of the Dhrabi River Catchment, while Ahmad et al. [16] reported annual soil loss rates of 9–26 t ha^{-1} in the Fateh Jang watershed (Attock) with a slope of 1–10%. Similarly, Nasir et al. [11] predicted an annual soil loss that ranged from 0.1 to 28 t ha^{-1} at the small mountainous watershed of Rawal Lake in Rawalpindi. The literature data findings indicated that the semi-mountainous region of Pothwar is rainfed and soil erosion is a serious issue due to steep slope and heavy rainfall. The region receive erratic rainfall during a short rainy season and almost 70% precipitation occurs during monsoon [75,76]. The comparative studies on sediment and soil loss confirmed the results of the current study conducted in small watersheds of the Pothwar region.

The soil and water conservation structures are effective measures in reducing the soil erosion problems in the Pothwar region that have varied land slopes. This study reveals that considering the topographic conditions, loose stone soil and water conservation structures should be installed in areas with a slope range of 0–10%, and wire-meshed stone structures should be installed in areas with a slope range of 6–10%. Proper energy dissipation arrangements should be implemented to prevent downstream erosion.

5. Conclusions

In this research, we performed SWAT watershed modeling to describe the driving hydrological and sediment transport related processes of small watersheds. The effectiveness of soil and water conservation structures for soil erosion control was assessed with a SWAT model. Stone bund-type structure interventions were done in the model through modification of the USLE support P-factor, the curve number, and the SLSUBBSN. The model results revealed that a 40–90% sediment yield reduction could be achieved using soil conservation structures. Thus, soil and water conservation structures are effective options for soil erosion control in rainfed areas. The land use change scenario results revealed that vegetation cover facilitated sediment yield reduction, in addition to soil conservation structures. An all-inclusive interpretation of the quantitative model results may be misleading because no model can fully simulate all the physical processes of soil and water interactions in a real sense. Some assumptions were made during modeling; however, based on the results, we suggest to policymakers and planners that more than 60% of the area in the Attock and Chakwal districts has potential for soil and water conservation structures.

Author Contributions: G.N. and F.H. contributed equally to this research. Conceptualization, G.N. and F.H.; data curation, F.H. and R.B.; formal analysis, F.H., R.-S.W., and V.N.; investigation, R.-S.W.; methodology, G.N.; project administration, R.B.; resources, V.N.; software, F.H.; supervision, G.N.; visualization, R.-S.W., V.N., and R.B.; writing—original draft, G.N. and F.H.; writing—review and editing, F.H. All authors have read and agreed to the published version of the manuscript.

Funding: This research received no external funding.

Acknowledgments: This study is part of a research project under the Consultative Group for International Agricultural Research-CGIAR Research Program (CRP) on Dryland Systems. University of Engineering and Technology, Centre for Excellence in Water Resources Engineering, Lahore, Pakistan, The International Center for Agriculture Research in the Dry Areas, Syria, country office Pakistan and Soil and Water Conservation Research Institute, Chakwal, Pakistan performed collaborative research work. The authors particularly thank all colleagues involved in the fieldwork.

Conflicts of Interest: The authors declare no conflict of interest.

References

1. Rodrigo-Comino, J.; Senciales, J.M.; Cerdà, A.; Brevik, E.C. The multidisciplinary origin of soil geography: A review. *Earth Sci. Rev.* **2018**, *177*, 114–123. [CrossRef]

2. Desruelles, S.; Fouache, E.; Eddargach, W.; Cammas, C.; Wattez, J.; Beuzen-Waller, T.; Martin, C.; Tengberg, M.; Cable, C.; Thornton, C.; et al. Evidence for early irrigation at Bat (Wadi Sharsah, northwestern Oman) before the advent of farming villages. *Quat. Sci. Rev.* **2016**, *150*, 42–54. [CrossRef]

3. Keesstra, S.D.; Bouma, J.; Wallinga, J.; Tittonell, P.; Smith, P.; Cerdà, A.; Montanarella, L.; Quinton, J.N.; Pachepsky, Y.; Van Der Putten, W.H.; et al. The significance of soils and soil science towards realization of the United Nations Sustainable Development Goals. *Soil* **2016**, *2*, 111–128. [CrossRef]

4. Scholes, R.; Montanarella, L.; Brainich, A.; Barger, N.; Ten Brink, B.; Cantele, M.; Erasmus, B.; Fisher, J.; Gardner, T.; Holland, T.G.; et al. *Summary for Policymakers of the Thematic Assessment Report on Land Degradation and Restoration of the Intergovernmental Science-Policy Platform on Biodiversity and Ecosystem Services*; IPBES Secretariat: Bonn, Germany, 2018; pp. 1–31.

5. Pimentel, D. Soil erosion: A food and environmental threat. *Environ. Dev. Sustain.* **2006**, *8*, 119–137. [CrossRef]

6. Ananda, J.; Herath, G. Soil erosion in developing countries: A socio-economic appraisal. *J. Environ. Manag.* **2003**, *68*, 343–353. [CrossRef]

7. Keesstra, S.; Mol, G.; de Leeuw, J.; Okx, J.; de Cleen, M.; Visser, S. Soil-related sustainable development goals: Four concepts to make land degradation neutrality and restoration work. *Land* **2018**, *7*, 133. [CrossRef]

8. KRIS, Watershed. Cumulative Watershed Effects. Klamath Resource Information System (KRIS). 2002. Available online: http://www.krisweb.com/watershd/impacts.htm (accessed on 1 August 2013).

9. Ontario Envirothon. [Chapter 7] Soil Erosion. Ontario Envirothon, a Program of Ontario Forestry Association. 2007. Available online: http://www.ontarioenvirothon.on.ca/files/soil/soil_Chapter7.pdf (accessed on 1 August 2013).

10. Mu, X. Trend and change-point analyses of streamflow and sediment discharge in the Yellow River during 1950 to 2005. *Hydrol. Sci. J.* **2010**, *55*, 275–285.

11. Nasir, A.; Uchida, K.; Ashraf, M. Estimation of soil erosion by using RUSLE and GIS for small mountainous watershed in Pakistan. *Pak. J. Water Resour.* **2006**, *10*, 11–21.

12. Ashraf, M.; Hassan, F.U.; Saleem, A.; Iqbal, M.M. Soil conservation and management: A prerequisite for sustainable agriculture in Pothwar. *Sci. Technol. Dev.* **2002**, *21*, 25–31.

13. Scherr, S.J.; Yadav, S. *Land Degradation in the Developing World: Implications for Food, Agriculture and the Environment to 2020*; Food, Agriculture and the Environment Discussion Paper 14; IFPRI: Washington, DC, USA, 1996.

14. Hussain, F.; Nabi, G.; Wu, R.-S.; Hussain, B.; Abbas, T. Parameter evaluation for soil erosion estimation on small watersheds using SWAT model. *Int. J. Agric. Biol. Eng.* **2019**, *12*, 96–108. [CrossRef]

15. Iqbal, M.N.; Jilani, G.; Ali, A.; Ali, S.; Ansar, M.; Aziz, I.; Sajjad, M.R. Soil and water loss from natural and cultivated slopes in Dharabi watershed. *J. Biodivers. Environ. Sci.* **2015**, *7*, 128–135.

16. Ahmad, S.; Ikram, M.A.M. Soil and water conservation and integration land use in Pothwar, Pakistan. In *Soil Physics: Applications Under Stress Environments*; Pakistan Agricultural Research Council: Islamabad, Pakistan, 1990; pp. 301–312.

17. Saleem, U.; Ali, A.; Iqbal, M.; Javid, M.; Imran, M. Geospatial assessment of soil erosion intensity and sediment yield: A case study of Potohar Region, Pakistan. *Environ. Earth Sci.* **2018**, *77*, 705. [CrossRef]

18. Bashir, S.; Baig, M.A.; Ashraf, M.; Anwar, M.M.; Bhalli, M.N.; Munawar, S. Risk Assessment of Soil Erosion in Rawal Watershed using Geoinformatics Techniques. *Sci. Int. (Lahore)* **2013**, *25*, 583–588.

19. Nabi, G.; Latif., M.; Ahsan, M.; Anwar, S. Soil erosion estimation of Soan river catchment using remote sensing and geographic information system. *Soil Environ.* **2008**, *27*, 36–42.

20. Rehman, O.; Rashid, M.; Kausar, R.; Alvi, S.; Hussain, R. Slope gradient and vegetation cover effects on the runoff and sediment yield in hillslope agriculture. *Turk. J. Agric. Food Sci. Technol.* **2015**, *3*, 478–483. [CrossRef]

21. Gordon, J.M.; Bennett, S.J.; Alfonso, C.V.; Bingner, R.L. Modeling long term soil losses on agricultural fields due to ephemeral gully erosion. *J. Soil Water Conserv.* **2008**, *63*, 173–181. [CrossRef]

22. Saco, P.M.; Willgoose, G.R.; Hancock, G.R. Eco-geomorphology of banded vegetation patterns in arid and semi-arid regions. *Hydrol. Earth Syst. Sci.* **2007**, *11*, 1717–1730. [CrossRef]

23. De Vente, J.; Poesen, J.; Verstraeten, G.; Govers, G.; Vanmaercke, M.; Van Rompaey, A.; Arabkhedri, M.; Boix-Fayos, C. Predicting soil erosion and sediment yield at regional scales: Where do we stand? *Earth Sci. Rev.* **2013**, *127*, 16–29. [CrossRef]

24. Wischmeier, H.; Smith, D.D. *Predicting Rainfall Erosion Losses*; Agriculture Handbook no 537; USDA Science and Education Administration: Hyattsville, MD, USA, 1978.

25. Young, R.A.; Onstad, C.A.; Bosch, D.D.; Anderson, W.P. AGNPS: A nonpoint-source pollution model for evaluating agricultural watersheds. *J. Soil Water Conserv.* **1989**, *44*, 168–173.

26. Ferro, V.; Porto, P. Sediment delivery distributed (SEDD) model. *ASCE J. Hydraul. Eng.* **2000**, *5*, 411–422. [CrossRef]

27. Arnold, J.G.; Srinivasan, R.; Muttiah, R.S.; Williams, J.R. Large-area hydrologic modeling and assessment: Part I. Model development. *J. Am. Water Res. Assoc.* **1998**, *34*, 73–89. [CrossRef]

28. Viney, N.R.; Sivapalan, M. A conceptual model of sediment transport: Application to the Avon River Basin in Western Australia. *Hydrol. Process.* **1999**, *13*, 727–743. [CrossRef]

29. Vertessey, R.A.; Watson, F.G.R.; Rahman, J.M.; Cuddy, S.D.; Seaton, S.P.; Chiew, F.H.; Scanlon, P.J.; Marston, F.M.; Lymbuner, L.; Jeanelle, S.; et al. New software to aid water quality management in the catchments and waterways of the south-east Queensland region. In Proceedings of the Third Australian Stream Management Conference: The Value of Healthy Streams, Brisbane, Queensland, 27–29 August 2001; pp. 611–616.

30. Beasley, D.B.; Huggins, L.F.; Monke, E.J. ANSWERS—A model for watershed planning. *Trans. Am. Soc. Agric. Eng.* **1980**, *23*, 938–944. [CrossRef]

31. Knisel, W.G. *CREAMS: A Field Scale Model for Chemicals, Runoff and Erosion from Agricultural Management Systems*; USDA: Hyattsville, MD, USA, 1980.

32. Lopes, V.L. A Numerical Model of Watershed Erosion and Sediment Yield. Ph.D. Thesis, The University of Arizona, Tucson, AZ, USA, 1987.

33. Wicks, J.M. Physically-Based Mathematical Modelling of Catchment Sediment Yield. Ph.D. Thesis, Department of Civil Engineering, University of Newcastle Upon Tyne, England, UK, 1988.

34. Morgan, R.P.C.; Quinton, J.N.; Smith, R.E.; Govers, G.; Poesen, J.W.A.; Auerswald, K.; Chisci, G.; Torri, D.; Styczen, M.E. The European soil erosion model (EUROSEM): A dynamic approach for predicting sediment transport from fields and small catchments. *Earth Surf. Process. Landf.* **1998**, *23*, 527–544. [CrossRef]

35. Klik, A.; Wakolbinger, S.; Obereder, E.; Strohmeier, S.; Melaku, N.D. *Impacts of stone bunds on soil loss and surface runoff: A case study from Gumara-Maksegnit Watershed, Northern Ethiopia*; Purdue University: West Lafayette, IN, USA, 2016.

36. Nyssen, J.; Poesen, J.; Gebremichael, D.; Vancampenhout, K.; Dáes, M.; Yihdego, G.; Deckers, J. Interdisciplinary on-site evaluation of stone bunds to control soil erosion on cropland in Northern Ethiopia. *Soil Tillage Res.* **2007**, *95*, 151–163. [CrossRef]

37. Klik, A.; Schürz, C.; Strohmeier, S.; Melaku, N.D.; Ziadat, F.; Schwen, A.; Zucca, C. Impact of stone bunds on temporal and spatial variability of soil physical properties: A field study from northern Ethiopia. *Land Degrad. Dev.* **2018**, *29*, 585–595. [CrossRef]

38. Melaku, N.D.; Renschler, C.S.; Holzmann, H.; Strohmeier, S.; Bayu, W.; Zucca, C.; Ziadat, F.; Klik, A. Prediction of soil and water conservation structure impacts on runoff and erosion processes using SWAT model in the northern Ethiopian highlands. *J. Soils Sediments* **2018**, *18*, 1743–1755. [CrossRef]

39. Arabi, M.; Frankenberger, J.R.; Engel, B.A.; Arnold, J.G. Representation of agricultural conservation practices with SWAT. *Hydrol. Process.* **2008**, *22*, 3042–3055. [CrossRef]

40. Ramos, M.C.; Benito, C.; Martínez-Casasnovas, J.A. Simulating soil conservation measures to control soil and nutrient losses in a small, vineyard dominated, basin. *Agric. Ecosyst. Environ.* **2015**, *213*, 194–208. [CrossRef]

41. Briak, H.; Mrabet, R.; Moussadek, R.; Aboumaria, K. Use of a calibrated SWAT model to evaluate the effects of agricultural BMPs on sediments of the Kalaya river basin (North of Morocco). *Int. Soil Water Conserv. Res.* **2019**, *2*, 176–183. [CrossRef]

42. Zakaullah Ashraf, M.; Afzal, M.; Yaseen, M.; Khan, K. Appraisal of Sediment Load in Rainfed Areas of Pothwar Region in Pakistan. *Glob. J. Res. Eng.* **2014**, *14*, 25–33.

43. Nizami, M.A.; Shafiq, M.; Rashid, M.; Aslam, M. *The Soils and Their Agricultural Development Potential in Pothwar*; WRRI-LRRP, National Agricultural Research Centre: Islamabad, Pakistan, 2004; p. 158.

44. Shafiq, M.; Rashid, A.; Mangrio, A.G. Agricultural potential soil resources of Pothwar Plateau. *Soil Environ.* **2005**, *24*, 109–119.

45. Khan, R.S. *Pothwar's Agricultural Potential*, Pakistan Agriculture Overview, Courtesy Daily Dawn, 24 May 2002.

46. Gassman, P.W.; Reyes, M.R.; Green, C.H.; Arnold, J.G. The Soil and Water Assessment Tool: Historical development, applications and future directions. *Trans. ASABE* **2007**, *50*, 1211–1250. [CrossRef]

47. Gassman, P.W.; Sadeghi, A.M.; Srinivasan, R. Applications of the SWAT model special section: Overview and insights. *J. Environ. Qual.* **2014**, *43*, 1–8. [CrossRef]

48. Douglas-Mankin, K.R.; Srinivasan, R.; Arnold, J.G. Soil and Water Assessment Tool (SWAT) model: Current developments and applications. *Trans. ASABE* **2010**, *53*, 1423–1431. [CrossRef]

49. Tuppad, P.; Douglas-Mankin, K.R.; Lee, T.; Srinivasan., R.; Arnold, J.G. Soil and Water Assessment Tool (SWAT) hydrologic/water quality model: Extended capability and wider adoption. *Trans. ASABE* **2011**, *54*, 1677–1684. [CrossRef]

50. Krysanova, V.; White, M. Advances in water resources assessment with SWAT—An overview. *Hydrol. Sci. J.* **2015**, *60*, 771–783. [CrossRef]

51. Bressiani, D.A.; Gassman, P.W.; Fernandes, J.G.; Garbossa, L.H.P.; Srinivasan, R.; Bonuma, N.B.; Mendiondo, E.M. A review of SWAT (Soil and Water Application Tool) applications in Brazil: Challenges and prospects. *Int. J. Agric. Biol. Eng.* **2015**, *8*, 9–35.

52. Tripathi, M.P.; Panda, R.K.; Raghuwanshi, N.S. Identification and Prioritization of Critical Sub-watersheds for Soil Conservation Management using the SWAT Model. *Bio Syst. Eng.* **2003**, *85*, 365–379. [CrossRef]

53. Zabaleta, A.; Meaurio, M.; Ruiz, E.; Antigüedad, I. Simulation climate change impact on runoff and sediment yield in a small watershed in the Basque Country, northern Spain. *J. Environ. Qual.* **2014**, *43*, 235–245. [CrossRef]

54. Lemann, T.; Zeleke, G.; Amsler, C.; Giovanoli, L.; Suter, H.; Roth, V. Modelling the effect of soil and water conservation on discharge and sediment yield in the upper Blue Nile basin, Ethiopia. *Appl. Geogr.* **2016**, *73*, 89–101. [CrossRef]

55. Roth, V.; Lemann, T. Comparing CFSR and conventional weather data for discharge and soil loss modelling with SWAT in small catchments in the Ethiopian Highlands. *Hydrol. Earth Syst. Sci.* **2016**, *20*, 921–934. [CrossRef]

56. Setegn, S.G.; Dargahi, B.; Srinivasan, R.; Melesse, A.M. Modeling of sediment yield from Anjeni-Gauged watershed, Ethiopia using SWAT model. *J. Am. Water Resour. Assoc.* **2010**, *46*, 514–526. [CrossRef]

57. USDA-SCS (U.S. Department of Agriculture-Soil Conservation Service). *National Engineering Handbook*; Section 4; Hydrology. U.S. Department of Agriculture-Soil Conservation Service: Washington, DC, USA, 1972.

58. Williams; Berndt, H.D. Sediment yield prediction based on watershed hydrology. *Trans. ASAE* **1977**, *20*, 1100–1104. [CrossRef]

59. Neitsch, S.L.; Arnold, J.G.; Kiniry, J.; Williams, J.R. *Soil and Water Assessment Tool Theoretical Documentation (Version 2005)*; USDA Agricultural Research Service and Texas A&M Blackland Research Center: Temple, TX, USA, 2005.

60. Williams, J.R. Sediment routing for agricultural watersheds. *JAWRA* **1975**, *11*, 965–974. [CrossRef]

61. Betrie, G.D.; Mohamed, Y.A.; van Griensven, A.; Srinivasan, R. Sediment management modelling in the Blue Nile Basin using SWAT model. *Hydrol. Earth Syst. Sci.* **2011**, *15*, 807–818. [CrossRef]

62. Addis, H.K.; Strohmeier, S.; Ziadat, F.; Melaku, N.D.; Klik, A. Modeling streamflow and sediment using SWAT in the Ethiopian Highlands. *Int. J. Agric. Biol. Eng.* **2016**, *9*, 51–66.

63. Herweg, K.; Ludi, E. The performance of selected soil and water conservation measures-case studies from Ethiopia and Eritrea. *Catena* **1999**, *36*, 99–114. [CrossRef]

64. Hurni, H. Erosion—Productivity—Conservation systems in Ethiopia. In Proceedings of the 4th International Conference on Soil Conservation, Maracay, Venezuela, 3–9 November 1985; pp. 654–674.

65. Bracmort, K.; Arabi, M.; Frankenberger, J.; Engel, B.; Arnold, J. Modeling long-term water quality impact of structural BMPs. *Trans. ASABE* **2006**, *49*, 367–374. [CrossRef]

66. Gebremichael, D.; Nyssen, J.; Poesen, J.; Deckers, J.; Haile, M.; Govers, G.; Moeyersons, J. Effectiveness of stone bunds in controlling soil erosion on cropland in the Tigray highlands, Northern Ethiopia. *Soil Use Manag.* **2005**, *21*, 287–297. [CrossRef]

67. Rieder, J.; Strohmeier, S.; Demelash, N.; Ziadat, F.; Klik, A. Investigation of the impact of stone bunds on water erosion in northern Ethiopia. In Proceedings of the EGU General Assembly Conference, Vienna, Austria, 27 April–2 May 2014; Volume 16, p. 3885.

68. Adimassu, Z.; Mekonnen, K.; Yirga, C.; Kessler, A. Effect of soil bunds on runoff, soil and nutrient losses, and crop yield in the central highlands of Ethiopia. *Land Degrad. Dev.* **2012**, *25*, 554–564. [CrossRef]

69. Oweis, T.; Ashraf, M. *Assessment and Options for Improved Productivity and Sustainability of Natural Resources in Dhrabi Watershed Pakistan*; ICARDA: Aleppo, Syria, 2012; p. 77.

70. Hejduk, S.; Kasprzak, K. A contribution to proposals of the width of protective grasslands strips. *Soil Water Conserv.* **2005**, *4*, 30–35.

71. Wei, M.; Bogaard, T.A.; Beek, R. Dynamic effects of vegetation on the long-term stability of slopes: Components of evaporation. *Geophys. Res. Abstr.* **2011**, *13*, 7720–7725.

72. Hofman, I.; Ries, R.F.; Gilley, G.E. Relationship of runoff and soil loss to ground cover of native and reclaimed grazing land. *Agron. J.* **1985**, *75*, 599–607. [CrossRef]

73. Schmidt, E.; Zemadim, B. Expanding sustainable land management in Ethiopia: Scenarios for improved agricultural water management in the Blue Nile. *Agric. Water Manag.* **2015**, *158*, 166–178. [CrossRef]

74. Yesuf, H.M.; Assen, M.; Alamirew, T.; Melesse, A.M. Modeling of sediment yield in Maybar gauged watershed using SWAT, northeast Ethiopia. *Catena* **2015**, *127*, 191–205. [CrossRef]

75. Baig, M.B.; Shahid, S.A.; Straquadine, G.S. Making rainfed agriculture sustainable through environmental friendly technologies in Pakistan: A review. *Int. Soil Water Conserv. Res.* **2013**, *1*, 36–52. [CrossRef]

76. Hussain, F.; Nabi, G.; Boota, M.W. Rainfall trend analysis by using the Mann-Kendall Test & Sen's Slope Estimates: A case study of district Chakwal rain gauge, Barani area, Northern Punjab Province, Pakistan. *Sci. Int. (Lahore)* **2015**, *27*, 3159–3165.

Article

Effects of Residue Cover on Infiltration Process of the Black Soil Under Rainfall Simulations

Yan Xin, Yun Xie * and Yuxin Liu

State Key Laboratory of Earth Surface Processes and Resource Ecology, Faculty of Geographical Science, Beijing Normal University, Beijing 100875, China; xinyanhao51@163.com (Y.X.); day_dream_lyx@163.com (Y.L.)
* Correspondence: xieyun@bnu.edu.cn; Tel.: +86-10-5880-7391

Received: 19 September 2019; Accepted: 5 December 2019; Published: 9 December 2019

Abstract: Residue cover is widely used for soil conservation after crop harvesting in the black soil region of the Northeastern China, which influences infiltration. It is necessary to optimize infiltration models for accurate predictions under bare and residue cover slope conditions. Rainfall simulation experiments were conducted to quantify the infiltration for the black soil under four rainfall intensities (30, 60, 90, and 120 mm/h), five residue coverage controls (15%, 35%, 55%, 75%, and bare slope), and two soil moisture (8% and approximately 30%) conditions. The observed data were used to fit and compare four infiltration models by Kostiakov, Mein and Larson (short for GAML, a modification of GreenAmpt model made by Mein and Larson), Horton, and Philip under the bare slope conditions. The residue cover infiltration factor (RCF_i) was derived to predict the infiltration under the residue cover slopes, which was defined as the ratio of infiltration from residue-covered soil to that from bare soil. The results showed that the newly derived equation coupling the Philip model with the RCF_i was the most accurate way of predicting the cumulative infiltration of black soil under various residue covers, and could be applied to the black soil region for residue cover infiltration predictions.

Keywords: infiltration estimation; black soil; residue cover; model validation

1. Introduction

Infiltration is the process of water entering soil from the soil surface [1,2]. This process is one of the most important components in the hydrological cycle and is related to many environmental problems and soil erosion [3]. It is influenced by many internal and external factors, such as rainfall characteristics, slope gradients, soil hydraulic properties, soil properties, and surface sealing, which makes infiltration hard to quantify [4].

Numerous models have been proposed for vertically homogeneous soils with constant initial soil water content and flow over horizontal surfaces for infiltration estimations [5,6], including physically-based models and empirical models [7–14]. However, these models have limited applicability under complex initial factors. Many studies have been conducted to modify the applications of these models under various scenarios and assumptions. The Green–Ampt model is one of the most widely used hydrological and erosion models, as it involves simple expression, uses few parameters, and has a specific physical meaning. The model was initially developed to simulate infiltration under ponding conditions in homogeneous soil [15]. Modifications have continuously been proposed to expand the scope of the model's application so that infiltration can be simulated under steady rainfall events [16], layered soil [17], or unsaturated soil with different slopes [18–20], as well as other conditions. The aforementioned models should be modified and improved based on a variety of scenarios to obtain the most accurate infiltration estimations.

Black soil, classified as a Haplic-Ustic sohumosol in the Chinese Soil Taxonomy and Udic Argiboroll in the U.S. Soil Taxonomy [21], is mainly distributed in Northeastern China. As this soil is rich in organic matter (OM) and appears black, it is referred to as "black soil" and the "black soil region" by local people and researchers alike [22]. The parent material of the soil contains mostly loess clay loam deposits. Due to the fine soil texture, black soil is poor in terms of permeability and is easily eroded. Coupled with the effect of seasonal variation in the region, i.e., freeze–thaw erosion in the winter and spring and rainstorm erosion in the flood season, black soil is impacted by severe runoff erosion. Erosion diminishes the OM content and soil thickness, and even induces decreasing soil productivity and environmental deterioration. It is essential to study black soil infiltration processes to prevent soil erosion.

Leaving field residues on the top of soil is a practical conservation tillage in the black soil region since there is a large annual production of crop residues. It was demonstrated that residue cover had a positive effect on soil infiltration, which increased infiltration into the soil and reduced surface runoff [19]. Although, researchers have paid much attention to the benefits of residue cover in terms of preserving soil and water and maintaining soil productivity, they have so far neglected infiltration estimation under residue cover [22,23]. Moreover, infiltration models have been mostly used for bare soils without the effects of residue cover. It is therefore necessary to modify the existing infiltration models to take into account the effects of residue cover for infiltration estimations of black soil on farmland.

In this study, the infiltration process of black soil under various rainfall intensities and residue coverages was studied with simulated rainfall. The objectives of the paper are:

(1)   To determine the optimal infiltration model for bare black soil;
(2)   To establish the infiltration model combined with the effect of residue cover for black soil.

2. Materials and Methods

2.1. Design of Rainfall Simulation Experiments

Rainfall simulation experiments were conducted to study infiltration using spray-nozzle rainfall simulators and soil flumes in the rainfall simulation laboratory of Beijing Normal University, China, in 2015 (Figure 1). Four levels of rainfall intensities (including 30, 60, 90, and 120 mm/h), two soil moistures (8% and 30%, measured by TDR soil moisture sensor (Time domain reflectometry, CAMPBELL TDR200, Campbell Scientific, Inc., U.S.), representing the extremely dry and extremely wet runs), and 7% slope (defined as "intense erosion", according to the grade scale of soil erosion intensity standard in the experimental field) were designed for the laboratory experiments. Five residue coverages (15%, 35%, 55%, 75%, and bare slope as a control), determined by the digital photograph method, were designed to quantify the residue cover effect. Corn stalks were collected to use in the experiments. Each rainfall event was performed in both dry and wet runs and lasted for 1 h under various rainfall intensities. The interval was 24 h between runs. In total, 80 runs were conducted. All treatments were performed twice for reproducibility and precision.

The experimental black soil used was the top 20 cm of surface soil from farmland and classified by light erosion intensity, collected from the Jiusan Soil Conservation Station. The characteristics of the black soil included an organic matter content of 5.0% and a particle size distribution of 32%, 33%, and 35% for sand, silt, and clay, respectively. The soil was filled into the flume with a depth of 0.25 m (2 m length × 1 m width × 0.35 m depth) and layered with a soil density of 1.07 to 1.25 g/cm^3 to model actual field conditions. The upper soil layer was 0.1 m deep and the sublayer was 0.15 m deep, and the contact surface of the two layers was roughened to reduce soil stratification which could affect infiltration [24]. A wooden board was used for the soil surface flatting to reduce the effect of soil roughness.

The precipitation of each rainfall event was controlled by simulators. The runoff was collected into bottles for measurement every 5 min during rainfall, and volumes were recorded after the rain. The infiltration amount was the difference between the rainfall amount and the runoff; evaporation and residue retention during the rainfall events were not measured. The infiltration data obtained were used to determine the hydrological parameters in the models.

Figure 1. Laboratory experimental setup.

2.2. The Residue Cover Factor for Model Validation

Once the optimal infiltration model was determined for the bare soil, the infiltration under the residue cover could be predicted using a ratio multiplying the function of the bare soil, according to Xin et al. [25]. Therefore, the optimal infiltration model under the residue cover was

$$f(i,t)_r = RCF_i \times f(i,t)_b,\tag{1}$$

where $f(i,t)_r$ is the infiltration model under residue cover (mm/min), i is the infiltration rate (mm/min), t is the corresponding time (min), RCF_i is the residue cover factor, and $f(i,t)_b$ is the optimal infiltration model of the bare soil. RCF_i is the ratio of infiltration amounts from the residue cover soil and the bare soil, which was described as

$$RCF_i = CI_r/CI_b,\tag{2}$$

where CI_r and CI_b are the cumulative infiltration amount (mm) of the residue cover and bare soil under the rainfall events, respectively. The relationship between the residue cover infiltration factor (RCF_i) and residue cover was established as

$$RCF_i = 0.94 \times RC + 1,\tag{3}$$

where RC is the residue cover (Figure 2).

The performances of four common infiltration models were compared to evaluate the bare black soil infiltration, including Kostiakov [8], Horton [9], Philip [11], and Mein and Larson (GAML) models (Table 1). The infiltration models were different in terms of mathematical structure and hydrological parameters, but their estimates were all based on the measured water infiltration data for bare soil conditions [26].

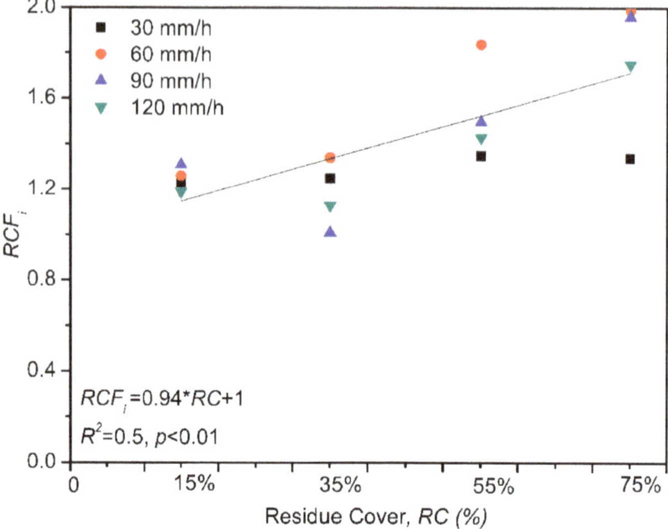

Figure 2. The relationship between the residue cover infiltration factor (RCF_i) to predict the infiltration and the residue cover.

Table 1. The four infiltration models.

Model	Year	Equation	Symbols
Kostiakov [8]	1932	$f(t) = Bt^{-n}$	B, fitting parameters
Horton [9]	1933	$f(t) = f_c + (f_0 - f_c)e^{-kt}$	f_0, initial rate f_c, constant rate k, decay constant
Philip [11]	1957	$f(t) = 0.5St^{-0.5} + K_s$	S, sorptivity K_s, saturated hydraulic conductivity
Mein and Larson (GAML) [16]	1973	$f(t) = K_e\left(\frac{1+S \cdot M}{F}\right)$	F, cumulative infiltration K_e, effective hydraulic conductivity S, suction at the wetting front M, hydrological parameter of difference between saturated and initial volumetric moisture content

Note: $f(t)$ is the infiltration rate and t represents the time.

2.3. Accuracy Assessment Methods

Nonlinear regression was used to determine the values of the parameters in the infiltration models with the rainfall data under the bare soil. The observed values beneath the residue cover and the corresponding predicted values were compared to evaluate the simulations of the models using the 1:1 line method. This method pertains to the t-test method to estimate whether the confidence interval of the slope and intercept of the regressed equation included the numbers 1 and 0, respectively [25]. If included, no difference existed between the regressed curve and the 1:1 line.

The root mean square error (RMSE), the Nash–Sutcliffe efficiency (NSE), and the determination coefficient (R^2) were used to evaluate the accuracy of the infiltration models [27,28]. The equations of the statistical indexes were as follows:

$$\text{RMSE} = \sqrt{\frac{\sum_{i=1}^{N}(Y_i - O_i)^2}{N}}, \tag{4}$$

$$\mathrm{NSE} = 1 - \frac{\sum_{i=1}^{N} (Y_i - O_i)^2}{\sum_{i=1}^{N} \left(O_i - \overline{O_i}\right)^2}, \tag{5}$$

$$\mathrm{R}^2 = \frac{\sum_{i=1}^{N} \left(Y_i - \overline{O_i}\right)^2}{\sum_{i=1}^{N} \left(O_i - \overline{O_i}\right)^2}, \tag{6}$$

where O_i is the ith observed value, $\overline{O_i}$ is the average observed value of all of the observed events, Y_i is the ith predicted value, and N is the total number of events. The higher the NSE values were, the better the model performed, as it represented the level of agreement between the observed and predicted values [29]. The values of the RMSE showed the opposite result, namely, the lower the RMSE values were, the better the model performed. The closer to 1 the determination coefficient R^2 was, the higher the correlation was. The range of the values of R^2 was 0–1, while the range of the values of NSE was $(-\infty)$–1. In the present study, O_i represented the observed values and Y_i represented the predicted values of the infiltration rates.

3. Results

3.1. Infiltration Rates of Bare Black Soil Under Various Rainfall Events

The average values of two repeated trials were used for the infiltration rate curves for different rainfall intensities and soil moistures of the black soil (Figure 3). The initial infiltration rates were equal to the rainfall intensities before ponding. When the water ponded on the surface, infiltration occurred at the potential infiltration rate and runoff generation began. The times to runoff in the dry runs were around 48.0, 16.0, 7.4, and 5.1 min, and the total infiltration amounts were 27.67, 38.30, 45.14, and 47.54 mm under the four rainfall intensities, respectively. The higher the rainfall intensities were, the earlier the runoff began. Then, infiltration rates decreased and tended to become steady as the rainfall continued due to the soil crust that formed on the soil surface. The final infiltration rates of the dry runs were 0.37, 0.40, 0.53, and 0.54 mm/min. In contrast, the runoff times of the wet runs were within 5 min and the infiltration rate became stable soon after runoff generation under various rainfall intensities. The final infiltration rates corresponding to the rainfall intensities were 0.21, 0.23, 0.43, and 0.51 mm/min, respectively. The total infiltration amounts were 44.43, 55.05, 72.81, and 80.18 mm under the four rainfall intensities.

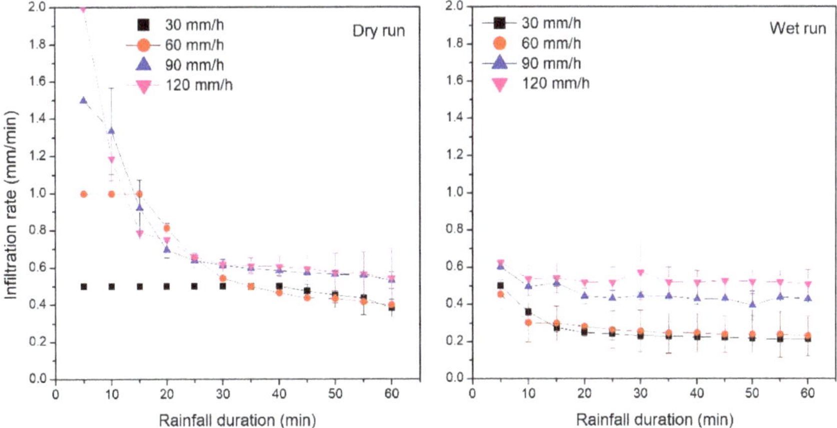

Figure 3. Infiltration rates under various rainfall events.

3.2. Estimation of the Infiltration Model Parameters

The infiltration rates under the various rainfall intensities on the bare soil were used to fit the four models. The values of each parameter were regressed; these are listed in Table 2. As shown, the S·M parameter in the GAML model was negative (equal to −0.572), which was an invalid value. Therefore, the other three models were used for the infiltration estimation of the black soil, but not the GAML model. The determination coefficient R^2 was approximately 0.5, and the regression results passed the significance test at the $p = 0.01$ level.

Table 2. The fitted parameters of different infiltration models.

Kostiakov			Horton				Philip			GAML		
B	n	R^2	f_c	f_0	k	R^2	S	K_s	R^2	K_e	S·M	R^2
1.511	0.352	0.494	0.421	1.199	0.096	0.511	3.185	0.290	0.509	0.507	−0.572	0.553

3.3. Cumulative Infiltration Amounts for the Residue-Covered Black Soil under Various Rainfall Events

The cumulative infiltration amounts of the residue cover on the black soil under different rainfall events are shown in Figure 4. It was indicated that the residue cover tillage was effective at promoting the infiltration of the black soil and delaying the runoff generation. Under the 30 mm/h scenario, the precipitation nearly seeped into the soil when the residue cover was more than 55%, which proved that the tillage was effective under the relatively small rainfall intensity events. The infiltration amounts did not show that the higher the residue coverages were, the higher the infiltration amounts were. The infiltration amounts under 35% residue cover were less than under 15% residue cover. This result might have been due to the fixed runoff flow path, which could have promoted runoff generation [30].

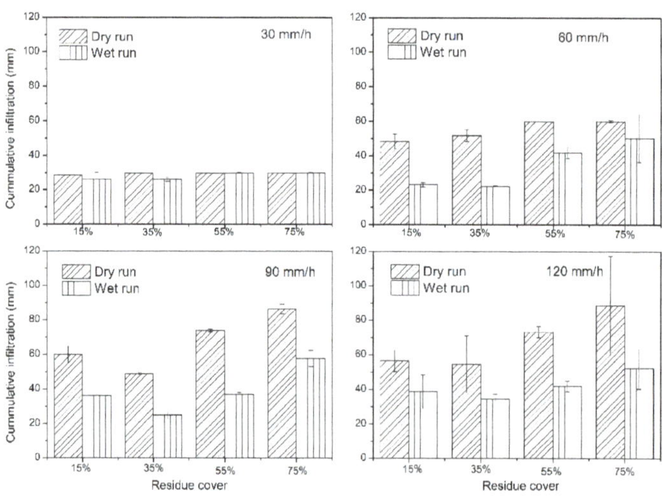

Figure 4. Cumulative infiltration amounts under various rainfall events.

3.4. Performance of the Models for the Residue-Covered Black Soil

According to Xin et al. [25], RCF_i was used to fit the infiltration rates of the residue-covered black soil. The performances of the three models were evaluated and the results showed that the Kostiakov model performed poorly. As it did not pass the 1:1 line test for the confidence interval of the slope and intercept of the regressed equation, excluding the numbers 1 and 0, respectively. The Horton and Philip models performed well (Figure 5). As is generally accepted, the performance of the Kostiakov model was robust for many soils over short time periods [31]. In our study, the performance of this model was good for the bare black soil, but it did not perform well after adding the effects of the residue cover in the multiplication form (Equation (1)). The multiplication form underestimated the initial infiltration rate under the high residue coverage scenario, which was in accordance with Almeida et al. [32] who used the Kostiakov–Lewis model for estimation, with the results indicating that the Kostiakov–Lewis model underestimated the infiltration rates at the beginning of the rainfall event and overestimated the rates at the end of the rainfall.

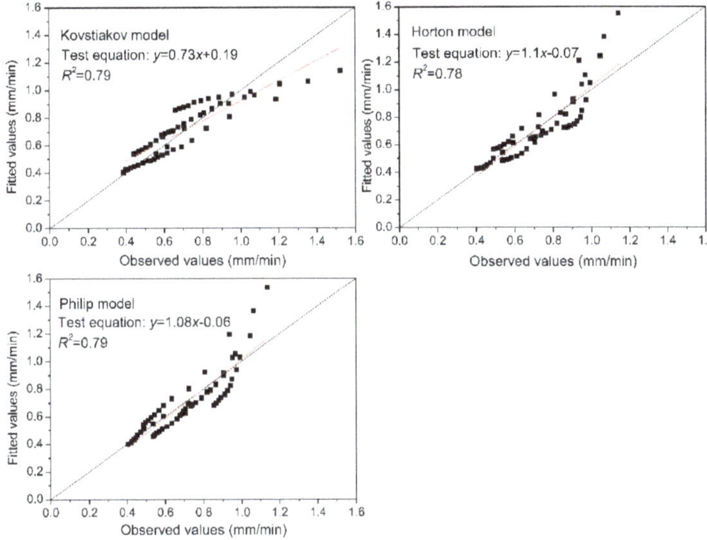

Figure 5. Comparison of the observed and fitted infiltration rates obtained with different residue covers (15%, 35%, 55%, and 75%) for the Kostiakov, Horton, and Philip models.

The statistical indexes NSE and RMSE were used for the comparison and are shown in Table 3. The performance of the Philip model was better than the Horton model, as the lower the values of the RMSE and the closer the NSE values were to 1, the better the fitting results were. From the above, the Philip equation was optimal for the infiltration estimation of the black soil under the residue cover conditions. The equation was

$$i(t)_r = RCF_i \times \left(1.59 \times t^{-0.5} + 0.290\right),\qquad(7)$$

where $i(t)_r$ is the infiltration rate under the residue cover (mm/min).

Table 3. The values of root mean square error (RMSE) and the Nash–Sutcliffe efficiency (NSE) under the Horton and Philip models.

	RMSE	NSE
Horton	0.112	0.666
Philip	0.105	0.699

4. Discussion

Four models were compared to evaluate the infiltration of black soil. The GAML model, derived from the Green-Ampt model, demonstrated that infiltration during a steady rainfall event could be simulated. However, the model was not suitable for infiltration estimations of the black soil because of the negative values observed from the hydrological parameters. It might be that the original form was usually applied to initially dry, uniform, coarse-textured soil, such as sands and sand-fraction media [33], whereas the main textural classes of the black soil are silt clay loam and clay loam, according to the USDA (U.S. Department of Agriculture) classification [34]. As with all fine-textured soils, the resistance of soil pores to water flow was higher than in the coarse-textured soils [35]. The permeability of the black soil was poor, which caused the inapplicability of the GAML model.

The infiltration estimations of the Horton and Philip models for the residue-covered black soil performed well, with the Philip model performing better regarding the comparison of the statistical indexes NSE and RMSE. It is worth noting that both models overestimated the initial infiltration rates, especially under high residue coverage.

The average infiltration rates under the four rainfall intensities were used for the model fitting to remove the effects of heavy rain, but only the residue cover was considered, which might have been the reason for the outliers. Considering the physical significance of the Philip model, this derived residue cover infiltration model was suggested for use in estimations of cumulative infiltration amounts.

5. Conclusions

In this study, we observed the infiltration processes of black soil slopes under bare and residue-covered conditions by simulated rainfall experiments. The optimal infiltration model for the residue-covered black soil was derived in combination with the Philip model and the residue cover infiltration factor (RCF_i) after comparisons. The model was suitable for the estimation of cumulative infiltration amounts under residue cover conditions for the black soil. The model was meaningful for the infiltration estimation, and thus provided effective governance for soil erosion management of the black soil.

Author Contributions: Conceptualization, Y.X. (Yun Xie); data curation, Y.X. (Yan Xin); formal analysis, Y.X. (Yan Xin); funding acquisition, Y.X. (Yun Xie); investigation, Y.L.; methodology, Y.X. (Yan Xin) and Y.L.; supervision, Y.X. (Yun Xie); validation, Y.X. (Yan Xin); writing—original draft, Y.X. (Yan Xin); writing—review and editing, Y.X. (Yun Xie).

Funding: This work was supported by the National Key R&D Program (No. 2018YFC0507000) and the China Postdoctoral Science Foundation (No. 2018M630102).

Acknowledgments: We are thankful for the valuable comments of anonymous reviewers.

Conflicts of Interest: The authors declare no conflict of interest.

References

1. Hillel, D. *Environmental Soil Physics*; Academic Press: New York, NY, USA, 1998.
2. Zhang, J.; Lei, T.; Yin, Z.; Hu, Y.; Xiusheng Yang, X. Effects of time step length and positioning location on ring-measured infiltration rate. *Catena* **2017**, *157*, 344–356. [CrossRef]
3. Philip, J.R. Hillslope infiltration: Divergent and convergent slopes. *Water Resour. Res.* **1991**, *27*, 1035–1040. [CrossRef]
4. Herrada, M.A.; Gutiérrez-Martin, A.; Montanero, J.M. Modeling infiltration rates in a saturated/unsaturated soil under the free draining condition. *J. Hydrol.* **2014**, *515*, 10–15. [CrossRef]
5. Morbidelli, R.; Saltalippi, C.; Flammini, A.; Govindaraju, R.S. Role of slope on infiltration: A review. *J. Hydrol.* **2018**, *557*, 878–886. [CrossRef]
6. Wang, K.; Yang, X.; Liu, X.; Liu, C. A simple analytical infiltration model for short-duration rainfall. *J. Hydrol.* **2017**, *555*, 141–154. [CrossRef]
7. Green, W.; Ampt, G. Studies on soil physics, part I: The flow of air and water through soils. *J. Agric. Sci.* **1991**, *4*, 1–24.
8. Kostiakov, A.N. On the dynamics of the coefficient of water percolation in soils and on the necessity of studying it from a dynamic point of view for purposes of amelioration. *Trans. Sixth Comm. Int. Soc. Soil Sci. Part A* **1932**, *1*, 7–21.
9. Horton, R. The role of infiltration in the hydrologic cycle. *Am. Geophys. Union Trans.* **1933**, *14*, 446–460. [CrossRef]

10. Holtan, H.N. *A Concept for Infiltration Estimates in Watershed Engineering*; USDA-ARS Bull. 41–51; United States Department of Agriculture, Agricultural Research Service: Washington, DC, USA, 1961.

11. Philip, J.R. The theory of infiltration: 1. The infiltration equation and its solution. *Soil Sci.* **1957**, *83*, 345–357. [CrossRef]

12. Soil Conservation Service. *SCS National Engineering Handbook*; Section 4, Hydrology; United States Department of Agriculture: Washington, DC, USA, 1972.

13. Smith, R.E.; Parlange, J.Y. A parameter-efficient hydrologic infiltration model. *Water Resour. Res.* **1978**, *14*, 533–538. [CrossRef]

14. Corradini, C.; Melone, F.; Smith, R.E. Modeling infiltration during complex rainfall sequences. *Water Resour. Res.* **1994**, *30*, 2777–2784. [CrossRef]

15. Van Den Putte, A.; Govers, G.; Leys, A.; Langhans, C.; Clymans, W.; Diels, J. Estimating the parameters of the Green–Ampt infiltration equation from rainfall simulation data: Why simpler is better. *J. Hydrol.* **2013**, *476*, 332–344. [CrossRef]

16. Mein, R.G.; Larson, C.L. Modeling infiltration during a steady rain. *Water Resour. Res.* **1973**, *9*, 384–394. [CrossRef]

17. Chu, X.; Mariño, M.A. Determination of ponding condition and infiltration into layered soils under unsteady rainfall. *J. Hydrol.* **2005**, *313*, 195–207. [CrossRef]

18. Chen, L.; Young, M.H. Green-Ampt infiltration model for sloping surfaces. *Water Resour. Res.* **2006**, *42*. [CrossRef]

19. Derpsch, R.; Franzluebbers, A.J.; Duiker, S.W.; Reicosky, D.C.; Koeller, K.; Friedrich, T.; Sturny, W.G.; Sá, J.C.M.; Weiss, K. Why do we need to standardize no-tillage research? *Soil Till. Res.* **2014**, *137*, 16–22. [CrossRef]

20. Gavin, K.; Xue, J. A simple method to analyze infiltration into unsaturated soil slopes. *Comput. Geotech.* **2008**, *35*, 223–230. [CrossRef]

21. Gong, Z.T.; Chen, Z.C.; Shi, X.Z. *Chinese Soil Taxonomy*; Science Press: Beijing, China, 1999.

22. Gu, Z.; Xie, Y.; Gao, Y.; Ren, X.; Cheng, C.; Wang, S. Quantitative assessment of soil productivity and predicted impacts of water erosion in the black soil region of northeastern China. *Sci Total Environ.* **2018**, *637-638*, 706–716. [CrossRef]

23. Wang, J.; Xie, Y.; Liu, G.; Zhao, Y.; Zhang, S. Soybean root development under water stress in eroded soils. *Acta Agric. Scand. Sect. B Soil Plant Sci.* **2015**, *65*, 374–382. [CrossRef]

24. Zhao, X.N.; Wu, P.; Gao, X.D.; Tian, L.; Li, H.C. Changes of soil hydraulic properties under early-stage natural vegetation recovering on the Loess Plateau of China. *Catena* **2014**, *113*, 386–391. [CrossRef]

25. Xin, Y.; Xie, Y.; Liu, Y.; Liu, H.; Ren, X. Residue cover effects on soil erosion and the infiltration in black soil under simulated rainfall experiments. *J. Hydrol.* **2016**, *543*, 651–658. [CrossRef]

26. Bamutaze, Y.; Tenywa, M.M.; Majaliwa, M.J.G.; Vanacker, V.; Bagoora, F.; Magunda, M.; Obando, J.; Wasige, J.E. Infiltration characteristics of volcanic sloping soils on Mt. Elgon, Eastern Uganda. *Catena* **2010**, *80*, 122–130. [CrossRef]

27. Nash, J.E.; Sutcliffe, J.V. River flow forecasting through conceptual models: Part 1. A discussion of principles. *J. Hydrol.* **1970**, *10*, 282–290. [CrossRef]

28. Moriasi, D.N.; Arnold, J.G.; van Liew, M.W.; Bingner, R.L.; Harmel, R.D.; Veith, T.L. Model evaluation guidelines for systematic quantification of accuracy in watershed simulations. *Trans. ASABE* **2007**, *50*, 885–900. [CrossRef]

29. Wu, L.Z.; Zhou, Y.; Sun, P.; Shi, J.S.; Liu, G.G.; Bai, L.Y. Laboratory characterization of rainfall-induced loess slope failure. *Catena* **2017**, *150*, 1–8. [CrossRef]

30. Meyer, L.D.; Wischimeier, W.H.; Foster, G.R. Mulch rates required for erosion control on steep slopes. *Soil Sci. Soc. Am. J.* **1970**, *34*, 928–931. [CrossRef]

31. Naeth, M.A.; Bailey, A.W.; Chanasyk, D.S.; Pluth, D.J. Water holding capacity of litter and soil organic matter in mixed prairie and fescue grassland ecosystems of Alberta. *J. Range Manag.* **1991**, *44*, 13–17. [CrossRef]

32. De Almeida, W.S.; Panachuki, E.; de Oliveira, P.T.S.; Da Silva Menezes, R.; Sobrinho, T.A.; de Carvalho, D.F. Effect of soil tillage and vegetal cover on soil water infiltration. *Soil Till. Res.* **2018**, *175*, 130–138. [CrossRef]

33. Mohammadzadeh-Habili, J.; Heidarpour, M. Application of the Green–Ampt model for infiltration into layered soil. *J. Hydrol.* **2015**, *527*, 824–832. [CrossRef]

34. Wu, Y.; Zheng, Q.; Zhang, Y.; Liu, B.; Cheng, H.; Wang, Y. Development of gullies and sediment production in the black soil region of northeastern China. *Geomorphology* **2008**, *101*, 683–691. [CrossRef]
35. Bouwer, H. Infiltration into increasingly permeable soils. *J. Irrig. Drain. Eng.* **1976**, *102*, 127–136.

Article

Validation of the EROSION-3D Model through Measured Bathymetric Sediments

Zuzana Németová [1,*], David Honek [2,3], Silvia Kohnová [1], Kamila Hlavčová [1], Monika Šulc Michalková [2], Valentín Sočuvka [4] and Yvetta Velísková [4]

[1] Department of Land and Water Resources Management, Faculty of Civil Engineering, Slovak University of Technology, Radlinského 11, 81005 Bratislava, Slovakia; silvia.kohnova@stuba.sk (S.K.); kamila.hlavcova@stuba.sk (K.H.)

[2] Department of Geography, Faculty of Science, Masaryk University, Kotlářská 2, 61137 Brno, Czech Republic; 324439@mail.muni.cz or david.honek@vuv.cz (D.H.); monika.michalko@gmail.com (M.Š.M.)

[3] T. G. Masaryk Water Research Institute, p. r. i., Podbabská 2582/30, 16000 Prague, Czech Republic

[4] Institute of Hydrology, Slovak Academy of Sciences, Dúbravská cesta 9, 84104 Bratislava, Slovakia; socuvka@uh.savba.sk (V.S.); veliskova@uh.savba.sk (Y.V.)

* Correspondence: zuzana.nemetova@stuba.sk; Tel.: +421-02-59-274-621

Received: 3 March 2020; Accepted: 7 April 2020; Published: 10 April 2020

Abstract: The testing of a model performance is important and is also a challenging part of scientific work. In this paper, the results of the physically-based EROSION-3D (Jürgen Schmidt, Berlin, Germany) model were compared with trapped sediments in a small reservoir. The model was applied to simulate runoff-erosion processes in the Svacenický Creek catchment in the western part of the Slovak Republic. The model is sufficient to identify the areas vulnerable to erosion and deposition within the catchment. The volume of sediments was measured by a bathymetric field survey during three terrain journeys (in 2015, 2016, and 2017). The results of the model point to an underestimation of the actual processes by 30% to 80%. The initial soil moisture played an important role, and the results also revealed that rainfall events are able to erode and contribute to a significant part of sediments.

Keywords: sediment budget; rainfall event; EROSION-3D; small water reservoir

---

1. Introduction

Soil erosion, as a process of the detachment, transport and accumulation of soil materials from any part of the Earth's surface, plays an important role in research since it assumes the most significant position among the individual degradation processes [1,2]. The importance of soil erosion research lies in the protection of soil as a fundamental resource for human food supplies; therefore, the understanding of soil erosion processes has important practical implications over large areas of the Earth [3,4].

Research on soil erosion and sediment transport has rapidly increased, thanks to technical developments and the increasing use of computer applications. Predictions of sediment yields can be carried out by taking advantage of an erosion model and a mathematical operator that expresses the efficiency of the sediment transport of a hill slope and channel network [5–9]. Mathematical simulation models have been developed, starting with empirical models such as the Universal Soil Loss Equation (USLE) by [10]. Empirical models are observation-based with a primary focus on the prediction of average soil loss, while conceptual models are based on the representation of catchments [11]. However, the USLE only generates annual mean soil losses but ignores the highly non-continuous character of erosion processes.

Nowadays, the development of models is based on the concept of equations for the conservation of mass, momentum, and energy [12], the use of equations describing streamflow or overland flows, and the equation for the conservation of mass for sediments. These models belong to the group of physically-based models and represent mathematical relationships derived from physical laws and the mechanisms controlling erosion, which means that the parameters used are measurable [13], i.e., they allow for the sufficient representation and quantitative estimation of the detachment, transport of soil, and deposition processes [14]. In comparison with empirical models, the main advantages of physically-based models are distinguished by their more appropriate representation of erosion, deposition processes and extrapolation to different land uses in a more satisfactory style, as well as a more precise estimation of erosion, deposition and sediment yields from an event-based rainfall or application to complex varying soil conditions and surface characteristics [15].

Despite the large number of soil erosion and sediment transport models, choosing a suitable model is still a very complex and complicated task. There are many models that suffer from a range of problems, such as overestimation due to the uncertainty of the models and the unsuitability of the assumptions and parameters in conformity with the local conditions [16]. It needs to be stated that the modelling of a natural system is always limited by many variations in terms of spatial and temporal scales, spatial heterogeneity, the transport media, and the lack of available data [17]. Different types of factors affecting water erosion, such as the climate, topography, soil structure, vegetation and anthropogenic activities, tillage systems and soil or conservation measures [18], can result in different values of sediment generation and deposition as well. Among the factors mentioned above, rainfall intensity and the runoff rate are the major triggers of splash and sheet erosion [19], together with the human activities in rivers, causing changes in the magnitude and nature of material inputs to estuaries, which can trigger erosion with consequences for populations and ecosystems [20].

The amount of sediment in a catchment is heterogeneous in space and over time, depending on the land use, vegetation cover, climate, and landscape characteristics, i.e., the soil type, topography, any slopes, and the drainage conditions [21,22]. The quantification of eroded material can be made by a variety of methods, and the selection of the method depends on the financial support, objectives, the size of the study area and the characteristics of the research group [23]. The bathymetric measurement of sediment deposited in a reservoir is a suitable method for assessing the volume of eroded material in a study area. Boyle et al. [24] noted that calculating the lake sediment is useful for quantifying the historical impact of agriculture on soil erosion and sediment yield, as well as a good approach for calibrating and testing the erosion models compared to the actual bathymetry measurements.

This paper presents a suitable approach as to how to validate an erosion model through the sedimentation in a small reservoir. The aims of the paper are as follows:

(a)   To model and measure the sedimentation of a small reservoir in a small rural catchment;
(b)   To evaluate the role of an intensive rainfall event in the erosion process;
(c)   To validate the results from the physically-based EROSION-3D model through the bathymetric measurement of the mass of sediment in a small reservoir.

2. Materials and Methods

2.1. EROSION-3D Model

The EROSION-3D model is a physically-based computer tool predominantly developed for simulating runoff and deposition processes on arable land. The model can be applied for predicting the amount of soil loss on agricultural land, sediments, depositions, the volume and concentration of eroded sediments, and the amount of surface runoff produced by intensive rainstorms [25,26]. The EROSION-2D model was initially developed by [25]. Based on that concept, the EROSION-3D model has been developed since 1995 by Michael von Werner at the Department of Geography at the Free University of Berlin [27]. The difference between EROSION-3D and EROSION-2D is that EROSION-2D simulates soil erosion on a slope profile and EROSION-3D model is raster-based and uses digital elevation model which determines connectivity processes between raster cells uses. Testing of the model was done during comprehensive rainfall simulation studies that were carried out on agricultural land in Saxony, Germany, from 1992 to 1996 [27].

The structure of the EROSION-3D model consists of two main submodels, i.e., the infiltration and the erosion models (Figure 1). The calculations of the soil erosion are conducted by the erosion submodel, which considers the processes such as rainfall infiltration, generation of surface runoff, detachment of soil particles through the kinetic energy of raindrops and surface runoff or long-term modifications of the land surface relief due to soil erosion [27]. Because the EROSION-3D model is raster based, it requires a grid cell representation of a catchment.

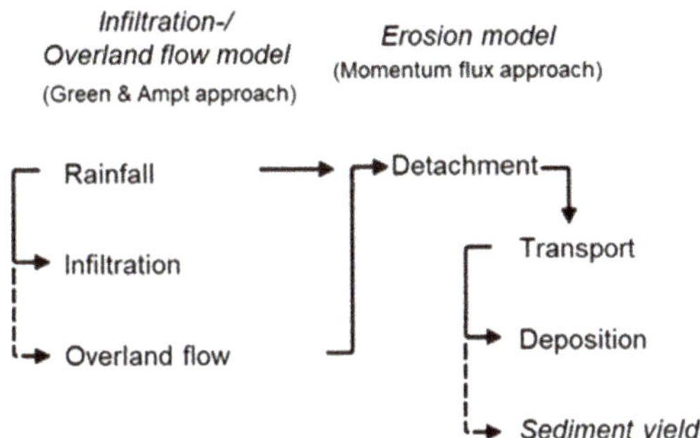

Figure 1. The EROSION-3D model structure [24].

The erosion submodel represents soil erosion processes in three steps, i.e., the detachment of soil particles from the impact of raindrops as well as their transport and deposition. The mathematical expression of the erosion submodule is based on the momentum flux approach [25], which involves an overland flow and is defined by the Equation:

$$\varphi_{qD} = \frac{q \times \rho_q \times v_q}{\Delta x},\tag{1}$$

where φ_{qD} is the momentum of the flux exerted by overland flow (N); q is the flow (m^3/(m·s)); ρ_q is the fluid density (kg/m^3); v_q is the mean velocity of the flow according to Manning (m/s); and Δx is the length of a specified slope segment (m). Because the infiltration process is complicated, the mathematical description of the infiltration process is divided into the gravitational component i_1 and the matrix component i_2 as described below.

The gravitational component (i_1) is defined by the gravitational potential as follows:

$$i_1 = k_s \times \frac{\Delta\psi_g}{x_{f1}} = k_s \times g,\tag{2}$$

where i_1 is the infiltration rate of the gravitational component (kg/(m^2·s)); k_s is the hydraulic conductivity of the transport zone ((kg·s)/m^3); $\Delta\psi_g$ is the gravitational potential ((N·m)/kg); x_{f1} is the depth of the wetting front of the gravitational component (m); and g is the gravitational constant (m/s^2).

The dynamic component of the matrix i_2 is described by a function of the matrix potential $\Delta\psi_m$:

$$i_2 = k_s \times \frac{\Delta\psi_m}{x_{f2}(t)},\tag{3}$$

where i_2 is the infiltration rate of the matrix component (kg/(m^2·s)); ks is the hydraulic conductivity of the transport zone ((kg·s)/m^3); $\Delta\psi_m$ is the matrix potential ((N·m)/kg); and $x_{f2}(t)$ is the depth of the wetting front of the gravitational component (m) in time t.

The saturated hydraulic conductivity depends on the soil structure, soil texture, and the occurrence of macropores. For determining the saturated hydraulic conductivity, an empirical equation according to [28] is used:

$$k_s = 4 \times 10^{-3} \left(1.3 \times 10^{-3}/\rho_b\right)^{1,3b} \exp\left(-0.069T - 0.037U\right), \tag{4}$$

$$b = (10 - 3\,D) - 0.5 + 0.2\,\delta p, \tag{5}$$

where k_s is the saturated hydraulic conductivity (kg·s/m^3); ρ_b is the bulk density (kg/m^3); T is the clay content (kg/kg); U is the silt content (kg/kg); D is the average particle size (m); δp is the standard deviation (-); and b is the local parameter (−).

2.2. Long-Term Simulation in the EROSION-3D Model

Although the EROSION-3D model has mainly been established as an event-based model, it contains the possibility of executing long-term simulations as an additional tool [29]. The long-term simulation submodel has a continuous character and can perform long-term simulations and run more calculations one after the other; after each event, the digital elevation model is adjusted to reflect the amount of erosion and deposition [25]. Long-term simulations can be performed in the following ways [30]:

(1) The simulations are based on iterations where one or more events have recurred. How often single events or sequences of events are repeated is determined according to the iterative value.
(2) A combination of individual single events is used to summarize the sequences of rainfall events. Based on this assumption, the model behaves like a continuous model and provides overall results.
(3) A long-term simulation based on a continuous rainfall series consists of a chronological series of single rainstorms that occur within the period evaluated. Each rainfall event needs its own soil data set whose parameters account for the individual soil conditions and the stages of crop growth as of that date.

3. The Case Study

The Svacenický Creek catchment, which has a total area of 6.3 km^2, is located in the Western Slovak Republic in the middle of the Myjava Hill Lands (Figure 2). The climate is continental, warm and moderately humid, with a mild winter and warm summer. The mean annual precipitation in the catchment is between 650 and 700 mm (1981–2015); the mean annual temperature is about 8.8 °C (1981–2013), as measured at the Myjava meteorological station. The elevation ranges from 311.4 m to 545.6 m a. s. l. The Svacenický Creek catchment is part of the flysch massif of the White Carpathians, and the dominant type of soil is Luvisols (68.0%), followed by Pararendzina (28.7%), and Cambisols (3.7%). The relief is composed of moderate slopes in the lower part of the catchment and narrow stream valleys with steep slopes in the upper part. The aerial-photograph from 2018 was used for creating of land use map and the arable land covers almost 66% of the catchment (Table 1).

The Svacenický reservoir (Figures 2 and 3) was constructed in 2010 at the bottom of the Svacenický Creek catchment as a flood-protection measure, and the permanent water level covers approximately 3 ha of the catchment. According to the project documentation [31], the height of the embankment dam is 10.25 m and the maximum retention capacity is 215,808 m^3 with a maximum water level of 8.5 m. The planned minimum water level height is 4.4 m (316.4 m about mean sea level (AMSL)). The modelled retention capacity during a 100-year flood is 207,330 m^3, with a peak discharge of 16.0 m^3·s^{-1}.

Table 1. The land use categories in the study area.

Land Use Category	Total	Path	Paved Area	Arable Land	Water Body	Forest	Shrubbery	Grassland	Orchard, Garden
Area (km^2)	6.26	0.03	0.11	4.11	0.49	0.55	0.06	0.54	0.37
Area (%)	100	0.5	1.8	65.7	7.8	8.8	1.0	8.6	5.9

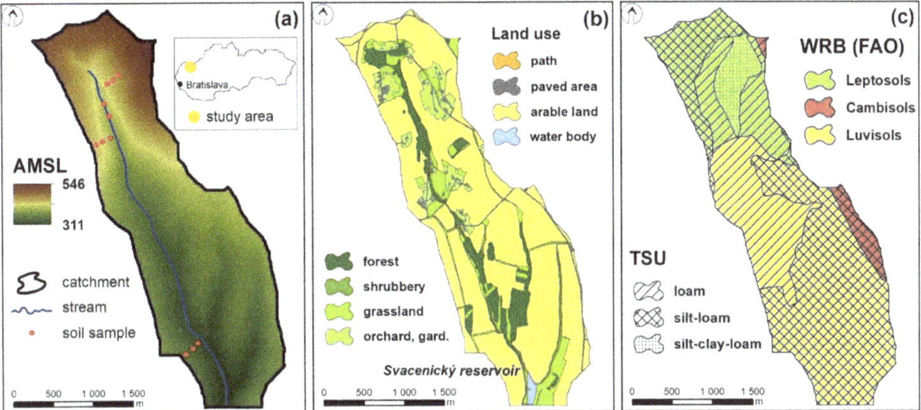

Figure 2. The characteristics of the Svacenický Creek catchment study area: (**a**) Relief, soil samples and location in the Slovak Republic; (**b**) land use; (**c**) textural soil unit (TSU) and World Reference Base (WRB(FAO)).

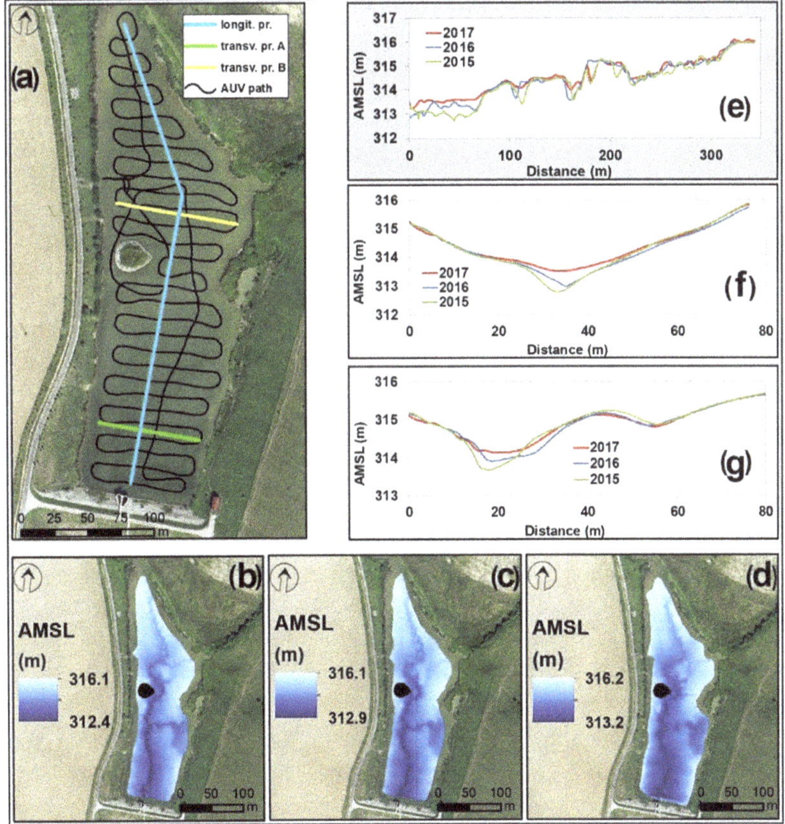

Figure 3. The Svacenický reservoir: (**a**) The autonomous underwater vehicle (AUV) EcoMapper path and longitudinal/transversal profiles; DEM of the reservoir in (**b**) 2015, (**c**) 2016, and (**d**) 2017; Graphs of the bottom's development—(**e**) longitudinal profile, (**f**) transversal profile A, (**g**) transversal profile B.

4. Data

An Autonomous Underwater Vehicle recorded the bathymetry in the Svacenický reservoir during field measurements carried out by the Slovak Academy of Science (SAS) in 2015 (22 September), 2016 (6 October) and 2017 (2 October). The EcoMapper is ideal for hydrographic spatial environmental monitoring in coastal and shallow-water applications. The speed of the AUV was 3.7 km/h and the sampling time interval was set at 1 s per sample. The total number of collected sample points is 2017 in 2015 and 2016, and 9211 sample points in 2017. The water level height was 4.45 m in 2015 and 2016, and 4.35 m in 2017. Figure 3 presents the results from the Digital Elevation Model (DEM) of the reservoir bottom with the 1 × 1 m spatial resolution. Post processing and data analysis were accomplished using Esri's ArcGIS software (Jack Dangermond, Redlands, CA, United States) (ArcMap 10.7). The DEM of the Svacenický Creek reservoir was created by geostatistical analyst tool through the Topo to Raster. Topo to Raster provides the functionality of incorporating other types of geographic features, which can assist in the creation of a DEM. Finally, the input DEM of the Svacenický Creek catchment (in grid cell size 10 × 10 m) is provided by the Esprit spol. s.r.o. (cartographic company).

In comparison with other physically-based models like EUROSEM (The European Soil Erosion Model) [32], WEPP (The Water Erosion Prediction Project) [33], and LISEM (The Limburg Soil Erosion Model) [34], the EROSION-3D model requires fewer soil input parameters (Table 2). Some of the input soil data were acquired during field measurements, and the laboratory processing of eleven soil samples followed (Figure 2a, Table 3). Also, the initial soil moisture (one of the most unstable and changeable parameters) was determined based on the field measurements in the study area. The importance of this model input parameter is due to its correlation with the previous precipitation index (Figure 4). The previous precipitation index was estimated as the 5-days sum of antecedent precipitation amount before each simulated event. The functions are derived based on rainfall experiments with known rainfall intensities and durations. The rainfall experiments took place at experimental plots in Slovakia (the Myjava region) and the Czech Republic (the Plzen region) in cooperation with the Czech Technical University in Prague and with the Research Institute for Soil and Water Conservation. A graphical representation of the functions is shown in Figure 4 (A—data from the Czech Republic, B—data from the Slovak Republic). The soil input parameters from the Parameter Catalogue were estimated for the specific crop and its growth phase in different months within the year.

Table 2. The input soil parameters required by the EROSION-3D model.

Input Parameter	Unit	Data Source
Altitude (DEM)	(m)	Esprit, s. r. o.
Rainfall intensity per time step	(mm/min)	Slovak Hydrometeorological Institute
Bulk density	(kg/m^3)	Field measurement
Organic carbon content	(%)	Field measurement
Grain size distribution	(%)	Field measurement
Skin factor	(−)	Parameter Catalogue
Surface roughness	$(s/m^{1/3})$	Parameter Catalogue
Initial soil moisture	(%)	Field measurement
Erosion resistance	(N/m^2)	Parameter Catalogue

The rainfall data covers the time period investigated between September 2015 and October 2017. The rainfall events were used in the model calculations and the rainfall series consist of effective erosive events which occurred within the periods selected. Each rainfall event requires its own soil data set, whose parameters account for the current soil conditions and stages of crop growth of that date. The model runs were done for 95 rainfall events with specific rainfall intensities higher than 2 mm (for the model calculations, intensive rainfall events involved in soil erosion processes were selected). The summarized characteristics of the rainfall events selected are shown in Table 4. Figure 5 presents the occurrence of the selected rainfall events during the time period investigated.

Table 3. The soil characteristics in the study area.

Soil ID	Soil Particle Size (%)			Organic Carbon Content (%)	Bulk Density (g/cm^3)
	Sand	**Silt**	**Clay**		
1	15.9	55.7	5.1	9.5	1.070
2	19.4	53.1	2.8	12.5	1.161
3	19.1	41.4	3.6	11.8	1.032
4	8.4	73.5	7.2	8.8	1.097
5	7.8	76.5	6.3	9.4	1.011
6	9.1	71.9	7.0	9.2	1.466
7	9.8	68.3	7.2	12.1	1.334
8	3.6	69.5	12.1	15.1	0.890
9	9.8	68.3	7.2	12.1	1.334
10	8.9	70.4	7.2	12.8	1.048
11	7.9	75.0	6.5	10.9	0.982

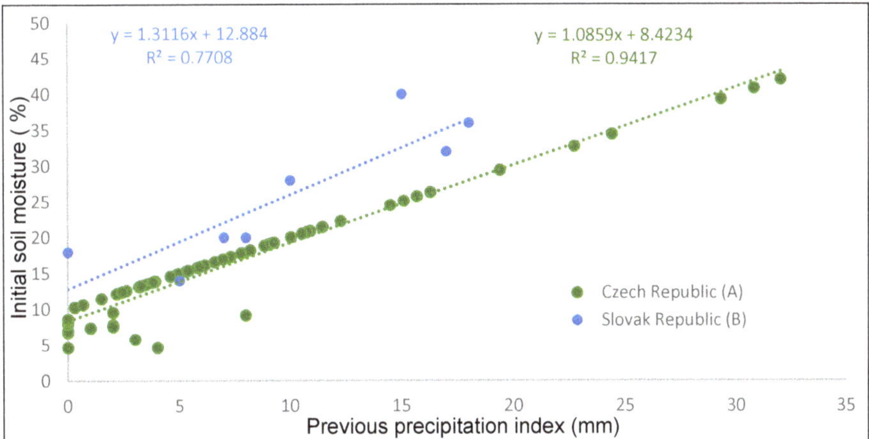

Figure 4. The functions (dependence) between the initial soil moisture and previous precipitation index determined according to the field measurements in the Slovak and Czech Republics.

Table 4. Characteristics (minimum, maximum and mean value ($\bar{x}$)) of the selected 95 rainfall events between September 2015 and October 2017.

Time Period	Duration (min)	Rainfall Amount (mm)	Rainfall Intensity (mm/min)
2015–2016	1–1041 ($\bar{x}$ 141)	0.57–26.11 ($\bar{x}$ 6.63)	0.02–0.57 ($\bar{x}$ 0.12)
2016–2017	10–504 ($\bar{x}$ 86)	0.28–52.30 ($\bar{x}$ 5.59)	0.02–0.71 ($\bar{x}$ 0.10)

According to the documentation of the local farmers, 49% of the arable land (from September 2015 to October 2016) was covered by wheat, which was followed by corn (19%), barley (14%), rye (13%), and lucerne (5%). In the next time period (from October 2016 to October 2017), corn covered 54% of the arable land, which was followed by wheat (31%), barley (10%), and lucerne (5%). Figure 6 shows the spatial distribution of the crops in the study area from September 2015 to October 2016.

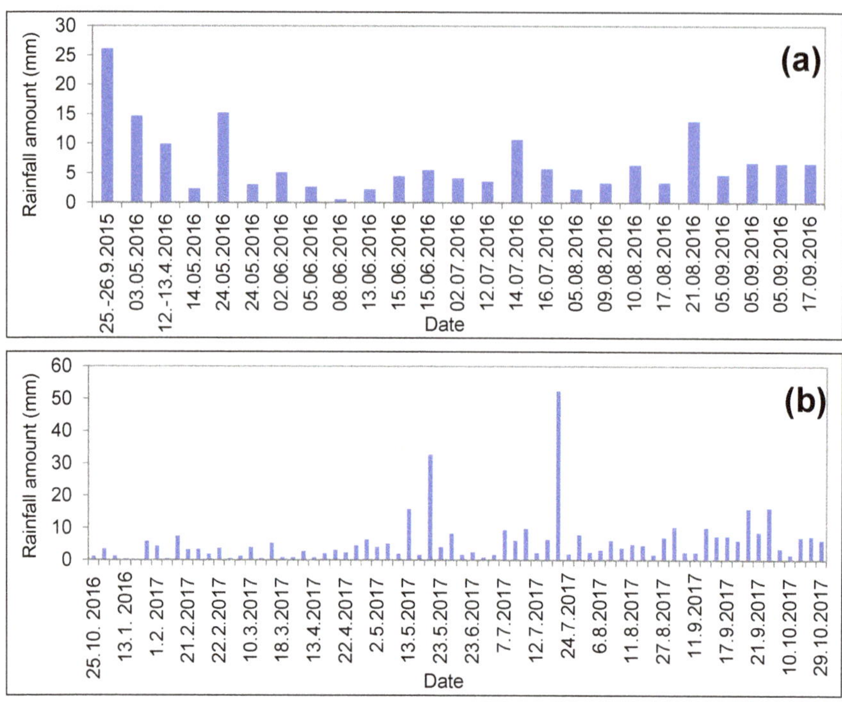

Figure 5. Total rainfall amounts and date of occurrence of selected rainfall events during the two time periods: (**a**) September 2015–October 2016; (**b**) October 2016–October 2017.

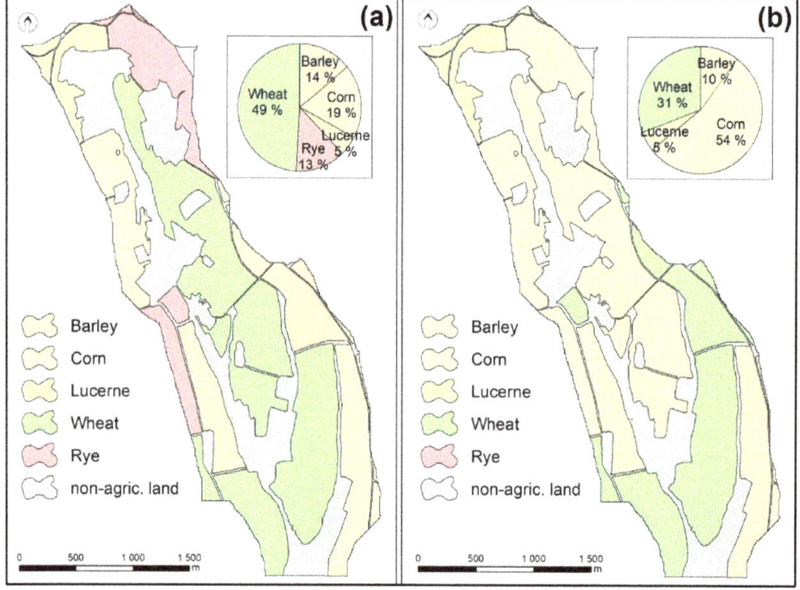

Figure 6. Annual crop distribution in the study area: (**a**) 2015–2016; (**b**) 2016–2017.

5. Results

The arable land is influenced by the intensive erosion processes in the Svacenický Creek catchment. The most vulnerable areas are located at the bottom of the slopes and in the steepest parts of particular slopes (Figure 7a). The intensity of the erosion is often higher than 5.0 t/ha/yr. Together with inappropriate crop management (corn), these localities are exposed to extreme intensities of erosion processes (>10 t/ha/yr). Wide-row crops such as corn do not represent such a threat on short slopes as in the right part of the Svacenický Creek catchment. Nevertheless, narrow-row crops such as wheat or lucerne result in a lower rate of soil degradation (3–5 t/ha/yr). The intensities of the erosion look similar in both periods of time (Figure 7a). Nevertheless, there is a difference, which is more notable in the case of the volume of sediment (Figure 7b).

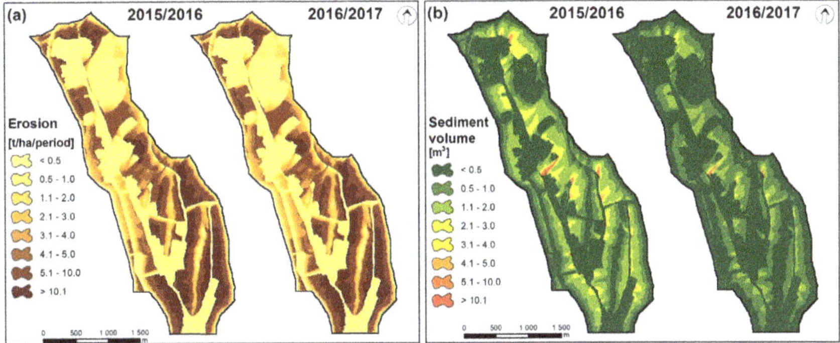

Figure 7. Results from the EROSION-3D model (calculated for a function A): (**a**) Erosion intensity; (**b**) Volume of sediment.

The intensive erosion processes are caused by intensive rainfall events. Table 5 presents the results for the rainfall-erosion events selected with a total rainfall amount higher than 12.5 mm [35–37]. In the first time period investigated (2015–2016), the rainfall-erosion events produced almost 23% of the eroded material compared to the total production of the time period in Table 6. On the other hand, only 2.4% of the sediments were produced by rainfall-erosion during the second time period investigated (2016–2017). The difference can be found in the total number of rainfall events, because the period 2016–2017 was richer in rainfall than the period 2015–2016, which made this period much drier and the soil more vulnerable to the erosion process.

Table 5. Results from the EROSION-3D model calculated for the erosion-rainfall events (total rainfall amount >12.5 mm; [35–37] and function A.

Date	Duration (min)	Rainfall Amount (mm)	Rainfall Intensity (mm/min)	Surface Runoff (m³)	Erosion/ Deposition Rate (t/ha)	Sediment Volume (m³)
25–26 September 2015	1041	26.11	0.03	10.20	2.75	12.27
3 May 2016	206	14.66	0.07	6.25	0.35	15.66
24 May 2016	161	15.22	0.09	6.23	0.00	13.54
21 August 2016	79	13.8	0.18	4.33	0.18	7.60
Total Sediment Production:						**49.07**
13 May 2017	51	15.8	0.31	4.20	1.36	1.19
23 May 2017	58	32.67	0.56	10.20	2.37	1.30
22 July 2017	74	52.3	0.71	22.30	2.36	1.41
21 September 2017	504	15.74	0.03	10.60	2.57	3.87
3 October 2017	226	16.07	0.07	0.50	0.21	1.23
Total Sediment Production:						**9.00**

Table 6. Comparison between the bathymetric measured data and modelled data of the sediment budget in the Svacenický reservoir (2015–2017).

Time Period	Predicted Sediment Volume [1] (m^3)	Predicted Sediment Volume [2] (m^3)	Observed Sediment Volume (m^3)	Relative Error (%)	
2015–2016	216.5	648.6	913.1	−76.3 [1]	−28.9 [2]
2016–2017	375.8	721.5	508.1	−26.0 [1]	41.9 [2]

[1] Calculated for the initial soil moistures from the function A in Figure 4. [2] Calculated for the initial soil moistures from the function B in Figure 4.

The production of the eroded material is connected with areas of intensive erosion processes (Figure 7a). The total predicted sediment volume was 216.5 m^3 in the period 2015–2016 and 375.8 m^3 in the period 2016–2017. The higher amount of sediment in the period 2016–2017 correlates to the higher total number of rainfall events compared to the period 2015–2016. Comparing the predicted (modelled) and observed (bathymetric measured) sediment volume in the Svacenický reservoir, the EROSION-3D predicted a lower amount of sediments (Table 6). More specifically, the predicted volume of sediment (calculated for a function A) in the period 2015–2016 is 76% lower and 26% lower in the period 2016–2017 than the observed volume of sediment in the reservoir. Based on the function B, the production of sediment increased, and the modelled volume of sediment is even higher in the period 2016–2017 compared to the bathymetry. It must be noted that the actual measured volume of sediment was reduced by 56.5% because of the assumed water content in the reservoir sediment [38].

The eroded material reached the catchment outlet and the amount was measured by the AUV EcoMapper (YSI Company, Yellow Springs, OH, USA). The measured sediment volume was reduced by considering a sediment water content of 56.5% according to [38].

6. Discussion

The characteristics such as land use, and crop management (type of crop, crop rotation) belong to temporal dynamic features, while attributes such as slope morphology or type of soil represent relatively constant catchment characteristics [19]. The crop distribution with a higher amount of winter wheat pointed to a lower volume of sediments and generally less endangered soil [39]. The probable causes are the lower values for the parameters skin factor, surface roughness, and erosion resistance in comparison with corn. The higher amounts of the surface runoff (m^3), the erosion and deposition rates (t/ha), and the volume of sediment (m^3) in the Svacenický Creek catchment detected for the second period were the result of two main causes. The first one is the higher total number of rainfall events. The second cause lies in the crop management of the study area, where corn represents 54% of the arable land. The combinations of the effects of the management systems and rain conditions resulted in the higher amounts of erosion processes in the catchment. The type of crop has a great influence on the erosive process and soil losses [40]. There is a direct connection between the crop management system and soil losses; therefore, the crop management system is one of the main factors affecting soil erosion by water [40]. Appropriate land management can significantly help reduce erosion processes [41–43]. In the case of model parameters, the biggest influence on runoff and erosion processes has bulk density, soil moisture and skin factor. The skin factor is the parameter used to reduce the prediction error resulting from the simplified assumptions of the model because the EROSION-3D model considers simplifications of homogenous soil matrix, but the process of infiltration is influenced by many different factors, i.e., soil compaction, soil crusting, surface soil pores, biological activity due to rodent burrows or worm and it is necessary to include correcting parameter.

Although erosion resistance and hydraulic roughness are considered as important factors, it has been found that they do not affect model results to a significant extent. However, the general trend of decreasing resistance of erosion with increasing erosion intensity has been confirmed by experimental results [44,45].

The results also pointed to the importance of rainfall-erosion events that are able to result in high erosion of soil [46]. In this case, the role of such events could increase in the future because of the predicted increase in extreme rainfall events due to climate change [4,46–50]. According to the Slovak Hydrometeorological Institute, the number of rainfall events with durations of 5 to 240 min has been increasing during the last decade, while rainfall-erosion events occur 2.5 to 4.7 times per year; this amount is expected to increase in the future, especially during the spring and summer months [51,52].

When testing the application of the EROSION-3D model, some difficulties were found during the appropriate determination of the input parameters and when verifying the model results through the actual measured data. A model dependability heavily depends on its calibration [53] and the testing of a physically-based erosion model is considered as a necessary part of any scientific work in order to develop an understanding of what the model will demonstrate [54].

Due to the testing process, the strengths and weaknesses of a model can be discovered. According to [55], long-term results are generally simulated in the best possible way. Here, the results presented pointed at a notable difference between the modelled and observed sediment yields, which indicates the difficulty posed by the model simplification of the natural phenomenon of erosion [18].

As it is mentioned in [56], mathematical simulation models consist of three sub-models: (a) Rainfall-runoff submodel, (b) a soil erosion sub-model and (c) a stream sediment transport sub-model. All three sub-models should be considered, but in this case, it was not done, because relevant data for stream sediment transport sub-model was not possible to obtain. In spite of a fact that small streams exist in the polder basin, significant flow occurs only in case of intense precipitation. This fact largely limits the acquirement and measurement of the relevant data needed to calculate the flow of streambed erosion. In addition to that, it can be seen (Figure 2a), that majority of these small streams flow in low slope area, so flow velocity is also low and so, frame velocity is a weak one as well. So, for this reason, we assume this type of erosion has not to be taken into account. Besides all, the used simulation model cannot simulate it as well.

However, this part of the available sediments in the catchment should be taken into account [56], but because of the conditions mentioned above, there was not done. On the other hand, during the three-years monitoring time there was found depositions of sediment in one inlet part of the reservoir (the longest contributing creek), that confirm flow stream erosion, but their volume is not significant. From all of those, it can be supposed that omission of streambed erosion could be a part of differences between calculated and measured /observed sediment volume data.

As is mentioned, based on the field measurements of the bottom bathymetry which started in 2015, the current status of the clogging of the reservoir was evaluated. The results confirmed our theories about the on-going sedimentation processes in the Svacenicky Creek reservoir. According to the analysis, we determined that during the years 2012–2019, over 10.4285 m^3 of sediment on the area of the Svacenicky Creek catchment have accumulated and the polder volume capacity has decreased 6%. The average sediment transport during the years was estimated to be 1400 t/year. The sediment transport has lowered since 2016, which can be e.g., due to lower precipitations as also flood protection measures adopted in this area. According to Table 6, the lifetime of the reservoir varies between 303.7 years for observed sedimentation, 315.0 years for predicted sedimentation B and 728.7 years for sedimentation A.

Possible sources of errors in the predicted sediment volume by the EROSION-3D model can mostly be associated with the model parameters, values of the initial soil moisture before each simulated event, and the grid size of the catchment area. The parameters of the EROSION-3D model as the resistance to erosion, surface roughness, and skin factor were chosen from the Parameter Catalogue for EROSION-3D, which contains their tabularized values by the type of soil, land use and the specific crop and its growth phase in different months within the year. Because of the attempt to continual modelling based on the modelling of a sequence of individual rainfall-erosion events, it was not possible to calibrate these parameters for each event. For the continual modelling, the crucial input parameters were the values of initial soil moisture before each simulated rainfall-erosion event. The values of initial soil moisture were estimated from the relationships between the soil moisture and the previous precipitation index, developed experimentally for the Plzen region in the Czech Republic (Figure 4, function A) and for the experimental plots in the Myjava region in Slovakia (Figure 4, function B). The data of initial soil moisture estimated from the function developed for the Myjava region in Slovakia, which is more representative for the case study area, improved the predictive sediment volumes in comparison with the measured sediment volumes—the relative error decreased to 40% for the period 2015–2016 and 30% for the period 2016–2017. On the other hand, the function B was developed only from seven pairs of experimental data, and we suppose that the extension of the data and improvement of this function would improve the performance of the model.

Also, the use of bathymetry to estimate trapped sediment in a reservoir brings out other uncertainties. According to [23], several problems with this method can be observed, e.g., the unknown effectiveness of the reservoir, the importance of a detailed analysis of the sediment cores; the precise location of sources of the sediment; the size of the reservoir and related discharges; and depreciation of the organic fraction of the sediment due to the measurements. Nevertheless, the bathymetric method is arguably the best way to determine the volume of sediment and can be carried out by several approaches [23,24,57]. The use of the autonomous underwater vehicle (AUV) to investigate the temporal evolution of changes in the bottom of the reservoir through the DEM was approved in this paper.

7. Conclusions

The physically-based EROSION-3D model was applied to simulate the runoff-erosion processes in the Slovak catchment. The model helps to identify the most sensitive erosion and deposition zones within a catchment. The long-term simulations were done for two periods (2015–2016 and 2016–2017), which were evaluated according to the time of the bathymetric measurements conducted. The results of the model pointed to an underestimation of the actual processes. The initial soil moisture plays an important role through all the input parameters of the EROSION-3D model. Finally, the rainfall-erosion events were able to erode and produce a significant part of the sediments, especially during a drier year or season.

Author Contributions: Conceptualization, Z.N. and D.H.; methodology, Z.N., D.H., S.K. and K.H.; software, Z.N., V.S. and Y.V.; validation, Z.N. and D.H.; formal analysis, Z.N. and D.H.; investigation, Z.N., D.H., V.S. and Y.V.; writing—original draft preparation, Z.N., D.H and M.Š.M.; writing—review and editing, S.K., K.H. and M.Š.M.; visualization, D.H. All authors have read and agreed to the published version of the manuscript.

Funding: This project was funded by Contracts No. APVV-15-0497 and VEGA Grant Agency No. 1/0632/19. This work was supported by the Specific Research project at Masaryk University (MUNI/A/1356/2019), the Slovak Research and Development Agency under Contracts No. APVV-15-0497, APVV-18-0347 and APVV-18-0205, the VEGA Grant Agency No. 1/0632/19 and No. 2/0025/19. The authors thank the agency for its research support

Conflicts of Interest: The authors declare no conflict of interest.

References

1.	Rawat, P.K.; Tiwari, P.C.; Pant, C.C.; Sharama, A.K.; Pant, P.D. Modelling of stream run-off and sediment output for erosion hazard assessment in Lesser Himalaya: Need for sustainable land use plan using remote sensing and GIS: A case study. *Nat. Hazards* **2011**, *59*, 1277–1297. [CrossRef]

2.	Markantonis, V.; Meyer, V.; Schwarze, R. Valuating the intangible effects of natural hazards—Review and analysis of the costing methods. *Nat. Hazards Earth Syst. Sci.* **2012**, *12*, 1633–1640. [CrossRef]

3.	Wainwright, J.; Parsons, A.J.; Michaelides, K.; Powell, D.M.; Brazier, R. Linking Short- and Long-Term Soil—Erosion Modelling. In *Long Term Hillslope and Fluvial System Modelling*, 1st ed.; Lang, A., Dikau, R., Eds.; Springer: Berlin/Heidelberg, Germany, 2003; pp. 37–51.

4.   Hlavčová, K.; Kohnová, S.; Borga, M.; Horvát, O.; Šťastný, P.; Pekárová, P.; Majerčáková, O.; Danáčová, Z. Post-event analysis and flash flood hydrology in Slovakia. *J. Hydr. Hydrom.* **2016**, *64*, 304–315. [CrossRef]

5.   Renfro, W.G. Use of erosion equation and sediment delivery ratios for predicting sediment yield. In *Present and Prospective Technology for Predicting Sediment Yields and Sources*, 1st ed.; US Department of Agriculture Publication ARS-S-40: Washington, DC, USA, 2018; pp. 33–45.

6.   Kirkby, M.J.; Morgan, R.P.C. *Soil Erosion*, 1st ed.; John Wiley and Sons: Chichester, UK, 1980; p. 306. [CrossRef]

7.   Walling, D.E. The sediment delivery problem. *J. Hydrol.* **1983**, *65*, 209–237. [CrossRef]

8.   Richards, K. Sediment delivery and drainage network. In *Channel Network Hydrology*, 1st ed.; Beven, K., Ed.; Wiley: New York, NY, USA, 1993; Volume 1, pp. 221–254.

9.   Atkinson, E. Methods for assessing sediment delivery in river systems. *Hydrol. Sci. J.* **1995**, *40*, 273–280. [CrossRef]

10.   Wischmeier, W.H.; Smith, H. *Predicting Rainfall Erosion Losses—A Guide to Conservation Planning*, 1st ed.; U.S. Dep. of Agriculture: Washington, DC, USA, 1978; p. 85.

11.   Nearing, M.A.; Lane, L.J.; Lopes, V.L. Modeling Soil Erosion. In *Soil Erosion Research Methods*, 2nd ed.; Lal, R., Ed.; Routledge: New York, NY, USA, 1994; p. 127.

12.   Kandel, D.D.; Western, A.W.; Grayson, R.B.; Turral, H.N. Process parameterization and temporal scaling in surface runoff and erosion modelling. *Hydrol. Process.* **2004**, *18*, 1423–1446. [CrossRef]

13.   Merritt, W.S.; Letcher, R.A.; Jakeman, A.J. A review of erosion and sediment transport models. *Environ. Mod. Soft.* **2003**, *18*, 761–799. [CrossRef]

14.   Kwong, F.A.L. Quantifying Soil Erosion for the Shihmen Reservoir Watershed, Taiwan. *Agric. Sys.* **1994**, *45*, 105–116. [CrossRef]

15.   Pandey, A.; Himanshu, S.K.; Mishra, S.K.; Singh, V.P. Physically based soil erosion and sediment yield models revisited. *Catena* **2016**, *147*, 595–620. [CrossRef]

16.   Hajigholizadeh, M.M.; Hector, A.F. Erosion and Sediment Transport Modelling in Shallow Waters: A Review on Approaches, Models and Applications. *Int. J. Environ. Res. Public Health* **2018**, *15*, 518. [CrossRef]

17.   Jakeman, A.J.; Green, T.R.; Beavis, S.G.; Zhang, L.; Dietrich, C.R.; Crapper, P.F. Modelling upland and instream erosion, sediment and phosphorus transport in a large catchment. *Hydrol. Process.* **1999**, *13*, 745–752. [CrossRef]

18.   Kuznetsov, M.S.; Gendugov, V.M.; Khalilov, M.S.; Ivanuta, A.A. An equation of soil detachment by flow. *Soil Tillage Res.* **1998**, *46*, 97–102. [CrossRef]

19.   Wei, H.; Nearing, M.A.; Stone, J.J.; Guertin, D.P.; Spaeth, K.E.; Pierson, F.B.; Nichols, M.H.; Moffet, C.A. A New Splash and Sheet Erosion Equation for Rangelands. *Soil Sci. Soc. Am. J.* **2009**, *73*, 1386–1392. [CrossRef]

20.   Iglesias, I.; Avilez-Valente, P.; Bio, A.; Bastos, L. Modelling the Main Hydrodynamic Patterns in Shallow Water Estuaries: The Minho Case Study. *Water* **2019**, *11*, 1040. [CrossRef]

21.   Marttila, H.; Klöve, B. Dynamics of erosion and suspended sediment transport from drained peatland forestry. *J. Hydrol.* **2010**, *388*, 414–425. [CrossRef]

22.   Ayele, G.T.; Teshale, E.Z.; Yu, B.; Rutherfurd, I.D.; Jeong, J. Streamflow and Sediment Yield Prediction for Watershed Prioritization in the Upper Blue Nile River Basin, Ethiopia. *Water* **2017**, *10*, 782. [CrossRef]

23.   García-Ruiz, J.M.; Beguería, S.; Nadal-Romero, E.; González-Hidalgo, J.C.; Lana-Renault, N.; Sanjuán, Y. A meta-analysis of soil erosion rates across the world. *Geomorphology* **2015**, *239*, 160–173. [CrossRef]

24.   Boyle, J.F.; Plater, A.J.; Mayers, C.; Turner, S.D.; Stroud, R.W.; Weber, J.E. Land use, soil erosion, and sediment yield at Pinto Lake, California: Comparison of a simplified USLE model with the lake sediment record. *J. Paleolimnol.* **2011**, *45*, 199–212. [CrossRef]

25.   Schmidt, J. A mathematical model to simulate rainfall erosion. *Catena* **1991**, *19*, 101–109.

26.   Green, W.H.; Ampt, G. Studies of soil physics: The flow of air and water through soils. *J. Agric. Sci.* **1911**, *4*, 1–24.

27.   Von Werner, M. *Erosion-3D: User Manual*, version. 3.1.1; Michael von Werner Berlin: Berlin, Germany, 2006; p. 54.

28.   Campbell, G.S. Soil physics with basic: Transport models for soil-plant systems. DEV. *Soil Sci.* **1985**, *14*, 252–254.

29.   Von Werner, M. *Erosion-3D Version 3.0 User Manual—Samples. 2003*; GeoGnostics Software: Berlin, Germany, 2003; p. 34.

30.   Schmidt, J. *Development and Application of a Physically Established Simulation Model for the Erosion of Agricultural Areas*; Institute of Geographic Sciences: Berlin, Germany, 1996; p. 148.

31. VODOTIKA. *Polder Svacenický Creek. Bratislava, 10-01-2008*; Project documentation; Vodotika: Bratislava, Slovak Republic, 2008; p. 35.

32. Morgan, R.; Quinton, J.; Smith, R.; Govers, G.; Poesen, J.; Auerswald, K.; Chisci, C.; Torri, D.; Styczen, M.E. The European soil erosion model (EUROSEM)—A process-based approach for predicting sediment transport from fields and small catchments. *Earth Surf. Proc. Land.* **1998**, *23*, 527–544. [CrossRef]

33. Laflen, J.M.; Elliot, W.J.; Flanagan, D.C.; Meyer, C.R.; Nearing, M.A. WEPP-predicting water erosion using a process-based model. *J. Soil Water Conserv.* **1997**, *52*, 96–102.

34. De Roo, A.P.J.; Wesseling, C.G.; Ritsema, C.J. LISEM: A single event physically based hydrological and soil erosion model for drainage basins. I: Theory, input and output. *Hydrol. Process.* **1996**, *10*, 1107–1117. [CrossRef]

35. Stocking, M.A.; Elwell, H.A. Rainfall erosivity over Rhodesia. *Trans. Inst. Brit. Geogr.* **1976**, *1*, 231–245. [CrossRef]

36. Pasák, V. *Soil Protection against Erosion*, 1st ed.; State Agricultural Publishing House Prague: Prague, Czech, 1984; p. 164.

37. Brychta, J.; Janeček, M. Determination of erosion rainfall criteria based on natural rainfall measurement and its impact on spatial distribution of rainfall erosivity in the Czech Republic. *Soil Water Res.* **2019**, *14*, 153–162. [CrossRef]

38. Hucko, P.; Šumná, J. *Verification of the System of Sediment Disposal from Water Management Reservoirs*, 1th ed.; VÚVH: Bratislava, Slovak, 2003.

39. Schindewolf, M.; Bornkampf, C.; von Werner, M.; Schmidt, J. Simulation of Reservoir Siltation with a Process-based Soil Loss and Deposition Model. In *Effects of Sediment Transport on Hydraulic Structures*; Vlassias, H., Ed.; Democritus University of Thrace: Xanthi, Greece, 2015; pp. 41–57. [CrossRef]

40. Carvalho, D.F.; Eduardo, E.N.; de Almeida, W.S.; Santos, L.A.F.; Sobrinho, T.A. Water erosion and soil water infiltration in different stages of corn development and tillage systems. *Rev. Bras. Eng. Agríc. Ambient. Agric.* **2015**, *19*, 1072–1078. [CrossRef]

41. Panagos, P.; Borelli, P.; Meusburger, K.; Alewell, C.; Lugato, E.; Montanarella, L. Estimating the soil erosion cover-management factor at the European scale. *Land Use Policy* **2015**, *48*, 38–50. [CrossRef]

42. Bo, M.; Xiaoling, Y.; Fan, M.; Zhanbin, L.; Faqi, W. Effects of Crop Canopies on Rain Splash Detachment. *PLoS ONE* **2014**, *9*, 10. [CrossRef]

43. Hugo, V.; Zuazo, D.; Rocío, C.; Pleguezuelo, R. Soil-erosion and runoff prevention by plant covers. A review. *Agric. Sustain. Dev.* **2008**, *1*, 65–86. [CrossRef]

44. Ebabu, K.; Tsunekawa, A.; Haregeweyn, N.; Adgo, E.; Meshesha, D.T.; Aklog, D.; Masunaga, T.; Tsubo, M.; Sultan, D.; Fenta, A.A.; et al. Effects of land use and sustainable land management practices on runoff and soil loss in the Upper Blue Nile basin. *Sci. Total Environ.* **2019**, *648*, 1462–1475. [CrossRef] [PubMed]

45. Knapen, A.; Poesen, J.; De Baets, S. Seasonal variations in soil erosion resistance during concentrated flow for a loess-derived soil under two contrasting tillage practices. *Soil Tillage Res.* **2007**, *94*, 425–440. [CrossRef]

46. Uhrová, J.; Bachan, R.; Štěpánková, P. Determination of soil loss from erosion rills by method of digital photogrammetry and method of volumetric quantification. *VTEI* **2018**, *6*, 35–38.

47. Uhrová, J.; Štěpánková, P.; Osičkovám, K. Complex system of natural water retention measures against erosion and flash floods. *VTEI* **2016**, *4*, 13–19.

48. Michael, A. Anwendung des Physikalisch Begründeten Erosionsprognosemodells EROSION 2D/3D—Empirische Ansätze zur Ableitung der Modellparameter. Ph.D. Thesis, Technische Universität Bergakademie Freiberg, Freiberg, Germany, 2000.

49. Zolina, O.; Simmer, C.; Kapala, A.; Shabanov, P.; Becker, P.; Mächel, H.; Gulev, S.; Groisman, P. Precipitation Variability and Extremes in Central Europe: New View from STAMMEX Results. *Bull. Am. Met. Soc.* **2014**, *99*, 995–1002. [CrossRef]

50. Dolák, L.; Řezníčková, L.; Dobrovolný, P.; Štěpánek, P.; Zahradníček, P. Extreme precipitation totals under present and future climatic conditions according to regional climate models. In *Climate Change Adaptation Pathways from Molecules to Society*, 1st ed.; Vačkář, D., Ed.; Global Change Research Institute, Czech Academy of Sciences: Brno, Czech, 2017; pp. 27–37.

51. Trnka, M.; Brázdil, R.; Vizina, A.; Dobrovolný, P.; Mikšovský, J.; Štěpánek, P.; Hlavinka, P.; Řezníčková, L.; Žalud, Z. Droughts and Drought Management in the Czech Republic in a Changing Climate. In *Drought and Water Crises: Integrating Science, Management, and Policy*, 1st ed.; Wilhite, D.A., Ed.; CRC Press, Taylor & Francis: Boca Raton, FL, USA, 2017; pp. 461–480.

52. Slovak Hydrometeorological Institute (Increase in the Extremes of Total Rainfall Events in Slovakia). Available online: http://www.shmu.sk/sk/?page=2049&id=965 (accessed on 20 March 2019).

53. Acuña, G.J.; Ávila, H.; Fausto, A.C. River Model Calibration Based on Design of Experiments Theory. A Case Study: Meta River, Colombia. *Water* **2019**, *11*, 1382. [CrossRef]

54. Quinton, J. The Validation of Physically-based Erosion Models with Particular Reference to EUROSEM. Ph.D. Thesis, Cranfield University, Wharley End, UK, 1994.

55. Favis-Mortlock, D.; Boardman, J.; Bell, M. Modelling long-term anthropogenic erosion of a loess cover: South Downs, UK. *Holocene* **1997**, *7*, 79. [CrossRef]

56. Hrissanthou, V. *Computation of Lake or Reservoir Sedimentation in Terms of Soil Erosion, In Sediment Transport in Aquatic Environments*, 3rd ed.; Manning, A.J., Ed.; IntechOpen: London, UK, 2011; pp. 233–262. [CrossRef]

57. Zhao, G.; Mu, X.; Jiao, J.; Gao, P.; Sun, W.; Li, E.; Wei, Y.; Huang, J. Assessing response of sediment load variation to climate change and human activities with six different approaches. *Sci. Total Environ.* **2018**, *639*, 773–784. [CrossRef]

Article

Effect of Sediment Load Boundary Conditions in Predicting Sediment Delta of Tarbela Reservoir in Pakistan

Zeeshan Riaz Tarar [1], Sajid Rashid Ahmad [1], Iftikhar Ahmad [1], Shabeh ul Hasson [2,3], Zahid Mahmood Khan [4], Rana Muhammad Ali Washakh [5,6], Sardar Ateeq-Ur-Rehman [4,7,*] and Minh Duc Bui [7]

[1]   College of Earth & Environmental Sciences (CEES), University of the Punjab, Punjab 54590, Pakistan
[2]   CEN, Institute of Geography, University of Hamburg, 20148 Hamburg, Germany
[3]   Department of Space Science, Institute of Space Technology, Islamabad 44000, Pakistan
[4]   Department of Agriculture Engineering, BZ University, Multan 60800, Pakistan
[5]   Key Laboratory of Mountain Hazards and Surface Process, Institute of Mountain Hazards and Environment, Chinese Academy of Sciences, Chengdu 610041, China
[6]   University of Chinese Academy of Sciences, Beijing 100049, China
[7]   Chair of Hydraulic and Water Resources Engineering, Technical University of Munich, 80333 Munich, Germany
*   Correspondence: sardar.ateeq@tum.de

Received: 16 June 2019; Accepted: 13 August 2019; Published: 18 August 2019

Abstract: Setting precise sediment load boundary conditions plays a central role in robust modeling of sedimentation in reservoirs. In the presented study, we modeled sediment transport in Tarbela Reservoir using sediment rating curves (SRC) and wavelet artificial neural networks (WA-ANNs) for setting sediment load boundary conditions in the HEC-RAS 1D numerical model. The reconstruction performance of SRC for finding the missing sediment sampling data was at $R^2 = 0.655$ and NSE = 0.635. The same performance using WA-ANNs was at $R^2 = 0.771$ and NSE = 0.771. As the WA-ANNs have better ability to model non-linear sediment transport behavior in the Upper Indus River, the reconstructed missing suspended sediment load data were more accurate. Therefore, using more accurately-reconstructed sediment load boundary conditions in HEC-RAS, the model was better morphodynamically calibrated with $R^2 = 0.980$ and NSE = 0.979. Using SRC-based sediment load boundary conditions, the HEC-RAS model was calibrated with $R^2 = 0.959$ and NSE = 0.943. Both models validated the delta movement in the Tarbela Reservoir with $R^2 = 0.968$, NSE = 0.959 and $R^2 = 0.950$, NSE = 0.893 using WA-ANN and SRC estimates, respectively. Unlike SRC, WA-ANN-based boundary conditions provided stable simulations in HEC-RAS. In addition, WA-ANN-predicted sediment load also suggested a decrease in supply of sediment significantly to the Tarbela Reservoir in the future due to intra-annual shifting of flows from summer to pre- and post-winter. Therefore, our future predictions also suggested the stability of the sediment delta. As the WA-ANN-based sediment load boundary conditions precisely represented the physics of sediment transport, the modeling concept could very likely be used to study bed level changes in reservoirs/rivers elsewhere in the world.

Keywords: Upper Indus Basin (UIB); Tarbela Reservoir; Besham Qila; sediment modeling; uncertainty; wavelet transform analysis-artificial neural network (WA-ANN); sediment rating curve (SRC); HEC-RAS

---

1. Introduction

The uncertainties in modeling reservoir sedimentation are due to: (a) both flow and sediment; (b) the distribution of sediment particle size; (c) the specific weights of sediment deposits; (d) reservoir geometry; and (e) the operational rules of reservoirs [1]. These uncertainties are propagated, particularly, due to the varying input of sediment loads as boundary conditions. Normally, sediment series, as input to the model, are estimated by utilizing sediment rating curves (SRCs), prepared by developing relationships through simple regression techniques, between flow and sediment, observed over a considerable period, adequately representing the complete hydrological cycle over decades [2,3]. It has been observed in various sediment studies of reservoirs around the world that SRCs, though a simple and convenient way to estimate missing values of sediment inflow, often overestimate and overshoot the sediment entry into the reservoirs against the actual conditions, up to 50% [4,5]. Tarbela Reservoir hydrographical/bathymetric surveys have been conducted since 1979 to observe the sediment entry and position/advancement of the delta in the reservoir. Each year, the reservoir authorities issue Sedimentation Reports based on the above conducted surveys. As per the Sedimentation Report of Tarbela Reservoir [6], the actual observed sediment deposits in the reservoir are about 171.3 Mt/year, which are about 53% of the average of the below-mentioned studies, i.e., 47% overestimation. Hence, precise hydro-morphodynamic boundary conditions play a principal role in modeling the transport processes in rivers and reservoirs.

The Tarbela Dam Project (TDP) was completed in the mid-1970s and is the backbone of the hydropower and water resources of Pakistan, with its 3478 MW of existing installed and 6298 MW of near future capacity. It is the world's largest earth-filled dam and also by structural volume [7]. The Tarbela Reservoir drains UIB and lies at its lowest point. The drainage area up to the dam is about 170,000 km^2, as shown and demarcated in Figure 1. The huge body of water created behind the dam, originally 11.620 million acre feet (MAF), has been reduced by sedimentation to 6.856 MAF in 2019 [8], meaning that it is only 59% of its original storage volume, and the rest has been consumed by sedimentation. The feasibility and engineering studies of Tarbela Dam that were conducted in the mid-1960s and 1970s took serious note of the potential sedimentation problems that were likely to arise after some years of dam construction. Various studies at the time and afterwards estimated sediment entering the reservoir to be substantially overestimated, based primarily on techniques in vogue and with less data. The Tarbela Dam Consultants (Tippets, Abbett, McCarthy, Stratton (TAMS)) used 235 million tons (Mt) annually as the sediments entering the reservoir [9]. The Kalabagh Dam Consultants estimated the annual sediment load entering Tarbela as 295.7 Mt using sediment rating curves. The same figure of 295.7 Mt was adopted for sediment studies of Tarbela by the Consultants of the Ghazi Barotha Hydropower Project located just 8 km downstream of Tarbela Dam. The Consultants for the Mega Diamer-Basha Dam, making use of additional data from 1962–2003 in sediment rating curves, calculated the load for Tarbela Reservoir as 233 Mt annually [10]. Future sedimentation scenarios fir Tarbela Reservoir hold a pivotal position for authorities and water managers alike, as a reduction in the storage capacity of Pakistan's largest water body and its implications for all related disciplines would be sensitive enough to provoke studies into alternative or preventive measures.

A list of studies also cited by [11], in addition to the ones mentioned above, calculating sediment entering Tarbela Reservoir/main Upper Indus Basin (UIB), is tabulated in Table 1:

Table 1. Published estimates of sediment load (SL) of the Upper Indus River.

Sediment Load (Mt year^{-1})	Estimate by
480	[12]
400	[13]
475	[14]
200	[15] as per the report by [16]
675	[17]
300	[18]
200	[19]
197	[20]
200	[21]

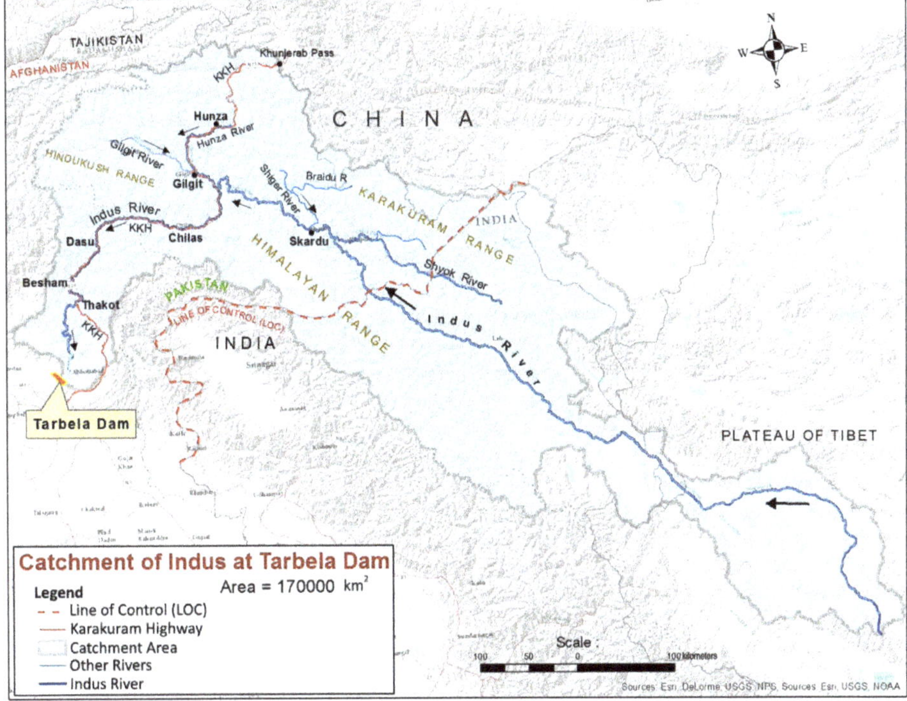

Figure 1. Demarcated Upper Indus Basin at Tarbela Dam.

All above estimates were based on sediment rating curve (SRC) method and varied in a wide range from around 200 Mt y^{-1} – 675 Mt y^{-1} over the last 50 years. Unfortunately, the accuracy of SRCs is limited, as they map all scattered data points of discharge and sediment loads using a single fitting line, which is more likely to be affected by data outliers [22–24]. Therefore, the single fitting line cannot handle sediment transport processes connected to the phenomenon of hysteresis and noticeable hydrological variations, such as: (a) fluvial erosion and transport processes, interacting with other sediment-production processes; (b) sediment temporary storage in the main channel of the river [25]; (c) landslide phases related to aggradation and degradation [26]; (d) on average, 5–10 waves of high flow of an average of 10–12 days' duration during the monsoon period; (e) different discharge and sediment conveyance times and their differing lag-times from sources to the gauge recording stations. Basically, all these processes cause different sediment concentrations on same magnitude of discharge during rising and falling limbs of flood events, which is referred to as the hysteresis phenomenon. As SRCs are mostly employed in the estimation process of sediment load boundary conditions due to their construction simplicity, a marked compromise could arise in the numerical or physical modeling outcomes.

Since the variations in sediment load boundary conditions affect the calculations of the morphodynamics, it is essential to model time-related changes in sediment supply more accurately, influenced by the above-mentioned phenomenon of hysteresis and noticeable hydrological changes. During recent years, artificial neural networks (ANNs) have gained increased reception as new analytical techniques due to their robustness and ability to model non-stationary data series. Therefore, ANNs have a clear advantage over other conceptual models as they do not need previous knowledge of the process because they build a relationship between data inputs and targets using non-linear activation functions. The ANNs have multiple inputs with dissimilar characteristics, making ANNs be able to represent time-space variation [1]. In spite of the adequate flexibility of ANNs in modeling time series, sometimes, ANNs have a weakness when signal alterations are highly non-stationary and physical hydrological processes operate under scales of large ranges, with variations of one day to several years. In such a situation, different methods have been proposed, among which are wavelet transforms. They have become a capable method for analyzing such changes and trends in hydrological time series [27–31]. A wavelet has been defined as a small wave whose energy is limited in a short period of time and is a logical method for signals that are non-stationary, having short-lived transient components, featuring at different scales, or singularities. A non-stationary signal can be broken up into a certain number of unvarying signals by wavelet transform. ANN is then combined with wavelet transform (WA-ANN). It is considered that WA-ANN models are more precise than the conventional methods since wavelet transforms provide effective break-ups of the original time series, and the wavelet transformation data improves the performance of conventional ANN models by catching effective information for various resolution levels [4,5,11].

In the present study, effort has been made to model the sediment delta of Tarbela Reservoir using the 1D HEC-RAS numerical model with the objective to reduce variations in its future prediction by employing first the conventionally-estimated sediment inflow based on SRC and then by the above elaborated innovative WA-ANN technique. The sediment series based on WA-ANN, as developed by [4,11], was further updated, calibrated, and validated by inclusion of sediment data up to 2014 and used as input to the model.

2. Methods

2.1. HEC-RAS Program System

The River Analysis System (HEC-RAS), a one-dimensional model, created by Hydrologic Engineering Centre, has been designed to carry out steady flow water surface profile computations of natural rivers and networks of natural and constructed channels, unsteady flow simulations, moving boundary sediment transport computations, and water quality analysis. All these components utilize a common geometric data representation and hydraulic computation procedures. The calculations of one-dimensional moving material of the river bed causing scour or deposition over a certain modeling period establish a base for sediment transport simulations. Generally, sediment transport in rivers, channels, and streams depends on two modes: bed load and suspended load, which in turn depend on sediment particle size, the velocity of water, and river bed slope. The basic idea of evaluating sediment transport capacity by HEC-RAS is by computing sediment capacity of each cross-section as a control volume and for all particle sizes. HEC-RAS requires boundary condition data of each type for making such calculations. The boundary conditions are necessary to get the solution to the differential equations set, describing the problem over the area of interest. There are a number of boundary conditions for steady flow and sediment analysis computations in HEC-RAS. Boundary conditions can be either external, which are specified at the ends of the simulated network at the upstream/downstream, or internal, which are to be used for connecting junctions. Background information regarding computational methods and equations used in modeling sediment transport is available in [32].

2.2. Data Collection

Owing to the noticeable global warming influence on the hydrological and river systems observed around the turn of the century, we considered to start the modeling process from 2005 onward [33–38]. For this, we collected the levels of Tarbela Reservoir and the flows of the Indus River at Besham Qila, the nearest station to the upper periphery of the reservoir located about 134 km upstream of the dam, from the project authorities for the 2005–2018 period. To hydrodynamically and morphodynamically initialize, calibrate, and validate the model, bathymetric surveys of the Tarbela Reservoir for the years 2005, 2013, and 2017 were also obtained.

To develop SRC and WA-ANN models, suspended sediment concentrations (ppm) and its gradational data at Besham Qila gauge recording station were collected for the 1969–2014 period from the Surface Water Hydrology Project (SWHP) of the Water and Power Development Authority (WAPDA), Pakistan. The raw data so collected are presented in Figure 2.

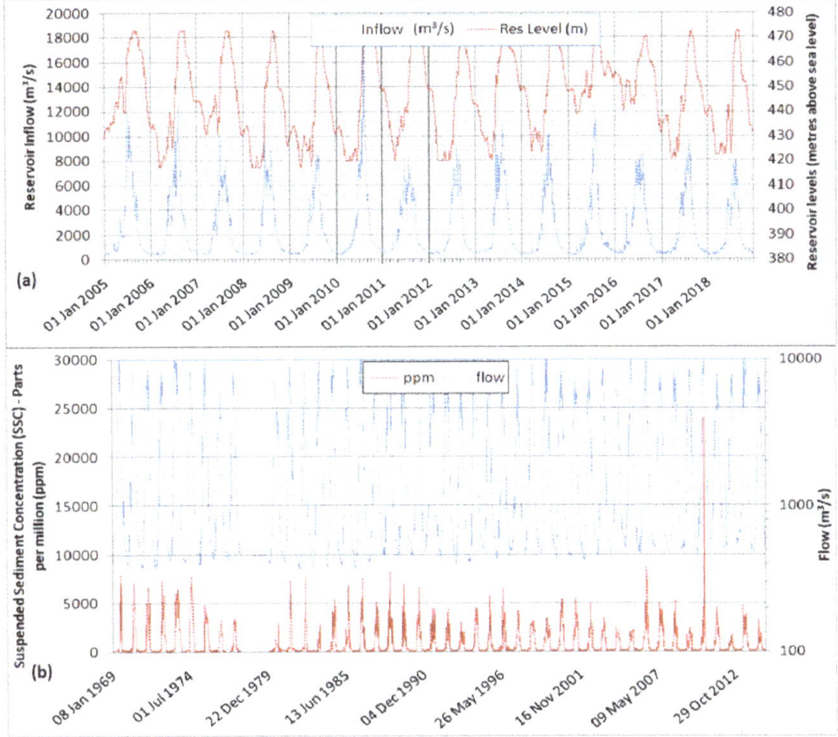

Figure 2. Data used in the study: (**a**) daily Tarbela Reservoir inflow and levels; (**b**) occasionally-collected suspended sediment concentration samples with observed flow.

The Tarbela Reservoir was cut into 73 cross-sections or range lines (R/Lines) to study the morphodynamics of the huge reservoir (see Figure 3).

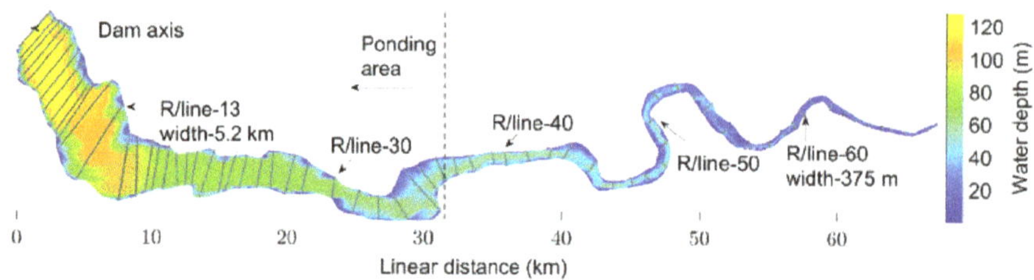

Figure 3. Range lines (R/Lines) of Tarbela Reservoir used from [5].

The first comprehensive reservoir survey after the dam's construction was in 1974, and since then, each year, hydro-graphic surveys of the Tarbela Reservoir have been conducted. To cover the whole reservoir area, i.e., 161 km^2, the hydro-graphic surveys were conducted using a systematic sounding method over the 73 cross-sectional range lines. Approximately 3500–4000 sounding measurements of the bed level alterations, reservoir depths, and water level elevations along these range lines are available, which were collected mostly during September–November. The distance between the cross-sections/range lines and the measured data points along these cross-sections are not identical. The average distance between each cross-section measured along River Thalweg was approximately 1000 m. However, compared to the upper periphery of the reservoir, the distances between the cross-sections nearer to the dam were smaller. The distance between measured data points along the cross-sections, i.e., lateral distance in y direction, was also variable with a mean of 39 m. The mean cross-sectional width near the dam axis was approximately 4000–5000 m, reducing to only 90–150 m near the upper periphery of the reservoir. Therefore, the major storage volume is near the dam axis, containing huge sediment deposits.

Water depths in the reservoir vary from a maximum 150 m near the dam to mostly 20 m at the reservoir inlet. To secure the stability of the dam and bank slopes along the reservoir, the maximum lowering and rising rate for the reservoir during operation is 4 m/day and 3 m/day, respectively, between reservoir levels 396 and 460 m and only 1 m/day up to the maximum conservation level of 472.5 m asl. The average slope of the river bed in 1979 was 0.0011211, which decreased in 2010 with an average slope of 0.0005988.

2.3. Performance Measures for Model Evaluation

To assess the performance of the models in terms of accuracy and consistency in simulating reservoir water depths and river bed levels, the following three statistical measures tests were made up of: (a) the coefficient of determination (R^2), an indication of the level of the relationship between the observed and simulated data, ranging from 0–1; (b) the observations' standard deviation ratio (RSR), the ratio of the root mean squared error (RMSE) to the standard deviation (STDEV) of the observed data; (c) the Nash–Sutcliffe efficiency (NSE), a statistical assessment to calculate the relative magnitude of residual variance compared to the measured data variance [39]. The formulas are shown in Table 2.

Table 2. Statistical performance parameters used to evaluate the modeling performance.

Parameters	Description	Ranges
Coefficient of determination	$R^2 = \left(\dfrac{\sum\limits_{i=1}^{P}(X_i^{obs} - \bar{X}^{obs})(X_i^{sim} - \bar{X}^{sim})}{\sqrt{\sum\limits_{i=1}^{P}(X_i^{obs} - \bar{X}^{obs})^2 \ \sum\limits_{i=1}^{P}(X_i^{sim} - \bar{X}^{sim})^2}} \right)^2$	0–1
Observations standard deviation ratio (RSR)	$RSR = \dfrac{RMSE}{STDEV_{obs}} = \dfrac{\sqrt{\frac{1}{P}\sum_{i=1}^{P}(X_i^{obs} - X_i^{sim})^2}}{\sqrt{\sum\limits_{i=1}^{P}(X_i^{obs} - \bar{X}^{obs})^2}}$	0–1
Nash–Sutcliffe efficiency (NSE)	$NSE = 1 - \dfrac{\sum\limits_{i=1}^{P}(X_i^{obs} - X_i^{sim})^2}{\sum\limits_{i=1}^{P}(X_i^{obs} - \bar{X}^{obs})^2}$	$-\infty$–1

X_i^{obs}, X_i^{sim} represent the ith observed and simulated value of parameters. $\bar{X}$ represents the mean.

2.4. Sediment Rating Curves

The SRC method is based on an empirical relationship between the discharge and the sediment concentration/load. Likewise, the collected suspended sediment concentration samples were converted into suspended sediment load (SSL) in t/day and related to their corresponding discharges in m^3/s to develop the rating curves, encompassing low and high flow conditions. Additionally, 10% bed load was added to the suspended load as recommended by [10]. Total load equations (Equations (1) and (2)) are expressed in the form $Q_T = a\,Q^b$, where Q_T is sediment discharge in t/day; Q is water discharge in m^3/s; and a and b are constants as solved on page 15 of [3]. They were entered as upper boundary conditions in the model and depicted graphically in Figure 4.

$$Q_T = 1.686 \times 10^{-4} Q^{2.627}, \text{ for } Q >= 481 \text{ m}^3/s \tag{1}$$

$$Q_T = 4.474 \times 10^{-32} Q^{12.868}, \text{ for } Q < 481 \text{ m}^3/s \tag{2}$$

where Q_T = total load (suspended + bed load) in t/day with respect to flow discharge Q in m^3/s.

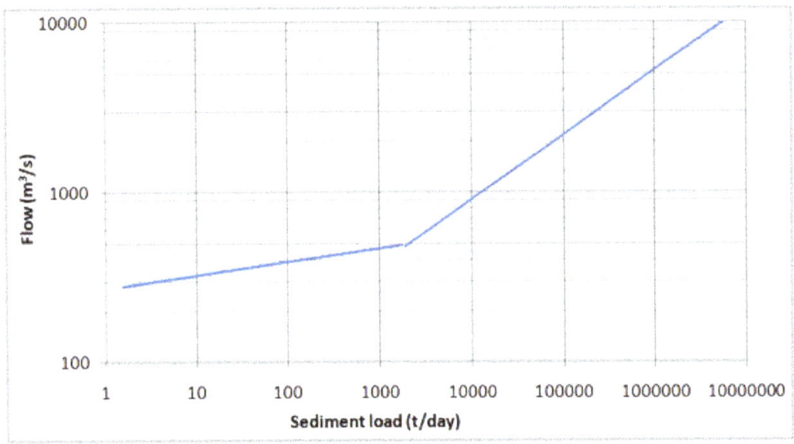

Figure 4. Sediment rating curve.

The annual load calculated by SRC was 212 million tons (Mt). The calculated monthly loads are shown in Table 3 and Figure 5. As can be seen, most of the sediment transport processes took place in the summer months. Against 84% of the annual flow, 98% of the sediment load transport occurred from May–September.

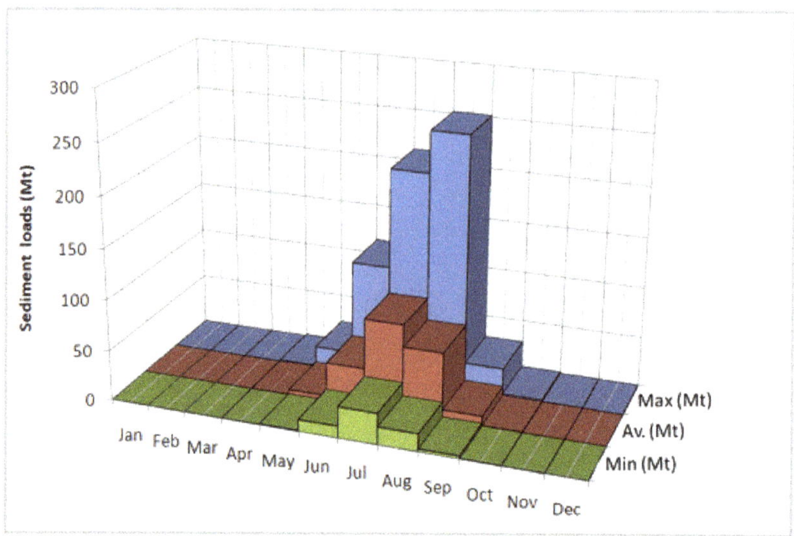

Figure 5. Monthly sediment load at Besham Qila with sediment rating curves (SRC).

Table 3. Average monthly load in Mt at Besham Qila with SRC.

Jan	Feb	Mar	Apr	May	Jun	Jul	Aug	Sep	Oct	Nov	Dec	Annual
0.037	0.026	0.068	0.374	5.171	38.613	87.884	66.285	9.800	0.606	0.154	0.077	212.068

2.5. Wavelet-Artificial Neural Network

2.5.1. Wavelet Transform

Recently, wavelet analysis has been widely accepted in a wide range of science and engineering applications. Some of the latest studies utilizing the wavelet analysis are [4,27–31]. The wavelet analysis technique has also been used in: data and image compression, partial differential equation solving, transient detection, pattern recognition, texture analysis, noise reduction, trend detection, etc. Wavelets have been identified as more effective tools than the Fourier transform (FT) in analyzing the non-stationary time series. Instead of FT, which analyses the data in two dimensions, i.e., time and frequency, wavelet transform was used, which analyses the data in three dimensions, i.e., time, space, and frequency. This provides a significant opportunity to examine the variation in the hydrological processes.

2.5.2. Continuous and Discrete Wavelet Analysis

Wavelet transform (WT) breaks down/separates data series into logically-ordered wave-like oscillations (wavelets) analogous to data vis-à-vis time within a range of frequencies. The original time series can be depicted with regard to a wavelet expansion that uses the coefficients of the wavelet functions. Several wavelets can be made from a function $\psi(t)$ known as a "mother wavelet", which is restricted in a finite/bound interval. That is, WT expresses/breaks a given signal into frequency bands and then analyses them in time. WT is widely categorized into the continuous wavelet transform (CWT) and discrete wavelet transform (DWT). CWT is defined as the sum over the whole time of the signal to be analyzed, multiplied by the scaled and shifted versions of the transforming function ψ. The CWT of a signal $f(t)$ is expressed as follows:

$$W_{a,b} = \frac{1}{\sqrt{a}} \int_{-\infty}^{\infty} f(t)\Psi^* \left(\frac{t-b}{a}\right) dt \tag{3}$$

where "*" denotes the complex conjugate. On the other hand, CWT looks for correlations/mutual relationships between the signal and wavelet function. This measurement is done at distinct scales of a and locally around the time of b. The result is a ripple/wavelet coefficient $W_{a,b}$ outline sketch. However, enumerating the wavelet/ripple coefficients at every likely scale (resolution level) demands a huge amount of data and calculation time. DWT analyzes a given time series with distinct resolutions for a distinct range of frequencies. This is done by decaying the data into coarse approximation and detail coefficients. For this, the scaling and wavelet/ripple functions are utilized. Choosing the scales a and the positions b based on the powers of two (binary scales and positions), DWT for a discontinuous time series f_i, becomes:

$$W_{m,n} = 2^{-\frac{m}{2}} \sum_{i=0}^{N-1} f_i \Psi^* \left(2^{-m}i - n\right) \tag{4}$$

where i is the integer time steps ($i = 0, 1, 2, ..., N-1$ and $N = 2^M$); m and n are integers that control, respectively, the scale and time; $W_{m,n}$ is the wavelet coefficient for the scale factor $a = 2^m$ and the time factor $b = 2^m n$. The original signal can be built back/recreated using the inverse discrete wavelet transform as follows:

$$f_i = A_{M,i} + \sum_{m=1}^{M} \sum_{n=0}^{(2^{M-m}-1)} W_{m,n} 2^{\frac{m}{2}} \Psi \left(2^{-m}i - n\right) \tag{5}$$

or in a simple form as:

$$f_i = A_{M,i} + \sum_{m=1}^{M} D_{m,i} \tag{6}$$

where $A_{M,i}$ is called an approximation sub-signal at level M and $D_{m,i}$ are the detail sub-signals at levels $m = 1, 2, ..., M$. The approximation coefficient $A_{M,i}$ represents the high-scale, low-frequency component of the signal, while the detailed coefficients $D_{m,i}$ represent the low-scale, high-frequency component of the signal.

There are a number of mother wavelets such as: Haar; Daubechies; Coiflet; and biorthogonal. Normally, Daubechies, belonging to the Haar wavelet, achieves improved results in sediment transport processes due to its inherent capacity to discover time localization information, such as dealing with the annual recurrence and hysteresis/lag phenomenon; the time localization information is beneficial in flow discharge and sediment processes. The different Daubechies wavelet families from [40] are shown in Figure 6. The Coiflet wavelet is more symmetrical than the Daubechies wavelet. Likewise, biorthogonal wavelets have the characteristic of the linear phase, which is required for signal rebuilding [29]. The appropriateness and selection of the mother wavelet are dependent on application type and characteristics of the data.

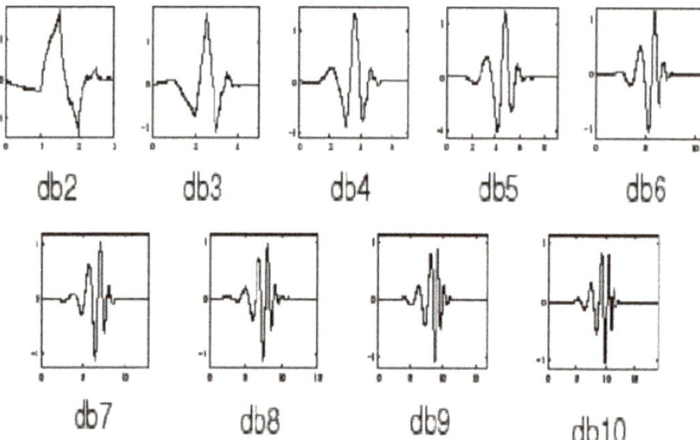

Figure 6. Daubechies wavelet families.

2.5.3. Combining Wavelet Analysis and Artificial Neural Networks

Wavelet transforms are mathematical tools that covert the one-dimensional time-domain signals into two-dimensional time-frequency-domain signals. The transformation separates significant changes in the time series in the form of high- and low-frequency signals. This property of wavelets is required for the identification of seasonality and hysteresis phenomenon in the data and helps ANNs to build a better relationship between inputs and sediment parameters. The level of transformation of signals depends on river properties, such as catchment, tributaries, lag-time, landslides, spatio-temporal sediment storage in tributaries, etc. Owing to the irregular and non-symmetric shape of the wavelets, their coupling with ANNs has been successful for filling missing sediment load data and for predictions in catchments where no land use/land cover changes occurred. There are many mother wavelets like Haar, Daubechies, or Coiflet. Application of the Daubechies wavelet using more than one decomposition level with a one-day lag-time has been proven more successful for the Upper Indus River [41]. We adopted the design of the WA-ANN model from [41], but extended the training period from 1969–2008 to 1969–2014.

3. Results and Discussion

In the numerical model, daily reservoir water levels (RWLs) of the Tarbela Reservoir were applied for the downstream boundary condition. At the upstream boundary, we specified daily inflows with corresponding sediment load. Modeled results were compared to observations and evaluated based on the statistical performance parameters like the coefficient of determination (R^2), the observations standard deviations ratio (RSR), and the Nash–Sutcliffe Efficiency (NSE).

Actual daily inflows of 14 years (2005–2018) were given as the upper flow boundary condition for running the model with the SRC-based sediment loads and were repeated thereafter up to 2030. For running the model with the WA-ANN-based sediment loads, actual daily inflows from 2005–2018 and thereafter futuristic flows from 2019–2030 as projected by [35] under plausible near-future climatic conditions were applied as upper boundary conditions. Actual daily RWLs of the Tarbela Reservoir were given as the downstream boundary condition up to 2018 and repeated thereafter for both SRC and WA-ANN runs of the model.

To check the performance of the SRC method (Equations (1) and (2)), sediment loads were generated and matched against observed sediment loads. The sediment equations output sediment load in t/day by the input of flow in m^3/s. The generated/estimated sediment load was matched against observed sediment load entering the reservoir for that particular day. The observed sediment load was calculated by converting observed sedimentation concentration in mg/L for that day into t/day by carrying out a dimensional analysis. The calculated values of NSE, R^2, and RSR were 0.635, 0.655, and 0.076, respectively, amply proving that the SRC technique, although in vogue, predicted output with an unacceptable level of certainty.

Applying the concept of data preprocessing on Besham Qila gauge station's data developed by [41], where he found the best relationship by selecting 70% of the input data for training, 15% for testing, and 15% for validation, we also obtained better results for the time period 1969–2014. The 70%, 15%, and 15% data from the entire available series was randomly selected for training, testing, and validation processes, respectively. It is also worthwhile to mention here that data pre-processing plays an important role where a short duration data series is available; however, our data series of more than 40 years also provided us the best results on even specifying 60% of data for training, 20% for testing, and the remaining 20% for validation. The coefficient of determination (R^2) for the training and testing datasets was 0.780 and 0.743, respectively. The Nash–Sutcliffe efficiency (NSE) was also 0.780 and 0.742 for training and testing, respectively. As our ANN trained best using single decomposition on Q(t), the inputs were only detailed and approximated coefficients of discharge without lag-time. The best trained WA-ANN used "tainsig" transfer functions in both the hidden and output layers. The number of hidden neurons in the single hidden layer of ANN was only five compared to seven for the same gauge station in [41]. As the Levenberg–Marquardt algorithm has fast convergence and also performed well for the Indus River [4], it also performed best in our training. The simulations stopped when the difference between the last and second to last simulation was less than 1/1000 or it reached maximum epochs of 1000 iterations. The work in [41] used the data series from 1969–2008 and reconstructed missing data for the Tarbela Reservoir with $R^2 = 0.773$ and 0.794 for testing and training, respectively. The statistical performance of our WA-ANN with a larger data series up to 2014 was slightly better for training data; however, it was slightly lower for testing data, which may be due to the inclusion of the exceptionally high flood of 2010. Similarly, increasing the decomposition levels slightly affected the model performance, which, interestingly, was significantly improved in [41]. In addition, the WA-ANN-generated sediment series showed an annual 160 Mt of suspended sediment load (excluding 10% bed load) entering the Tarbela Reservoir, which was similar to the estimate of [41].

3.1. Model Calibration

The model was calibrated for a period of nine years (2005–2013). The gradational analysis showed that on average, the Indus River transported silt (56.68%) as compared to sand (33.94%) and clay (9.78%).

Further, an extensive analysis of available particle size data of Besham Qila gauging station for 1983, 1989, 1991, 1994, and from 2002–2012 was conducted to calculate its variations with flow. Firstly, as mentioned in the previous paragraph, the average percentages for sand-, silt-, and clay-sized particles were calculated for all flow conditions. Then, the data were segregated into different sets corresponding to the indicated flow ranges in Table 4, and average percentages for sand-, silt-, and clay-sized particles were calculated for those particular flow ranges/bands. The analysis showed conclusively that the percentages of gradations across the sediment classes changed significantly with changing flow bands and were liable to affect sediment transport behavior as the flows increase/decrease. This analysis was important to study and model the morphodynamics across changing low and high flow bands accurately. The results are shown in Table 4 and entered in the sediment module of HEC-RAS as an adjunct to SRC and WA-ANN load series.

Table 4. Gradation percentages vis-à-vis increasing flow bands.

Flow Ranges (m^3/s)	Clay (%)	Silt (%)	Sand (%)
up to 1416	5.5	51.3	43.1
up to 2832	10.3	49.8	39.8
up to 4248	9.4	54.4	36.2
up to 5663	7.1	50.0	42.9
up to 7079	8.6	56.8	34.6
up to 8495	8.8	57.2	34.0
up to 9911	10.9	66.0	23.1
up to 11,327 and above	17.5	68.0	14.5

First, only hydrodynamic calibration was carried out up to 2013 by changing the value of Manning's roughness (n) throughout the length of the reservoir and comparing the calculated water levels with the observed water levels at different locations along the 66 available cross-sections. Initially, a uniform hydraulic roughness n = 0.04 from the literature [42,43] was adopted and subsequently adjusted in a plausible range of 0.035–0.04, throughout the 73 R/Lines of the reservoir and by comparing with available observed water levels, achieving an NSE and R^2 of 0.916 and 0.940, respectively. Next, hydro-morphodynamic calibration was attempted by varying both bed roughness and sediment parameters in the model.

Applying SRC sediment load at the inlet, it was noticed that the Ackers–White transport formula with the sorting method of Exner (7) was producing somewhat higher values of NSE and R^2. Exner (5) and Exner (7) are common bed sorting methods (sometimes called the mixing or armoring methods), which keep track of the bed gradation used by HEC-RAS to compute grain size-specific capacities and also to simulate armoring processes. Exner (5) uses a three-layer bed model that forms an independent coarse armor layer, which limits the erosion of deeper layers, whereas Exner (7) is an alternate version of Exner (5) designed for sand bed rivers as it forms armor layers more slowly and computes more erosion.

Hence, by keeping the combination of Ackers–White + Exner (7) constant, different fall velocities were tested to better the results. Amongst provisions to input commonly-used fall velocity methods like van Rijn, Ruby, and Tofaletti, HEC-RAS has an option to input the Report 12 fall velocity method, which finds solution iteratively by using the same curves as van Rijn, but using the computed fall velocity to compute the new Reynolds number until the assumed velocity matches with the computed velocity within tolerable limits. Consequently, a third tier calibration effort was attempted by varying scaling factors for transport and mobility functions of the transport formula as allowed by the HEC-RAS model for calibration fine-tuning, the result of which emerged with NSE and R^2 of 0.943 and 0.959, respectively. The default value of scaling factors was one, which was manipulated to achieve the maximum hydrodynamic calibration of NSE and R^2 of 0.996. It is worth mentioning here that for the sediment simulation and management study in Tarbela Reservoir in 1998 [44], the Ackers–White transport formula [45] was selected. The work in [43] also suggested the adoption of the Ackers–White formula, for the total load transport capacity of sand-sized fractions. However, other formulas were also tested in the calibration process as detailed in Table 5. A comparison with observed bed levels of 2013 was made and presented in Figure 7.

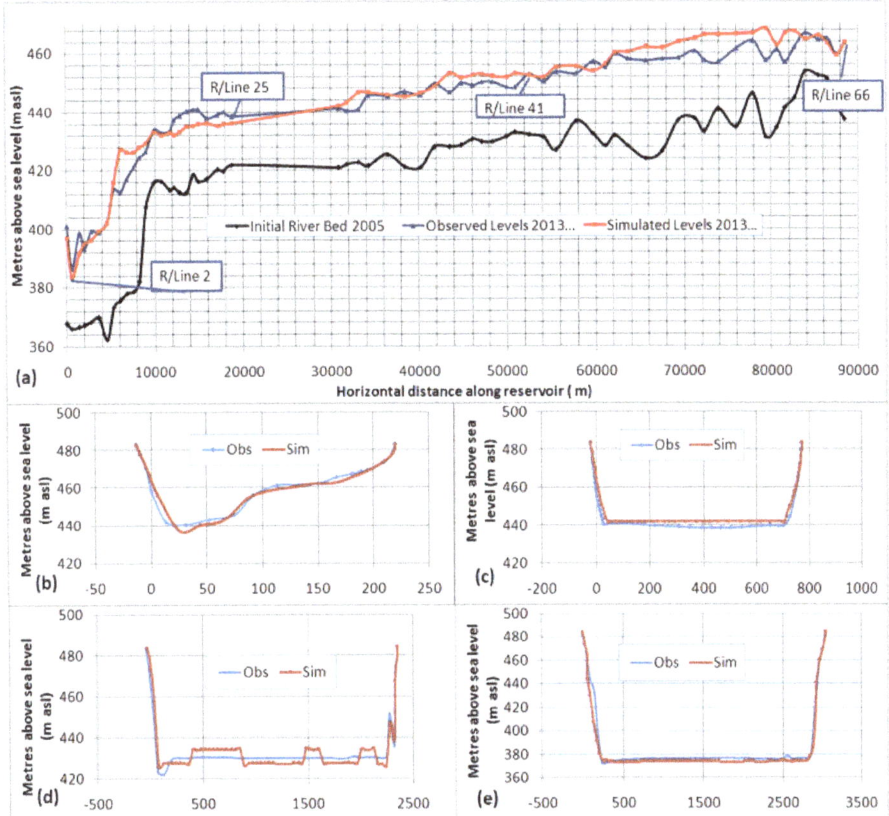

Figure 7. Comparison of observed and SRC simulated bed levels during calibration for 2013: (**a**) along the Tarbela Reservoir; (**b**) R/Line 66; (**c**) R/Line 41; (**d**) R/Line 25; (**e**) R/Line 2.

Table 5. Statistical performance of HEC-RAS with SRC sediment series by the input of different transport formulae and varying parameters.

Sed Transport Formulae	Sorting Method	Fall Velocity	Scaling Factors Applied	NSE	R^2
Yang	Exner (5)	van Rijn	No	0.817	0.943
Laursen-Copeland	Exner (5)	van Rijn	No	0.859	0.948
Engelund-Hansen	Exner (5)	van Rijn	No	0.867	0.950
Ackers–White	Exner (5)	van Rijn	No	0.869	0.952
Ackers–White	Exner (7)	van Rijn	No	0.896	0.956
Ackers–White	Exner (7)	Ruby	No	0.897	0.955
Ackers–White	Exner (7)	Report 12	No	0.898	0.956
Ackers–White	Exner (7)	Tofaletti	No	0.908	0.964
Ackers–White	Exner (7)	Tofaletti	Yes	0.943	0.959

Further, another extensive calibration exercise was carried out applying WA-ANN-based boundary conditions. Again, the Ackers–White transport formula with the sorting method of Exner (5) showed better results. Next, the above combination (Ackers–White + Exner-5) was evaluated by changing the fall velocity equations. Similar to the SRC case, the Tofaletti technique showed the best results hitherto, prior to application of scaling factors. Consequently, the best combination of input parameters (Ackers–White + Exner-7 + Tofaletti) was subjected to rigorous scaling of transport formula parameters. Hence, the highest NSE of 0.979 was achieved during calibration, and the results of the exercise tabulated in Table 6 in increasing order of NSE values. A comparison with observed bed levels of 2013 was made and presented in Figure 8.

Table 6. Statistical performance of HEC-RAS with WA-ANN sediment series by the input of different transport formulae and varying parameters.

Sed Transport Formulae	Sorting Method	Fall Velocity	Scaling Factors Applied	NSE	R^2
Ackers–White	Exner (7)	van Rijn	No	0.829	0.975
Laursen-Copeland	Exner (7)	van Rijn	No	0.830	0.975
Yang	Exner (7)	van Rijn	No	0.830	0.974
Engelund-Hansen	Exner (7)	van Rijn	No	0.831	0.976
Yang	Exner (5)	van Rijn	No	0.832	0.966
Engelund-Hansen	Exner (5)	van Rijn	No	0.855	0.969
Laursen-Copeland	Exner (5)	van Rijn	No	0.863	0.970
Ackers–White	Exner (5)	Report 12	No	0.869	0.971
Ackers–White	Exner (5)	Ruby	No	0.869	0.970
Ackers–White	Exner (5)	van Rijn	No	0.870	0.970
Ackers–White	Exner (5)	Tofaletti	No	0.876	0.972
Ackers–White	Exner (5)	Tofaletti	Yes	0.979	0.980

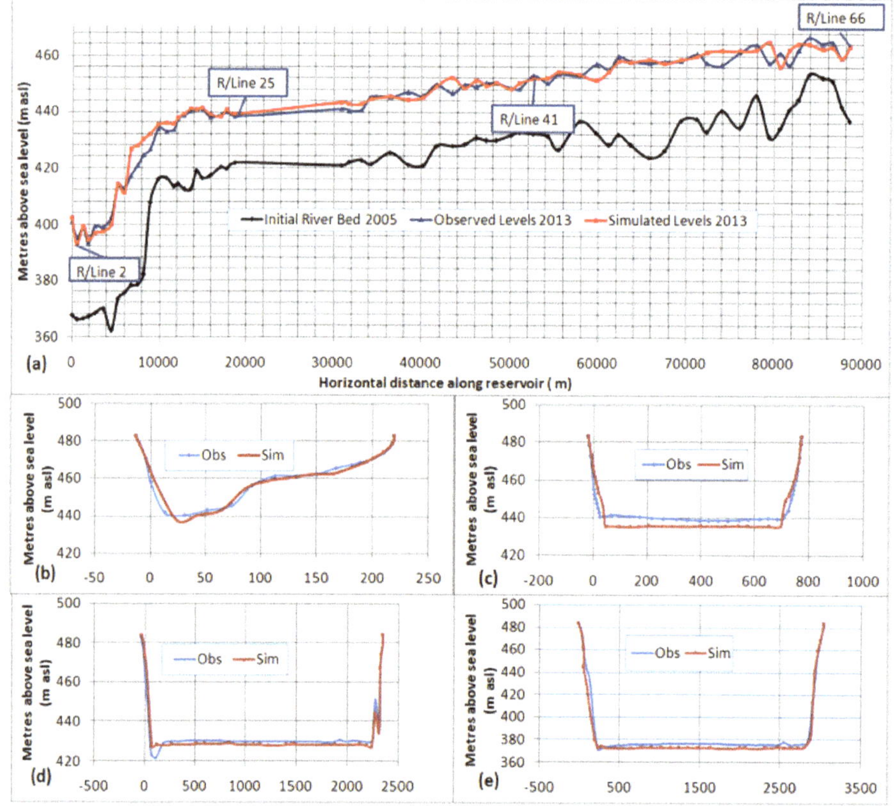

Figure 8. Comparison of observed and WA-ANN-simulated bed levels during calibration for 2013: (**a**) along the Tarbela Reservoir; (**b**) R/Line 66; (**c**) R/Line 41; (**d**) R/Line 25; (**e**) R/Line 2.

3.2. Model Validation

To validate the HEC-RAS model with the SRC technique, it was run for another four years up to 2017. The output was compared with observed sediment deposits of 2017 and is presented in Figure 9.

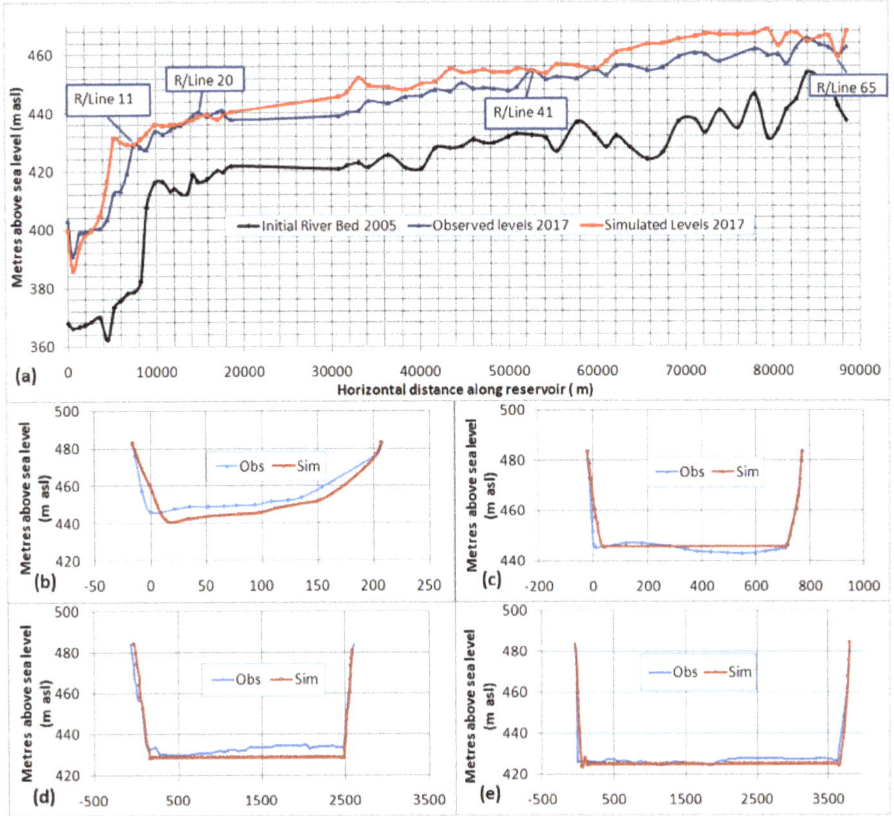

Figure 9. Comparison of observed and SRC simulated bed levels during validation for 2017: (**a**) along the Tarbela Reservoir; (**b**) R/Line 65; (**c**) R/Line 41; (**d**) R/Line 20; (**e**) R/Line 11.

The R^2 and NSE in the validation process were 0.950 and 0.893, respectively. The observed standard deviation was at 0.041. In a recent study [46], the HEC-RAS model was validated for the Tarbela Reservoir by simulating it only for one year, and an approximately 20-m difference between the observed and simulated river beds for the sediment delta in the year 2000 was found. However, in the present study, the difference of four years of simulation was only 4–5 m in the whole longitudinal profile (Figure 9). A better modeling performance might be due to more accurate sediment load boundary conditions generated using a long-term data series, i.e., 1969–2014, whereas [46] used only a 28-year data series, i.e., 1979–2006.

To validate the HEC-RAS model with the above calibrated WA-ANN sediment series, it was run for another four years up to 2017, similar to the SRC model. The output was compared with observed sediment deposits of 2017 and presented in Figure 10. The R^2 and NSE in the validation process were 0.968 and 0.959, respectively. The observed standard deviation was at 0.025.

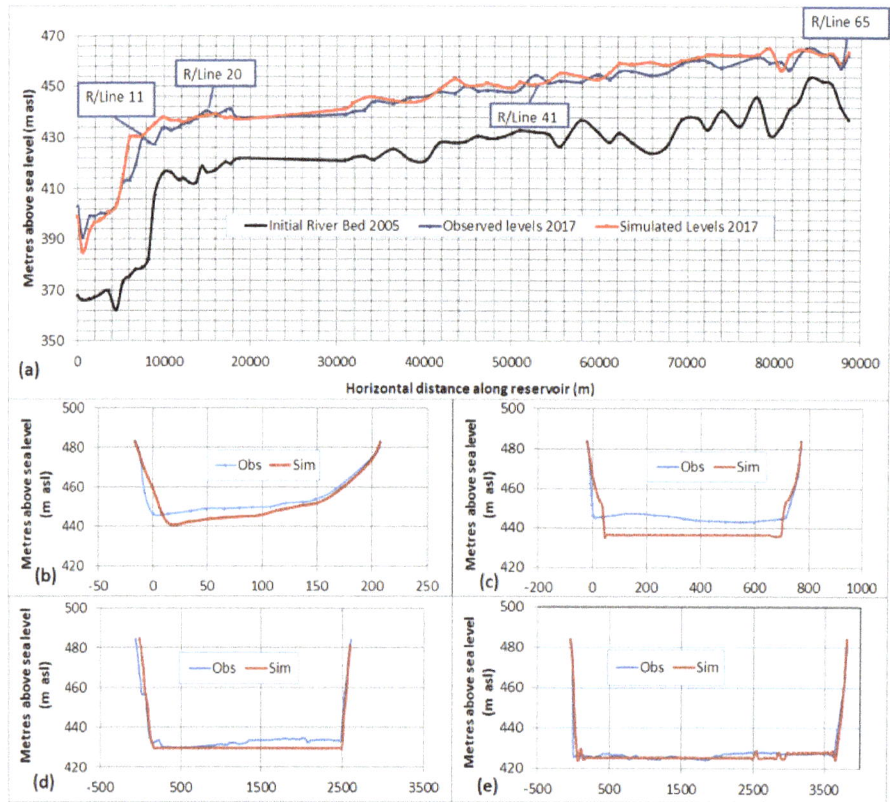

Figure 10. Comparison of observed and WA-ANN-simulated bed levels during validation for 2017:
(**a**) along Tarbela Reservoir; (**b**) R/Line 65; (**c**) R/Line 41; (**d**) R/Line 20; (**e**) R/Line 11.

3.3. HEC-RAS Model Performance with the SRC and WA-ANN Techniques

To sum up the above-elaborated calibration and validation exercises using SRC and WA-ANN-based boundary conditions, their statistical performance was compared and tabulated in Table 7. The statistical results (Table 7) clearly indicated a preferable performance of the model using WA-ANN-based sediment load boundary conditions. As SRC reconstructed the missing sediment load data with R^2 and NSE at 0.635 and 0.655, respectively, the model calibration took a long time to adjust transport parameters for attaining stability. However, due to better recondition accuracy using WA-ANN ($R^2 = 0.771$ and NSE = 0.771), the HEC-RAS model simulated the bed-levels changes with great stability. As the SRC overestimated sediment load, therefore to flush extra sediments, we needed to adjust the transport parameters that might not represent the correct physics of the transport processes in the reservoir. Therefore, more accurate boundary conditions played a vital role in precise modeling of the transport processes by keeping transport parameters within the physical limits.

Table 7. Statistical performance of HEC-RAS model with the SRC and WA-ANN techniques during the calibration and validation periods.

Process	Duration	R^2		RSR		NSE	
		SRC	WA-ANN	SRC	WA-ANN	SRC	WA-ANN
Calibration	2005–2013	0.959	0.980	0.030	0.018	0.943	0.979
Validation	2014–2017	0.950	0.968	0.041	0.025	0.893	0.959

3.4. Models' Application for Sediment Delta Prediction

More than 200 million people of Pakistan directly or indirectly depend on the irrigation supply and power generation from the Tarbela Dam. Therefore, it is very important to assess the future delta movement and sedimentation scenarios in the reservoir. It is pertinent to mention here that SRC-generated sediment load boundary conditions are being used for all types of sedimentation modeling in the upper Indus River projects [21,42]. Therefore, to check and ascertain the long-term application of the SRC and WA-ANN techniques, the HEC-RAS model was run up to the year 2030 using future discharges calculated by [35] employing the University of British Columbia (UBC) watershed model. UBC is a less data-extensive semi-distributed watershed model developed by the University of British Columbia. As discharge alone with one level of decomposition represents more accurately the transport processes at Besham Qila, calculated future discharges by [35] were used in the trained WA-ANN model for obtaining future sediment loads. Reservoir water levels from 2005–2018 were repeated for 2019–2030. The simulated/forecasted levels of the Tarbela Reservoir for 2022 and 2030 along with observed levels of 2013 and 2017 showed a huge volume lost due to sedimentation (see Figure 11). As SRC showed overestimation (190 Mt of suspended sediment load (SSL) compared to 160 Mt SSL using WA-ANN) for the Indus River (Table 1), therefore, using SRC as the boundary condition in the modeling process also overestimated the bed level variations in the major ponding area of the reservoir near the dam. As SRC has been used for sedimentation modeling of all studies of the Upper Indus River, and it has been predicting similar results. For example, the 4320-MW Dasu Hydropower Project, which is under construction upstream of Tarbela Dam, will be silted up just 20–25 years after its commissioning without conducting yearly flushing operations [21,41]. The predicted short life of the Dasu project could very likely be a result of the overestimation of sediment load using SRCs. Initially, the work in [13] in 1970 also estimated 400 Mt of sediment load using SRC for the Tarbela Reservoir, which showed a shorter life of the reservoir. However, later studies estimated 50% lower sediment load for the Indus River at Tarbela Dam (see Table 1). Due to less sediments entering the reservoir, it is still operational and not silted-up. It might be possible that in 1970, very limited sediment concentration data were available, which might have consisted of high-flow hydrological years. However, the availability of long data series of sediment sampling cannot help SRC to model the hysteresis phenomenon and hydrological variations related to shifting in high flows from summer to post- and pre-summer months at the Upper Indus Basin [41]. Therefore, the WA-ANN-generated sediment load boundary condition, using future projected discharges, can more precisely represent the sedimentation modeling processes.

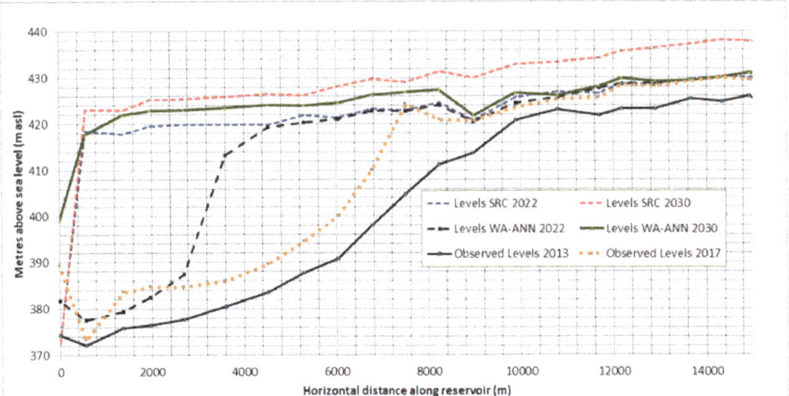

Figure 11. Comparison of simulated bed levels for 2022 and 2030 with the SRC and WA-ANN techniques, along with observed levels of 2013 and 2017. The longitudinal profile is only showing the sediment delta region of Tarbela Dam.

4. Conclusions

In the present study, the performance of HEC-RAS 1D model for modeling morphodynamic processes in the Tarbela Reservoir was tested using sediment rating curves and WA-ANN-based sediment load boundary conditions. A data series from 2005–2013 was used in the calibration, while a data series from 2014–2017 was used in the validation process. Based on the study results, the following conclusions can be drawn:

1. Compared to sediment rating curves, the WA-ANN model better represented the hysteresis phenomenon for the Indus River and could reconstruct the missing sediment load more accurately.
2. More accurate sediment load boundary conditions enabled the numerical model to calculate the bed level changes more precisely and also to provide stability in the calculation process. By comparing Figures 7d and 8d, it is evident that the simulated bed with WA-ANN showed stability against the SRC-simulated bed.
3. A de-synchronization between glacier melt and rainfall in the upper Indus catchment will cause a decrease in sediment to the Tarbela Reservoir and will decrease the sedimentation rate.

On the basis of the above conclusions, the following recommendations are being put forward:

1. Sediment rating curves should not be utilized to design the reservoir sediment management rules for the existing, under construction, or planned projects in the upper Indus Basin, as they cannot adjust the hysteresis phenomenon and contribute variations in the sediment load boundary conditions.
2. As we have repeated 2005–2018 reservoir operational rules for 2019–2030 in the modeling process, the future reservoir operational rules should be optimized to keep sediment delta stable.

Author Contributions: Z.R.T. defined the problem, outlined the research plan, carried out the research, and interpreted and drafted the outcome. S.R.A. calculated and analyzed flow and sediment concentration data, prepared SRCs, and made significant contributions to improve the draft. S.u.H. contributed the hydrological model for future flows of the Indus River. S.A.-U.-R. and M.D.B. developed WA-ANN models for setting sediment load boundary conditions and helped in the preparation of this paper with proof reading and corrections. R.M.A.W. collected/collated all the relevant data from different sources. Z.M.K. and I.A. supervised and directed the research study and reviewed the end result critically.

Funding: This research received no external funding.

Acknowledgments: The study data were provided by the Tarbela Dam Organisation of WAPDA upon specific request of College of Earth and Environmental Sciences (CEES). We appreciate their help and cooperation. Shabeh ul Hasson also acknowledges the support and funding from the Deutsche Forschungsgemeinschaft (DFG, German Research Foundation) through the Cluster of Excellence 'CliSAP' (EXC177), and under Germany's Excellence Strategy –EXC 2037 'CLICCS - Climate, Climatic Change, and Society' –Project Number: 390683824, contribution to the Center for Earth System Research and Sustainability (CEN) of Universität Hamburg. The corresponding author is also grateful to Rohit Kumar, DNZE Munich, for his help.

Conflicts of Interest: The authors declare no conflicts of interest.

References

1. Hassan, M.; Shamim, M.A.; Sikandar, A.; Mehmood, I.; Ahmed, I.; Ashiq, S.Z.; Khitab, A. Development of sediment load estimation models by using artificial neural networking techniques. *Environ. Monit. Assess.* **2015**, *187*, 1–13. [CrossRef] [PubMed]
2. Colby, B. *Relationship of Sediment Discharge to Streamflow*; United States Geological Survey: Reston, VA, USA, 1956.
3. Glysson, G. *Sediment-Transport Curves*; United States Geological Survey: Reston, VA, USA, 1987.
4. Ateeq-Ur-Rehman, S.; Bui, M.D.; Rutschmann, P. Variability and trend detection in the sediment load of the Upper Indus River. *Water* **2018**, *10*, 1–16. [CrossRef]
5. Ateeq-Ur-Rehman, S.; Bui, M.D.; Hasson, S.U.; Rutschmann, P. An Innovative Approach to Minimizing Uncertainty in Sediment Load Boundary Conditions for Modeling Sedimentation in Reservoirs. *Water* **2018**, *10*, 1411. [CrossRef]

6. Survey and Hydrology. *Annual Report on Tarbela Reservoir Sedimentation—2011*; Water and Power Development Authority (WAPDA): Lahar, Pakistan, 2012.

7. Asianics Agro-Dev. International (Pvt) Ltd. *Tarbela Dam and Related Aspects of the Indus River Basin, Pakistan, A WCD Case Study Prepared as an Input to the World Commission on Dams*; Asianics Agro-Dev. International (Pvt) Ltd.: Islamabad, Pakistan, 2000.

8. Office of Superintending Engineer, Survey & Hydrology Residency. *Tarbela Reservoir Capacity Table—2019*; Wapda: Lahore, Pakistan, 2019.

9. Consultants, G.G.H. *Technical Report No. 3, Sedimentology of Ghazi-Garriala Hydropower Project, Feasibility Report*; Technical Report; Water and Power Development Authority: Lahore, Pakistan, 1991.

10. Consultants, D.B. *Reservoir Sedimentation Studies, Appendix C to Reservoir Operation and Sediment Transport*; Technical Report; Water and Power Development Authority, Lahar, Pakistan, 2007.

11. Ateeq-Ur-Rehman, S.; Bui, M.D.; Rutschmann, P. Detection and estimation of sediment transport trends in the upper Indus River during the last 50 years. *Hydropower: A Vital Source of Sustainable Energy for Pakistan*; Sarwar, K., Ahmad, I., Eds.; Center of Excellence in Water Resources Engineering: Lahore, Pakistan, 2017; pp. 1–6.

12. Holeman, J.N. The sediment yield of major rivers of the world. *Water Resour. Res.* **1968**, *4*, 737–747. [CrossRef]

13. Peshawar University. *The Sediment Load and Measurements for Their Control in Rivers of West Pakistan*; Department of Water Resources, Peshawar, Pakistan, 1970.

14. Meybeck, M. Total mineral dissolved transport by world major rivers/Transport en sels dissous des plus grands fleuves mondiaux. *Hydrol. Sci. J.* **1976**, *21*, 265–284. [CrossRef]

15. Lowe, J.; Fox, I. Sedimentation in Tarbela reservoir. In *Commission Internationale des Grandes Barrages (International Commission on Large Dams-ICOLD, Paris, France)*. Quatorzieme Congres des Grands Barrages: Rio de Janeiro, Brazil, 1982.

16. Roca, M.; Tarbela Dam in Pakistan. Case study of reservoir sedimentation. In *Proceedings of River Flow*; Munoz, R.E.M., Ed.; CRC Press: London, UK, 2012.

17. Milliman, J.; Quraishee, G.; Beg, M. Sediment discharge from the Indus River to the ocean: Past, present and future. In *Marine Geology and Oceanography of Arabian Sea and Coastal Pakistan*; Haq, B.U., Milliman, J.D., Eds.; van Nostrand Reinhold/Scientific and Academic Editions: New York, NY, USA, 1984; pp. 65–70.

18. Summerfield, M.A.; Hulton, N.J. Natural controls of fluvial denudation rates in major world drainage basins. *J. Geophys. Res.-Solid Earth* **1994**, *99*, 13871–13883. [CrossRef]

19. Collins, D. *Hydrology of Glacierised Basins in the Karakoram: Report on Snow and Ice Hydrology Project in Pakistan with Overseas Development Administration, UK [ODA] and Water and Power Development Authority, Pakistan [WAPDA]*; University of Manchester: Manchester, UK, 1994.

20. Ali, K.F. Construction of Sediment Budgets in Large Scale Drainage Basins: The Case of the Upper Indus River. Ph.D. Thesis, Department of Geography and Planning, University of Saskatchewan, Saskatoon, SK, Canada, 2009.

21. Dasu Hydropower Consultants. *Detailed Engineering Design Report, Part A; Engineering Design*; Water and Power Development Authority: Lahar, Pakistan, 2013.

22. Graf, W.H. *Hydraulics of Sediment Transport*; Water Resources Publication: Littleton, CO, USA, 1984.

23. McBean, E.A.; Al-Nassri, S. Uncertainty in suspended sediment transport curves. *J. Hydraul. Eng.* **1988**, *114*, 63–74. [CrossRef]

24. Morris, G.L.; Fan, J. *Reservoir Sedimentation Handbook: Design and Management of Dams, Reservoirs, and Watersheds for Sustainable Use*; McGraw Hill Professional: New York, NY, USA, 1998.

25. Bogen, J. The hysteresis effect of sediment transport systems. *Nor. Geogr. Tidsskr.-Nor. J. Geogr.* **1980**, *34*, 45–54. [CrossRef]

26. Hewitt, K. Gifts and perils of landslides: Catastrophic rockslides and related landscape developments are an integral part of human settlement along upper Indus streams. *Am. Sci.* **2010**, *98*, 410–419.

27. Partal, T.; Kucuk, M. Long-term trend analysis using discrete wavelet components of annual precipitations measurements in Marmara region (Turkey). *Phys. Chem. Earth* **2006**, *31*, 1189–1200. [CrossRef]

28. Adamowski, K.; Prokoph, A.; Adamowski, J. Development of a new method of wavelet aided trend detection and estimation. *Hydrol. Process.* **2009**, *23*, 2686–2696. [CrossRef]

29. Railean, I.; Moga, S.; Borda, M.; Stolojescu, C.L. A wavelet based prediction method for time series. In *Stochastic Modeling Techniques and Data Analysis (SMTDA2010)*; Janssen, J., Ed.; ASMDA International: Chania, Greece, 2010.

30. Tan, C.; Huang, B.S.; Liu, K.S.; Chen, H.; Liu, F.; Qiu, J.; Yang, J.X. Using the wavelet transform to detect temporal variations in hydrological processes in the Pearl River, China. *Quat. Int.* **2017**, *440*, 52–63. [CrossRef]

31. Tarar, Z.R.; Ahmad, S.R.; Ahmad, I.; Majid, Z. Detection of Sediment Trends Using Wavelet Transforms in the Upper Indus River. *Water* **2018**, *10*, 918. [CrossRef]

32. Brunner, G.W. *HEC-RAS River Analysis System, Hydraulic Reference Manual*; US Army Corps of Engineers: Washington, DC, USA, 2016.

33. Hewitt, K. The Karakoram Anomaly? Glacier Expansion and the 'Elevation Effect', Karakoram Himalaya. *Mt. Res. Dev.* **2005**, *25*, 332–340. [CrossRef]

34. Lutz, A.F.; Immerzeel, W.W.; Kraaijenbrink, P.D.A.; Shrestha, A.B.; Bierkens, M.F.P. Climate Change Impacts on the Upper Indus Hydrology: Sources, Shifts and Extremes. *PLoS ONE* **2016**, *11*, e0165630. [CrossRef] [PubMed]

35. Hasson, S.U. Future Water Availability from Hindukush-Karakoram-Himalaya upper Indus Basin under Conflicting Climate Change Scenarios. *Climate* **2016**, *4*, 40. [CrossRef]

36. Wijngaard, R.R.; Lutz, A.F.; Nepal, S.; Khanal, S.; Pradhananga, S.; Shrestha, A.B.; Immerzeel, W.W. Future changes in hydro-climatic extremes in the Upper Indus, Ganges, and Brahmaputra River basins. *PLoS ONE* **2017**, *12*, e0190224. [CrossRef]

37. Bonekamp, P.N.J.; de Kok, R.J.; Collier, E.; Immerzeel, W.W. Contrasting Meteorological Drivers of the Glacier Mass Balance Between the Karakoram and Central Himalaya. *Front. Earth Sci.* **2019**, *7*, 107. [CrossRef]

38. Organization, W.M. *WMO Statement on the State of the Global Climate in 2018*; World Meteorological Organization (WMO): Geneva, Switzerland, 2019.

39. Nash, J.; Sutcliffe, J. River flow forecasting through conceptual models, part I: A discussion of principles. *J. Hydrol.* **1970**, *10*, 282–290. [CrossRef]

40. Dixit, A.; Majumdar, S. Comparative analysis of Coiflet and Daubechies wavelets using global threshold for image denoising. *Int. J. Adv. Eng. Technol.* **2013**, *6*, 2247–2252.

41. Ateeq-Ur-Rehman, S. Numerical Modeling of Sediment Transport in Dasu-Tarbela Reservoir Using Neural Networks and TELEMAC Model System. Ph.D. Thesis, Chair of Hydraulic and Water Resources Engineering, Technical University of Munich (TUM), Munich, Germany, 2019.

42. Dasu Hydropower Project. *Dasu Hydropower Project, Feasibility Report*; Water and Power Development Authority: Lahore, Pakistan, 2009.

43. HR Wallingford. *Sedimentation Study, Dasu Hydropower Project, Pakistan*; Report EX 6801 R1; Dasu Hydropower Consultants: Pakhtunkhwa, Pakistan, 2012.

44. Wallingford, T.C.I.H. *Tarbela Dam Sediment Management Study*; Wapda: Lahore, Pakistan, 1998; Volume 2.

45. Ackers, P.; White, W.R. Sediment Transport: New Approach and Analysis. *ASCE Hydr. Div. J.* **1973**, *99*, 2041–2060.

46. Rehman, H.U.; Rehman, M.A.; Naeem, U.A.; Hashmi, H.N.; Shakir, A.S. Possible options to slow down the advancement rate of Tarbela delta. *Environ. Monit. Assess.* **2018**, *190*, 39. [CrossRef] [PubMed]

Article

Seasonal Precipitation Variability and Gully Erosion in Southeastern USA

Ingrid Luffman * and Arpita Nandi

Department of Geosciences, East Tennessee State University, Johnson City, TN 37614, USA; nandi@etsu.edu
* Correspondence: luffman@etsu.edu

Received: 11 February 2020; Accepted: 20 March 2020; Published: 25 March 2020

Abstract: This study examines the relationship between gully erosion in channels, sidewalls, and interfluves, and precipitation parameters (duration, total accumulation, average intensity, and maximum intensity) annually and seasonally to determine seasonal drivers for precipitation-related erosion. Ordinary Least Square regression models of erosion using precipitation and antecedent precipitation at weekly lags of up to twelve weeks were developed for three erosion variables for each of three geomorphic areas: channels, interfluves, and sidewalls (nine models in total). Erosion was most pronounced in winter months, followed by spring, indicating the influence of high-intensity precipitation from frontal systems and repeated freeze-thaw cycles in winter; erosion in summer was driven by high-intensity precipitation from convectional storms. Annually, duration was the most important driver for erosion, however, during winter and summer months, precipitation intensity was dominant. Seasonal models retained average and maximum precipitation as drivers for erosion in winter months (dominated by frontal systems), and retained maximum precipitation intensity as a driver for erosion in summer months (dominated by convectional storms). In channels, precipitation duration was the dominant driver for erosion due to runoff-related erosion, while in sidewalls and interfluves intensity parameters were equally important as duration, likely related to rain splash erosion. These results show that the character of precipitation, which varies seasonally, is an important driver for gully erosion and that studies of precipitation-driven erosion should consider partitioning data by season to identify these drivers.

Keywords: gully erosion; seasonality; precipitation; statistical modeling; precipitation intensity

---

1. Introduction

Gully erosion is a global problem, particularly in the southeastern United States, where erodible soils, high relief, and climatic and meteorological factors encourage soil erosion. Gully erosion is one of the most dangerous forms of soil degradation, which is caused by natural and anthropogenic activities. Gullies are composed of several continuous or discontinuous channels and rills with varying slopes, which may later develop into deep trenches, inhibiting effective remediation by tillage. Gully erosion can initiate from anthropogenic factors like farming or grazing on susceptible soils, increased runoff from land-use changes due to logging or construction, and poor vegetative cover from wildfire or high soil salinity. Additionally, natural drivers for soil erosion are meteorological variables, topography, and soil type and texture [1,2].

Changes in land use can increase soil erosion. Vast regions of the United States experienced soil erosion when forested lands were converted to croplands in the late 19th century and the early 20th century [3]. Estimates of the volume of soil erosion in the United States caused by both sheet and rill erosion combined is 6.7 Mg/ha/y in cultivated cropland, 0.90 Mg/ha/y on federal lands, and 1.55 Mg/ha/y in pasture lands [3]. Considerable land area in the southeastern US was converted from forest to agriculture to support cotton farming in the 1800s and pasture for animal grazing [4,5].

Land cover change due to logging and conversion of forest to crop and pasture was linked to nineteenth-century European settlement in the southern Blue Ridge Mountains and Appalachian hillslopes [6–9]. Harvesting on the steep Appalachian hillslopes has been identified as one potential cause of soil erosion [10]. After recognizing the problem as early as 1933, soil conservation programs were implemented in the United States. As part of present soil conservation efforts, afforestation on the reclaimed land has partially halted erosion, but severe erosional areas from the past cotton farming era are still prominent [4]. Some researchers have described a multi-stage formation of severe soil erosion [11,12], i.e., a process for gully development in the Appalachian Piedmont: (i) rills and gullies are initiated along existing paths, tracks, ditches, or animal burrows, where runoff is concentrated due to reduced infiltration; (ii) head scarp erosion begins as runoff gains energy and is concentrated in steeply sloped land; (iii) gully downcutting eventually stabilizes when weathered bedrock and the shallow groundwater zone are encountered; (iv) erosion continues laterally along channel sidewalls and headwalls by slumping and under caving, inhibiting effective control by tillage. Hence, reclamation can be expensive.

Sidewalls (or midslopes), gully channels (or valleys), and interfluves (or gully divides) are major topographical factors that influence soil erosion [7,8,13]. Soil erosion increases with slope steepness, which is more relevant to gully sidewall erosion and less relevant to interfluves. Gully channels are dynamic and can serve as intermittent sediment sinks and sources, transporting sediment to the gully outlet [14].

In addition to land cover change and topographic variation, water-induced soil erosion from severe precipitation events erodes fertile soil, mainly in areas with poor agricultural management, land degradation from mining, road construction, or wild fires [2]. Unique climatic conditions in the humid subtropical climate (Köppen Cfa) of the southeastern United States are a major contributing factor in gully erosion [15]. Cold periods in the south are short and winters are mild, inhibiting deeper ground freezing. The thin surface layer (5–10 cm) of frost-heaved soil becomes loose after a few freeze-thaw cycles, and can erode easily from subsequent heavy rain or snow-melt runoff [13]. During warm periods, intensive rainfall that falls on steep, sparsely vegetated slopes contributes to erosion. General precipitation trends in the Appalachian hillslopes indicate that high-intensity events occur more during summer months, while higher accumulation low-intensity storms are more prevalent in winter months. Seasonal variability in precipitation characteristics impacts erosion, but the extent and nature of this relationship are not well understood in this region.

A short-term study of hillslope erosion in the Appalachians found that duration and accumulation of precipitation were more important than storm intensity as drivers for gully erosion [13]. The same study also found antecedent precipitation is a stronger predictor of erosion and discrete precipitation events alone may not result in measurable erosion. Antecedent precipitation along with successive precipitation events can saturate the soil, reduce shear strength, and cause erosion. To examine inter-annual variability and longer-term effects from antecedent precipitation, as well as the influence of seasonal events on soil erosion, a more extensive time series of precipitation and corresponding erosion data is necessary [16], however, it will be important to retain a high temporal resolution in the data to assess seasonal scale patterns.

In this context, the Appalachian hillslopes in the southern US are representative of a region of historic and modern land degradation from unique meteorological conditions, variable topography, and land use/land cover change. Therefore, the objective of the present study is to examine the effect of meteorological parameters, specifically precipitation, on soil erosion through long term high-resolution monitoring. This paper summarizes six years of comprehensive weekly monitoring of precipitation events and soil erosion in an Appalachian hillslope paying particular attention to seasonal effect. An understanding of the seasonal pattern of soil erosion with respect to precipitation-related drivers of erosion will improve the potential to achieve conservation measures.

2. Materials and Methods

The study site was a system of branching gullies located on a hillslope at the East Tennessee State University Valleybrook research facility in northeast Tennessee, USA (+36°25′36.77″, −82°32′10.63″) at an elevation of 530 m (Figure 1). The site was within the Appalachian Valley and Ridge physiographic province and consisted of northeast-southwest trending parallel limestone valleys (Maynardville Formation) and sandstone or shale ridges (Nolichucky Formation) [17]. The 1.5 ha study area was located on a grass and shrub hillslope surrounded by forest (on the ridges) and pasture (in the valleys). Soils were highly erodible fine-grained silt and clay Ultisols (Collegedale-Etowah complex (CeD3)) with an average erodibility factor (RUSLE K-factor) of 0.28, indicating susceptibility to raindrop impact and transport by surface runoff [18]. The region has a humid subtropical climate (Köppen Cfa) with year-round precipitation of 1070 mm (42 in) annually and an average annual temperature range from 1.1 °C (34 °F) in January to 23.3 °C (74 °F) in July. The National Oceanic and Atmospheric Administration describe Tennessee's winter precipitation as dominated by the polar front and summer precipitation that results from convectional systems. September and October are the driest months.

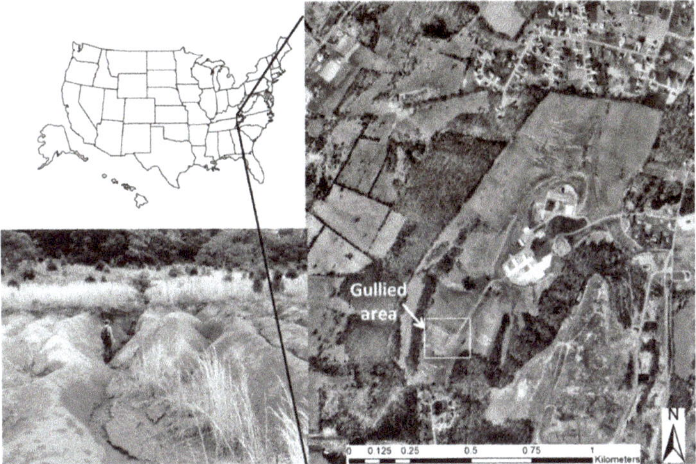

Figure 1. The study area was located in northeast Tennessee, USA on an actively eroding hillslope.

A detailed description of the site setup can be found in [13,19] and is summarized as following. Steel erosion pins were installed in transects throughout the 100 m × 100 m gullied zone. Each transect spanned interfluves, sidewalls, and the gully channel to assess erosion in these three morphological settings. In total, 105 erosion pins were installed, 34 (1 m × 5 mm) pins in channels, and the remaining (0.5 m × 5 mm) pins in interfluves (29 pins) and sidewalls (42 pins). From 23 May 2012 to 22 August 2018, pin length was recorded approximately weekly for each pin using a folding ruler. Pin attrition occurred periodically over the study period, such that some pins were eroded, damaged, or dislodged by animals. Therefore, in May 2015, 43 new pins were installed and 3 damaged pins were replaced, bringing the total number of pins to 105. The nature of the site surface limited access during and immediately after rain events, and over the six-year period, pin length was recorded 294 times. The difference between the exposed lengths of each pin was calculated between one measurement period and the next, and this dataset of pin change was compared to precipitation data to identify important drivers for erosion in each morphological setting.

For each setting, we created three erosion variables: (1) average of the absolute value of change (Avg|Ch|); (2) average of only positive changes in pin lengths (deposition) from one measurement period to the next (AvgDep), and; (3) average of only negative changes in pin lengths (erosion) from one measurement period to the next (AvgErosion). In prior research, a fourth variable, average change, was generated, however, because of a balance of erosion and deposition, especially in channels, the average change remained near zero and was not a useful parameter to capture weekly and longer-term erosion on-site [13,14,19–21]. Therefore, in this study, we have retained the three variables described above.

A Davis Vantage Pro wireless weather station (KTNJONES12, data available at https://www.wunderground.com/dashboard/pws/KTNJONES12) was located 350 m from the research site, and recorded precipitation, pressure, temperature, and wind data at five-minute intervals. Occasional data gaps were filled with data from a neighboring station 1.6 km away (KTNJONES7, data available at https://www.wunderground.com/dashboard/pws/KTNJONES7), with only 21 of 2282 study days missing weather data. See [19] for a detailed list of weather data gaps and coverage.

From these data, four precipitation parameters were generated for each measurement period: (1) Duration (total minutes of rainfall); (2) Total Accumulation (total precipitation in mm); (3) Average Intensity in mm/min (Total Accumulation/Duration), and; (4) Maximum Intensity in mm/min (the greatest station-reported rain rate during the measurement period). The rain rate is a smoothed function of rain accumulation over time that is calculated using the ratio of the tipping bucket depth-adjusted volume to the time between tips. As rainfall tapers off, the rate drops but does not reach zero immediately upon cessation of precipitation. Instead, it smooths the rate to more accurately represent how precipitation naturally tapers over an area at the end of a rain storm [22].

Prior research has shown that antecedent precipitation may be an important factor in erosion, and therefore a series of antecedent precipitation parameters were generated for the prior eleven measurement periods, for each of Duration, Total Accumulation (TotAcc), Average Intensity (AvgInt), and Maximum Intensity (MaxInt). These antecedent lagged variables were named Duration-1, Duration-2 ... Duration-11, TotAcc-1, TotAcc-2 ... and so-on, a total of 48 precipitation parameters, which we refer to as lagged precipitation parameters.

The relationship between erosion variables and all precipitation parameters was assessed with Spearman correlation coefficients. Ordinary Least Squares (OLS) regression models were created for the nine erosion variables using the set of current and lagged precipitation parameters. Further, because seasonal variability in erosion was observed in prior studies [13,19], the data were partitioned by season: winter (December, January, February); spring (March, April, May); summer (June, July, August); and autumn (September, October, November). OLS regression models were generated for the erosion variables using the precipitation parameters for each of the seasonal datasets.

3. Results

3.1. Precipitation

Precipitation accumulation for each measurement period had an annual mean of 22.2 mm, with the highest seasonal mean accumulation in winter (26.3 mm) and spring (24.3 mm), and the lowest in autumn (15.3 mm) (Table 1). Likewise, the duration of precipitation had an annual mean of 278.7 min, but the longest seasonal mean duration was received in winter (424.7 min), and the shortest in autumn (192.5 min). Both average and maximum precipitation intensity were higher in summer months (0.1 mm/min and 108.1 mm/min, respectively) compared to the annual values of these parameters (0.08 and 71.5 mm/min, respectively).

The study area experienced year-round precipitation, however, most of the accumulation was in winter (frontal systems) and summer (convectional storms) (Figure 2). September and October were the driest months, most notably in 2012, 2013, and 2016. The most intense rains occurred in summer months, for example, see high values for Average Intensity (AvgInt) in the summer of 2012, 2014, 2016, 2017, 2018, and to a lesser degree 2013 and 2015. One may also notice that when Total Accumulation (TotAccum) was high and Duration was low, Maximum Intensity (MaxInt) was also high because it follows that higher intensity rainfall occurred when high rainfall totals were received in a short time-period.

Table 1. Descriptive statistics of precipitation parameters by measurement period.

Parameter	Season	Mean	Median	Standard Deviation	Min.	Max.	Skewness	Kurtosis
Total Accumulation (mm)	All	22.2	17.0	21.1	0.0	132.6	1.6	4.0
	Spring	24.3	17.5	21.7	0.25	93.2	1.1	0.8
	Summer	22.7	19.6	18.6	0.0	72.1	0.7	−0.2
	Autumn	15.3	11.2	16.2	0.0	78.0	1.5	2.8
	Winter	26.3	19.1	25.9	0.25	132.6	2.1	5.7
Duration (min)	All	278.7	220.0	285.6	0	2600	3.0	16.5
	Spring	299.4	265.0	243.4	5	1185	1.3	2.0
	Summer	209.9	190.0	178.3	0	995	1.5	3.6
	Autumn	192.5	130.0	218.2	0	1175	2.3	7.1
	Winter	424.7	305.0	409.1	5	2600	2.8	11.6
Average Intensity (mm/min)	All	0.1	0.06	0.05	0	0.3	1.9	4.5
	Spring	0.1	0.07	0.04	0.03	0.2	1.5	1.5
	Summer	0.1	0.1	0.06	0	0.3	1.1	1.4
	Autumn	0.1	0.06	0.05	0	0.3	1.9	4.8
	Winter	0.1	0.05	0.01	0.05	0.1	2.2	5.6
Maximum Intensity (mm/min)	All	71.5	15.5	220.2	0	2090.2	6.4	45.2
	Spring	49.1	16.3	120.9	0	975.4	6.7	50.7
	Summer	108.1	48.5	276.8	0	2090.2	5.7	35.8
	Autumn	88.1	13.0	288.0	0	1625.6	5.0	24.6
	Winter	34.6	6.6	121.3	0	975.4	7.3	56.9

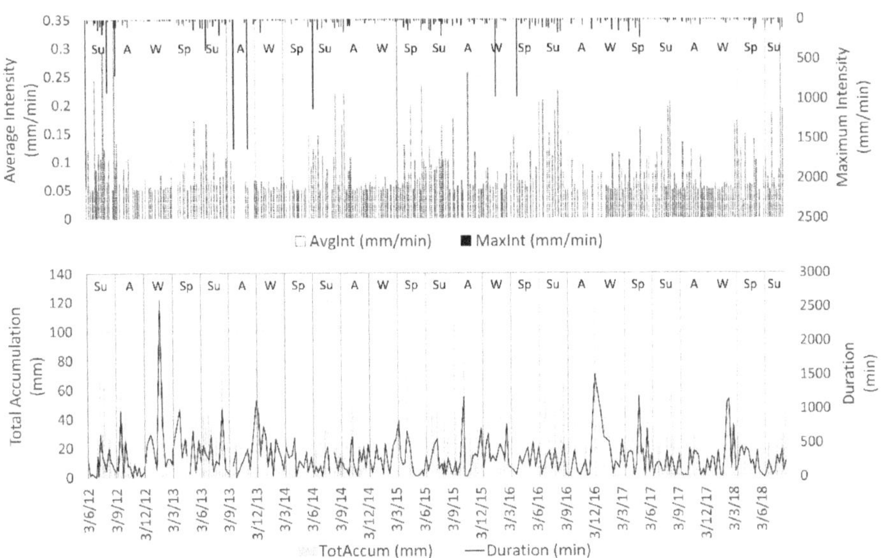

Figure 2. Time series of precipitation parameters. AvgInt and MaxInt refer to average and maximum precipitation intensity, respectively. TotAccum is the total depth of precipitation received during each weekly measurement period, and Duration is the total minutes during which precipitation was measured, for each measurement period. Columns delineate seasons (Su = summer, A = autumn, W = winter, and Sp = spring).

3.2. Erosion

Mean erosion by measurement period (assessed using the average absolute change variables CAvg|Ch|, IAvg|Ch|, and SAvg|Ch|, where C, I, and S, refer to channels, interfluves, and sidewalls, respectively) was greater in winter and spring than the overall mean for all three geomorphic areas (Table 2). Notably, in winter months, CAvg|Ch| was 16.8 mm compared to 9.9 mm overall and SAvg|Ch| was 8.0 mm compared to 5.0 mm overall. Seasonal effects on interfluves were less pronounced, with IAvg|Ch| in winter at 4.8 mm compared to the overall mean of 3.5 mm. As with precipitation parameters, autumn was the season with the lowest mean erosion by measurement period for all geomorphic areas at 4.8 mm for channels, 3.5 mm for sidewalls, and 2.8 mm for interfluves.

Table 2. Descriptive statistics for erosion variables by measurement period. All values measured in millimeters. C, channel I, interfluve; S, sidewall.

Variable	Season	Mean	Median	Standard Deviation	Min.	Max.	Skewness	Kurtosis
CAvg|Ch|	All	9.9	5.9	10.3	0.9	82.4	2.6	10.2
	Spring	11.1	6.9	12.2	1.4	82.4	3.4	16.1
	Summer	7.4	4.8	6.3	0.9	31.2	1.8	2.9
	Autumn	4.8	3.2	4.6	0.9	24.4	2.6	7.7
	Winter	16.8	12.8	12.2	1.4	53.9	1.1	0.7
CDep	All	10.6	6.2	12.2	1.0	79.4	2.9	10.5
	Spring	10.0	7.0	9.2	1.0	45.2	2.2	5.2
	Summer	8.4	5.3	9.7	1.0	72.3	4.1	23.4
	Autumn	4.8	3.5	4.2	1.0	25.8	2.7	9.5
	Winter	19.3	16.2	17.3	2.0	79.4	1.8	3.0
CErosion	All	−9.4	−5.2	10.7	−78.2	0	−2.7	9.5
	Spring	−11.4	−5.6	13.9	−78.2	−1.4	−2.7	8.7
	Summer	−7.1	−5.0	5.9	−28.6	−1.0	−2.0	3.9
	Autumn	−5.4	−3.4	5.7	−34.0	−1.0	−3.0	10.9
	Winter	−14.2	−10.6	12.8	−50.3	0	−1.4	1.4
IAvg|Ch|	All	3.5	3.2	1.8	0.6	14.1	2.2	8.1
	Spring	3.6	3.2	1.8	1.1	14.1	3.3	16.1
	Summer	3.0	2.9	1.0	1.0	6.7	0.8	1.9
	Autumn	2.8	2.6	1.1	0.6	6.9	0.9	2.4
	Winter	4.8	4.7	2.3	0.8	13.1	1.2	2.1
IDep	All	3.7	3.4	2.0	1.0	15.7	2.2	8.9
	Spring	3.8	3.5	1.6	1.0	9.6	1.0	1.6
	Summer	3.3	3.1	1.8	1.0	15.7	3.9	25.2
	Autumn	3.2	2.8	1.6	1.0	9.7	1.6	4.1
	Winter	4.7	4.3	2.5	1.4	14.1	1.8	4.4
IErosion	All	−4.1	−3.6	2.3	−19.3	0	−2.3	9.0
	Spring	−4.2	−3.8	2.4	−14.9	−1.3	−2.1	6.3
	Summer	−3.4	−3.3	1.4	−10.8	0	−2.0	10.3
	Autumn	−3.2	−3.0	1.4	−7.8	−1.0	−0.8	1.0
	Winter	−5.6	−4.9	3.1	−19.3	−1.2	−1.7	5.0
SAvg|Ch|	All	5.0	4.1	3.2	0.6	18.2	1.7	3.0
	Spring	5.0	4.3	2.9	1.7	15.2	1.9	3.9
	Summer	3.8	3.4	1.6	1.2	8.1	0.8	0.2
	Autumn	3.5	3.0	1.9	0.6	10.9	1.9	5.1
	Winter	8.0	7.6	3.8	1.6	18.2	0.8	0.1
SDep	All	5.2	4.2	3.3	1.0	20.2	1.8	4.2
	Spring	5.1	4.5	2.9	1.0	18.5	2.1	6.4
	Summer	3.9	3.6	1.8	1.3	9.3	0.8	0.2
	Autumn	3.9	3.4	2.2	1.1	12.3	1.7	4.0
	Winter	8.1	6.9	4.3	2.3	20.2	1.0	0.5
SErosion	All	−5.6	−4.5	3.7	−23.3	−1.0	−2.0	4.6
	Spring	−5.7	−4.6	3.8	−23.3	−2.2	−2.6	7.6
	Summer	−4.5	−3.7	2.5	−14.4	−1.5	−1.9	4.1
	Autumn	−4.0	−3.6	2.1	−13.7	−1.0	−2.2	7.2
	Winter	−8.4	−7.7	4.6	−22.7	−1.3	−0.9	0.9

Seasonally, erosion variables show the most variability during winter months (Figure 3). Winter of 2016–2017 experienced less erosion than other years for all geomorphic areas, however, the study area received high rainfall accumulation during two weekly measurement periods.

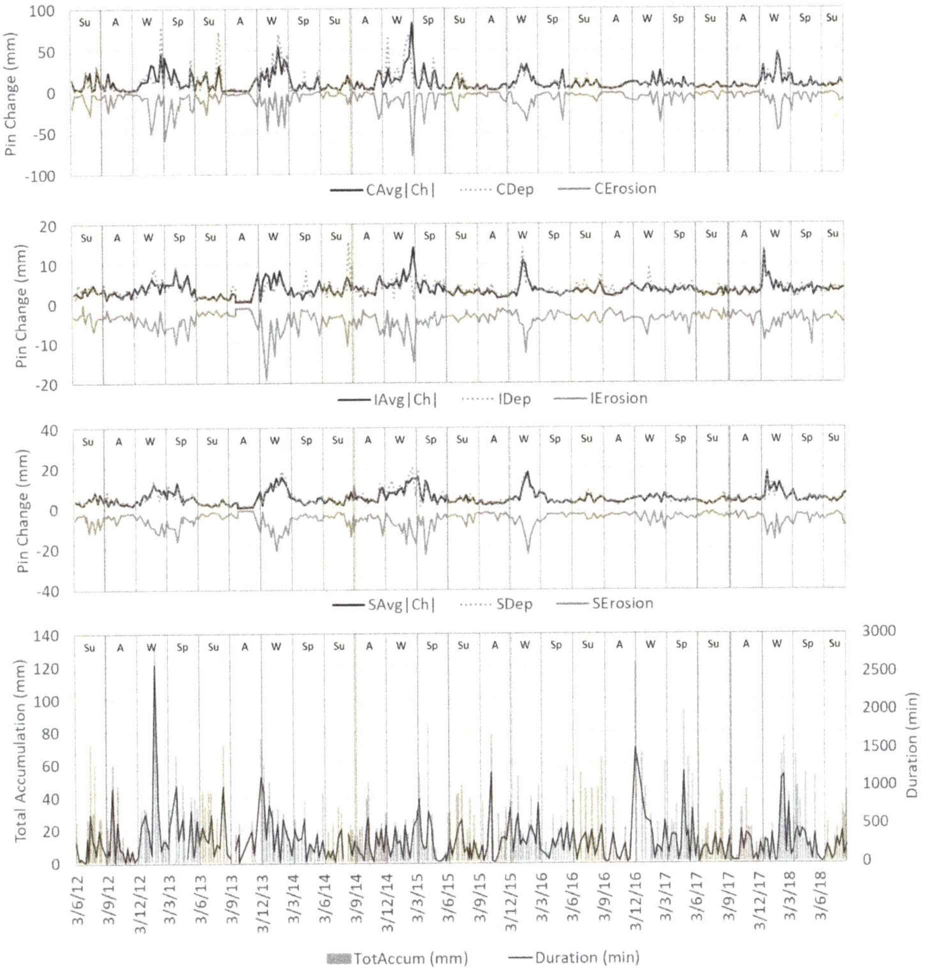

Figure 3. Comparison of erosion variables by geomorphic area. The top three graphs show erosion in channels (C, top), interfluves (I, middle), and sidewalls (S, lower), bottom graph shows precipitation. Columns mark seasons (Su = summer, A = autumn, W = winter, and Sp = spring).

3.3. Statistical Modeling

Erosion variables were significantly correlated with total accumulation and duration parameters for all variables except interfluve erosion (IErosion) (Table 3). Concordant with prior studies, erosion in channels was most strongly correlated with total accumulation ($r = 0.467$, $r = 0.352$, and $r = -0.469$ for CAvg|Ch|, CDep, and CErosion, respectively) and duration ($r = 0.470$, $r = 0.367$, and $r = -0.447$ for CAvg|Ch|, CDep, and CErosion, respectively). Note that all correlation coefficients for erosion variables (CErosion, IErosion, and SErosion) are negative because these variables are values below zero.

Spearman's correlation between the four precipitation parameters was compared to assess the potential for multicollinearity in statistical models, and total accumulation shows a very strong positive correlation with duration (r = 0.903) and a moderately strong positive correlation with average intensity (r = 0.591) and maximum intensity (r = 0.657). Likewise, average and maximum intensity were strongly and positively correlated (r = 0.794).

Table 3. Spearman's correlation coefficients for erosion variables and precipitation parameters. C, channel; I, interfluve; S, sidewall. Only significant correlations are shown (* significant at $\alpha = 0.05$, ** significant at $\alpha = 0.01$).

Variable Name	Total Accumulation (mm)	Duration (min)	Avg. Intensity (mm/min)	Max. Intensity (mm/min)
CAvg|Ch|	0.467 **	0.470 **	0.116 *	0.184 **
CDep	0.352 **	0.367 **	-	0.132 *
CErosion	−0.469 **	−0.447 **	−0.155 **	−0.230 **
IAvg|Ch|	0.130 *	0.178 **	-	-
IDep	0.138 *	0.146 *	-	-
IErosion	-	−0.156 **	-	-
SAvg|Ch|	0.238 **	0.279 **	-	-
SDep	0.240 **	0.265 **	-	-
SErosion	−0.199 **	−0.248 **	-	-
Total Accumulation (mm)	1.000	0.903 **	0.591 **	0.657 **
Duration (min)		1.000	0.278 **	0.461 **
Average Intensity (mm/min)			1.000	0.794 **
Maximum Intensity (mm/min)				1.000

Before modeling erosion by season, OLS regression models were developed for the annual dataset (all measurement periods) using the four precipitation parameters from the current period, plus lagged variables for up to 11 prior periods (weeks). Table 4 summarizes output from models for each erosion variable in columns, with the variable name and R^2 value at the head of the column, and retained parameters marked by *. Retained parameters (independent variables) were those with statistically significant coefficients in each OLS model output. Nine models are represented in Table 4, one for each erosion variable. Model coefficients are not presented (only significance) here because the purpose of the modeling was to identify the precipitation parameters that were universally important, which was completed through frequency analyses. All model linear equations are, however, presented in Table A1 in Appendix A. Duration and total accumulation were the most important variables for channel erosion, while average intensity was important for erosion in interfluves and sidewalls. Also notable is the influence of antecedent precipitation at lags of up to 11 weeks for some variables.

Table 4. Precipitation parameters retained (indicated by *) in Ordinary Least Squares regression models of erosion variables (dependent variables) using lagged precipitation parameters (independent variables). C, channel; I, interfluve; S, sidewall. Each column represents a different model.

Parameters Retained		CAvg|Ch|	CDep	CErosion	IAvg|Ch|	IDep	IErosion	SAvg|Ch|	SDep	SErosion
	R^2	0.297	0.191	0.354	0.119	0.093	0.120	0.174	0.137	0.205
Duration (min)	Current	*	*	*	*	*	*	*	*	
	Lag1			*						*
	Lag3									*
	Lag4	*	*			*		*		*
	Lag5								*	
	Lag6	*		*						*
	Lag8	*		*	*		*			
TotAcc (mm)	Current									
	Lag1									*
	Lag4	*	*							*
	Lag6	*		*						
	Lag8				*					
AvgInt (mm/min)	Lag2							*	*	*
	Lag4				*		*	*		
	Lag5						*			
	Lag7							*		
	Lag8				*	*		*	*	
	Lag9									*
	Lag11		*							
MaxInt (mm/min)	Lag4					*				
	Lag9		*							
	Lag10		*				*			
	Lag11									*

Seasonal OLS regression models clearly indicate the importance of precipitation intensity, which was, in prior studies, not retained in annual models of erosion (Table 5). Note that seasonal models for IAvg|Ch| were omitted from Table 5 because only one viable model was generated, and its coefficient of determination was extremely low ($R^2 = 0.064$).

Interestingly, in summer and winter, average and maximum intensity were important explanatory parameters both during the current period, but also in prior periods. Precipitation intensity was not often retained in models of erosion during spring and autumn. It is also important to note that viable OLS regression models were generated for all erosion variables for summer and winter, with coefficients of determination ranging from $R^2 = 0.245$ to $R^2 = 0.49$ (except for IErosion in summer at $R^2 = 0.131$ and SDep in winter at $R^2 = 0.087$), suggesting that precipitation is an important driver for erosion in these months, no matter the metric used. Moreover, these results show that the character of the precipitation is an important driver for erosion; antecedent precipitation has an influence on erosion in the following weeks and months and it varies with season.

Table 5. Parameters retained (indicated by *) in seasonal Ordinary Least Squares regression models of erosion variables (dependent variables) using lagged precipitation parameters (independent variables) Duration (min), Total Accumulation (TotAcc (mm)), and Average and Maximum Intensity (AvgInt and MaxInt, respectively (mm/min)). C, channel; I, interfluve; S, sidewall. Each column represents a separate model.

Season	Parameters Retained		CAvg|Ch|	CDep	CErosion	IDep	IErosion	SAvg|Ch|	SDep	SErosion
Spring		R^2	0.078	No Model	0.429	No Model	0.045	0.113	No Model	0.144
	Duration	Current						*		
		Lag3								*
		Lag6			*					
		Lag7			*					
		Lag8			*		*			*
	TotAcc	Current	*		*					
		Lag2			*					
		Lag8			*					
	AvgInt	Lag1			*				*	*
Summer		R^2	0.49	0.389	0.344	0.257	0.131	0.373	0.372	0.32
	Duration	Current	*	*						
		Lag2								*
		Lag7	*							
		Lag11	*					*	*	
	TotAcc	Current			*				*	
		Lag6					*			
		Lag8				*				
	AvgInt	Current				*		*	*	
		Lag5			*					
	MaxInt	Lag1	*					*		*
		Lag2				*				
		Lag4				*				
		Lag8						*	*	
		Lag10	*							
		Lag11					*	*	*	*
Autumn		R^2	0.364	0.19	0.258	0.125	no model	0.07	0.093	0.147
	Duration	Current	*	*	*					
		Lag3								*
		Lag7		*						
	TotAcc	Lag9								*
	AvgInt	Lag10	*		*	*		*		
	MaxInt	Current							*	
		Lag2	*		*					
Winter		R^2	0.374	0.324	0.472	0.273	0.251	0.245	0.087	0.347
	Duration	Current	*		*					
		Lag4		*						
		Lag6	*							
	TotAcc	Lag4		*		*				
		Lag6	*							
	AvgInt	Lag1						*		*
		Lag4				*				
		Lag9				*				
		Lag10				*				
		Lag11						*		*
	MaxInt	Current					*	*		*
		Lag1			*					
		Lag7					*			*
		Lag9	*	*	*				*	
		Lag10		*				*		
		Lag11			*					

4. Discussion

4.1. Erosion Variability

Variability exists in erosion statistics between the three geomorphic areas, such that channels had the highest variability and interfluves the lowest (Table 2), with sidewalls having intermediate variability. In particular, for both the overall annual dataset and for each seasonal partition, the mean and standard deviation were of similar magnitudes. Similar behavior was observed in a previous study in the same study area [13] and a study of gully erosion in the Karoo region of Africa [23]. Channels were dynamic and acted as both source and sink for sediment loads. Slugs of sediments gathered intermittently in the channel areas and were transported with channel flow following precipitation. Soil erosion was dominant in the gully sidewalls, however, the variability was moderate compared to channel erosion data, implying that sidewalls were less responsive with regard to erosion. In contrast, in the interfluve, the lesser amount of erosion and variability reflected the limited sediment yield, which may be due to the presence of vegetation that retarded erosion and lower gradient. Additionally, differences in soil cover thickness, soil types, moisture content, slope aspect and angle within the different geomorphic settings may explain the range of variability, however, that is beyond the scope of this paper and will be studied in the future.

4.2. Erosion-Precipitation Relationships

Seasonally, a comparison of erosion variables and precipitation parameters shows the same trend. Ordering seasonal precipitation parameters (Duration and TotAcc) and erosion variables from greatest to least, winter was greatest, followed by spring, summer and lastly, autumn. We see in Table 2 that winter months were the most dynamic, with the greatest mean erosion and the largest standard deviation of all seasons, and this pattern was consistent across channels, interfluves, and sidewalls for all erosion variables. This may be explained by the character of the winter precipitation: greater total accumulation and duration during these months associated with frontal precipitation events. Prior research has also demonstrated that freeze-thaw events are significant drivers of erosion in winter months at this site [14,19]. A similar pattern existed for spring, likely influenced by precipitation accumulation and duration as well as antecedent winter freeze-thaw activity [19]. Next, summer erosion and precipitation (Duration and TotAcc) ranked third, but interestingly, summer experienced the highest precipitation intensity of all seasons (both for AvgInt and MaxInt) (Table 1). This reflected the dominance of convectional precipitation events in summer. Autumn experienced the minimum erosion and precipitation accumulation and duration, but greater maximum precipitation intensity than the annual average. This suggests that autumn precipitation events were short duration, high-intensity events that did not produce much precipitation depth and had little erosive power.

During winter 2016–2017, precipitation variables were near normal levels for the winter season, however, erosion for all geomorphic areas was very low (Figure 3). We examined temperature during this time period to determine whether the reduced freeze-thaw activity may have played a part, but, while winter 2016–2017 had less intense freeze-thaw activity than other winters during the study period, freeze-thaw events occurred. The timing of the greatest precipitation accumulation and duration was late autumn/early winter, and because these events were coincident, they indicate a period of low-intensity precipitation that may have encouraged more infiltration and less runoff, leading potentially to less erosion during this period. Lower hydrostatic pressure in unsaturated soils increases cohesion [24] which may be a significant factor associated with reduced erosion in late autumn and early winter of that year.

Average Intensity and Maximum Intensity of precipitation were very different, with approximately three orders of magnitude between the generally low average precipitation intensities and maximum intensity for the full dataset and each seasonal partition (Table 1). Future research at this site should assess the soil's infiltration capacity and explore different metrics that may better capture the relation between precipitation intensity and erosion. For example, measuring the rainfall duration when the rain rate exceeds the soil's infiltration capacity would generate a metric of the length of time during which there was a high probability of runoff generation.

4.3. Precipitation as a Driver for Erosion

Prior research at this site using 14 months of data found that Duration and TotAcc were the drivers for erosion, most strongly in channels. With six years of data, the present study confirmed the earlier result when erosion and precipitation data were lumped without regard for season. OLS regression models of annual erosion for the nine erosion variables, using the set of lagged precipitation parameters as independent variables, and overwhelmingly retained Duration parameters most frequently (24 times) (Table 6). This means that overall nine OLS models of erosion outlined in Table 4, Duration and lagged Duration independent variables had significant coefficients 24 times. Despite the high correlation between TotAcc and erosion variables (Table 3), TotAcc was retained less frequently in the models (7 times) due to the high correlation between Duration and TotAcc (r = 0.903, p = 0.001) (Table 3), indicating multicollinearity. Lagged intensity parameters were likewise retained fewer times; AvgInt parameters were retained 14 times, while MaxInt parameters were retained only 5 times. Therefore, using lumped annual data, Duration was the most important predictor of erosion, indicating that over the long term, prolonged precipitation is key.

Table 6. Retention frequency of lagged precipitation parameters (Duration, Total Accumulation (TotAcc), and Average and Maximum Intensity (AvgInt and MaxInt, respectively) in OLS regression models of erosion annually and seasonally for the full study area and for each geomorphic area: Channels, Interfluves, and Sidewalls.

Geomorphic Area	Parameter	All Seasons	Spring	Summer	Autumn	Winter
Study area	Duration	24	7	7	5	4
	TotAcc	7	4	4	1	3
	AvgInt	14	3	4	4	7
	MaxInt	5	0	12	3	6
Channels	Duration	10	3	4	4	4
	TotAcc	4	4	1	0	2
	AvgInt	1	1	1	2	0
	MaxInt	2	0	2	2	6
Interfluves	Duration	6	1	0	0	0
	TotAcc	1	0	2	0	1
	AvgInt	5	0	1	1	3
	MaxInt	2	0	3	0	2
Sidewalls	Duration	8	3	3	1	0
	TotAcc	2	0	1	1	0
	AvgInt	8	2	2	1	4
	MaxInt	1	0	7	1	5

When erosion data were partitioned by geomorphic areas (Table 6), channel models overwhelmingly retained Duration most often. In contrast, sidewall and interfluve models retained Duration and AvgInt at approximately the same frequency (retained in 6 and 5 interfluve models and 8 sidewall models, respectively). This shows the importance of precipitation intensity as a driver for erosion in these two geomorphic areas. This may occur because interfluves and sidewalls may be more exposed to rain splash erosion, which is associated with higher intensity precipitation. Channels are not as steeply sloped as sidewalls and gully channel erosion is associated with the flow within the channel, which occurs after long-duration events that result in saturation-related runoff.

When erosion data were partitioned by season, the influence of precipitation intensity became apparent, especially during summer and to a lesser degree winter. This may be observed in Table 6, where MaxInt lagged parameters were retained 12 and 6 times in summer and winter erosion models, respectively, but only 0 and 3 times in spring and autumn models, respectively. This indicates that, while over the long term, Duration was the most important driver, during certain individual seasons intensity became important. This emphasizes the importance of the mechanics of convectional storms (summer) and frontal storms (winter) as an additional factor in seasonal erosion patterns. These patterns are also apparent when model results are partitioned by both season and geomorphic area (Table 6).

Partitioning the data by season, therefore, produces additional knowledge that was not previously captured. We conclude that different drivers may be more effective agents of erosion in different seasons and, therefore, we recommend that studies of precipitation driven erosion should, wherever possible, partition data by season.

5. Conclusions

This study examined the effect of precipitation parameters on soil erosion through six years of high-resolution weekly monitoring in an Appalachian hillslope paying particular attention to seasonal effect. The long-term data provided an understanding of the seasonal pattern of soil erosion in a humid sub-tropical environment, which was not noticeable in other studies in the region using an annual dataset.

Different gully morphologies responded differently to long-term erosion. Channels were most active, showed a wide range of variability, and responded most dynamically, whereas the interfluves were least disturbed by erosion. Sidewalls were prone to erosion but were not as dynamic as channels. To explore the reason behind varied gully erosion patterns in the different geomorphic settings, further studies are recommended to evaluate how erosion fluctuates with soil cover thickness, soil types, moisture contents, slope aspect, and slope angle.

Precipitation duration was the most important factor in initiating and continuing erosion year-round, yet seasonality played a significant role in the severity of gully erosion. Erosion was most pronounced in winter months, followed by spring, indicating the influence of high-intensity precipitation from frontal systems and repeated freeze-thaw cycles. Erosion in summer was driven by high-intensity precipitation from convectional storms. Soils in the study area were least prone to erosion during the moderate months of autumn. In channels, precipitation duration was the dominant driver for erosion due to runoff-related erosion, while in sidewalls and interfluves, intensity parameters were equally important as duration, likely related to rain splash erosion. This research shows that soil erosion is seasonally variable and an understanding of the seasonal pattern of soil erosion with respect to precipitation-related drivers improves the potential to achieve strategic conservation measures.

Author Contributions: Conceptualization, I.L. and A.N.; methodology, I.L. and A.N.; data collection, I.L. and A.N.; formal analysis, I.L. and A.N.; funding acquisition, I.L. and A.N.; data curation, I.L.; writing—original draft preparation, I.L. and A.N.; writing—review and editing, I.L. and A.N. All authors have read and agreed to the published version of the manuscript.

Funding: This research received funding support from East Tennessee State University's Honors College Federal Work Study program for collection of field data.

Acknowledgments: The authors gratefully acknowledge the assistance in data collection provided by Tim Spiegel, Nicholas Barnes, Tim Land, Jamie Kincheloe, Nicholas McConnell, and Jennifer Grant. The authors are grateful for the valuable contribution of the anonymous reviewers.

Conflicts of Interest: The authors declare no conflict of interest.

Appendix A

OLS Regression models of erosion are presented in Table A1. While model equations are useful for prediction when determination coefficients are high, even when they are relatively low, useful information can be revealed with respect to the importance of independent variables. Standardized coefficients can likewise provide information about the relative importance of independent variables within each model. For channels, Duration and TotAcc during the current and prior measurement periods were retained most often, and these variables had the highest standardized coefficients compared to the intensity parameters (AvgInt and MaxInt) (standardized coefficients are not shown in the table). For interfluves, AvgInt and MaxInt were also retained in the models, and for the IDep and IErosion models, standardized coefficients for all retained variables were of similar magnitudes. For sidewalls, a similar pattern was generally noted, with retention of the intensity variables. For the SErosion model, Duration and TotAcc parameters had the largest standardized coefficients.

Table A1. Regression equations for erosion variables (dependent variables) using lagged precipitation parameters (independent variables). Lagged variable names are appended with "LagN", where N indicates the number of measurement periods of antecedent lag. Duration_Lag1 indicates precipitation duration in prior measurement period (Lag of 1 period).

Dependent Variable	R^2	Equation
CAvg|Ch|	0.297	= 3.177 + 0.013 × Duration + 0.016 × Duration_Lag4 + 0.019 × Duration_Lag6 + 0.004 × Duration_Lag8 − 0.146 × TotAcc_Lag4 − 0.193 × TotAcc_Lag6
CDep	0.191	= 8.078 + 0.008 × Duration + 0.026 × Duration_Lag4 − 0.236 × TotAcc_Lag4 − 34.139 × AvgInt_Lag11 + 0.013 × MaxInt_Lag9 + 0.010 × MaxInt_Lag10
Cerosion	0.354	= −0.315 − 0.017 × Duration − 0.004 × Duration_Lag1 − 0.020 × Duration_Lag6 − 0.005 × Duration_Lag8 + 0.173 × TotAcc_Lag6
IAvg|Ch|	0.119	= 3.502 + 0.001 × Duration + 0.003 × Duration_Lag8 − 5.070 × AvgInt_Lag4 + 0.452 × AvgInt_Lag8 − 0.035 × TotAcc_Lag8
IDep	0.093	= 3.800 + 0.001 × Duration + 0.001 × Duration_Lag4 − 6.238 × AvgInt_Lag8 − 0.002 × MaxInt_Lag4
IErosion	0.120	= −4.426 − 0.001 × Duration − 0.001 × Duration_Lag8 + 5.831 × AvgInt_Lag4 + AvgInt_Lag5 − 0.002*MaxInt_Lag10
SAvg|Ch|	0.174	= 7.115 + 0.002 × Duration + 0.002 × Duration_Lag4 − 10.517 × AvgInt_Lag2 − 9.828 × AvgInt_Lag4 − 8.345 × AvgInt_Lag7 − 9.446 × AvgInt_Lag8
SDep	0.137	= 5.996 + 0.003 × Duration = 0.002 × Duration_Lag5 − 11.568 × AvgInt_Lag2 − 12.272 × AvgInt_Lag8
SErosion	0.205	= −5.623 − 0.006 × Duration_Lag1 − 0.002 × Duration_Lag3 − 0.005 × Duration_Lag4 − 0.002 × Duration_Lag6 + 0.062 × TotAcc_Lag1 + 0.052 × TotAcc_Lag4 + 7.750 × AvgInt_Lag2 + 11.889 × AvgInt_Lag9 − 0.002 × MaxInt_Lag11

References

1. Ionita, I.; Fullen, M.A.; Zgłobicki, W.; Poesen, J. Gully erosion as a natural and human-induced hazard. *Nat. Hazards* **2015**, *79*, 1–5. [CrossRef]
2. Poesen, J.; Nachtergaele, J.; Verstraeten, G.; Valentin, C. Gully erosion and environmental change: Importance and research needs. *Catena* **2003**, *50*, 91–133. [CrossRef]
3. United States Department of Agriculture Natural Resources. *Conservation Service 2012 National Resources Inventory Summary Report*; Natural Resources Conservation Service: Washington, DC, USA, 2015.
4. Bussell, P.; Galang, M.A.; Markewitz, D.; Morris, L.A. Land use change and gully erosion in the Piedmont region of South Carolina. *J. Soil Water Conserv.* **2007**, *62*, 122.
5. Trimble, S.W. *Man-Induced Soil Erosion on the Southern Piedmont, 1700–1970*; Soil and Water Conservation Society: Ankeny, IA, USA, 2008.
6. Reusser, L.; Bierman, P.; Rood, D. Quantifying human impacts on rates of erosion and sediment transport at a landscape scale. *Geology* **2015**, *43*, 171–174. [CrossRef]
7. Royall, D.; Kennedy, L. Historical erosion and sedimentation in two small watersheds of the southern Blue Ridge Mountains, North Carolina, USA. *Catena* **2016**, *143*, 174–186. [CrossRef]

8. Leigh, D.S.; Webb, P.A. Holocene erosion, sedimentation, and stratigraphy at Raven Fork, Southern Blue Ridge Mountains, USA. *Geomorphology* **2006**, *78*, 161–177. [CrossRef]

9. Price, K.; Leigh, D.S. Morphological and sedimentological responses of streams to human impact in the southern Blue Ridge Mountains, USA. *Geomorphology* **2006**, *78*, 142–160. [CrossRef]

10. Wang, J.; Edwards, P.J.; Hamons, G.W.; Goff, W.A. *Assessing RUSLE and Hill-slope Soil Movement Modeling in the Central Appalachians*; Transactions of the ASABE Annual International Meeting: Pittsburgh, PA, USA, 2010.

11. Ireland, H.A.; Sharpe, C.F.S.; Eargle, D.H. *Principles of gully erosion in the Piedmont of South Carolina No. 633*; US Department of Agriculture: Washington, DC, USA, 1939.

12. Morgan, R.P.C. *Soil Erosion and Conservation*; Blackwell Publishing: Oxford, UK, 2005.

13. Luffman, I.; Nandi, A.; Spiegel, T. Gully morphology, hillslope erosion, and precipitation characteristics in the Appalachian Valley and Ridge province, southeastern USA. *Catena* **2015**, *133*, 221–232. [CrossRef]

14. Barnes, N.; Luffman, I.; Nandi, A. Gully erosion and freeze-thaw processes in clay-rich soils, northeast Tennessee, USA. *GeoResJ* **2016**, *9*, 67–76. [CrossRef]

15. Sevon, W.; Potter, N.; Crowl, G. Appalachian peneplains: An historical review. *Earth Sci. Hist.* **1983**, *2*, 156–164.

16. Stroosnijder, L. Measurement of erosion: Is it possible? *Catena* **2005**, *64*, 162–173. [CrossRef]

17. Moore, H.L. Drainage problems in carbonate terrain of East Tennessee. In Proceedings of the 27th Annual Highway Geology Symposium, Orlando, FL, USA, 19–21 May 1976; pp. 112–131.

18. Soil Survey Staff, Natural Resources Conservation Service, United States Department of Agriculture, Web Soil Survey. Available online: https://websoilsurvey.sc.egov.usda.gov/ (accessed on 14 March 2020).

19. Luffman, I.; Nandi, A. Freeze-thaw induced gully erosion: A long-term high-resolution analysis. *Agronomy* **2019**, *9*, 549. [CrossRef]

20. Couper, P.; Stott, T.I.M.; Maddock, I.A.N. Insights into river bank erosion processes derived from analysis of negative erosion-pin recordings: Observations from three recent UK studies. *J. Br. Geomorphol. Res. Group* **2002**, *79*, 59–79. [CrossRef]

21. Kearney, S.; Fonte, S.; Garcia, E.; Smukler, S. Improving the utility of erosion pins: Absolute value of pin height change as an indicator of relative erosion. *Catena* **2018**, *163*, 427–432. [CrossRef]

22. *National Oceanic and Atmospheric Administration Surface Weather Observations and Reports*; National Climatic Data Center: Asheville, NC, USA, 1998.

23. Keay-Bright, J.; Boardman, J. Evidence from field-based studies of rates of soil erosion on degraded land in the central Karoo, South Africa. *Geomorphology* **2009**, *103*, 455–465. [CrossRef]

24. Winterwerp, J.C.; Van Kesteren, W.G.M.; Van Prooijen, B.; Jacobs, W. A conceptual framework for shear flow-induced erosion of soft cohesive sediment beds. *J. Geophys. Res. Space Phys.* **2012**, *117*. [CrossRef]

Article

Influences of Catchment and River Channel Characteristics on the Magnitude and Dynamics of Storage and Re-Suspension of Fine Sediments in River Beds

Jungsu Park [1,*], Ramon J. Batalla [2,3,4], Francois Birgand [5], Michel Esteves [6], Francesco Gentile [7], Joseph R. Harrington [8], Oldrich Navratil [9], Jose Andres López-Tarazón [2,10,11,12] and Damià Vericat [2,13]

[1] K-water Research Institute, Korea Water Resources Corporation (K-water), 200 Sintanjin-Ro, Daedeok-Gu, Daejeon 34350, Korea
[2] Fluvial Dynamics Research Group, Department of Environment and Soil Sciences, University of Lleida, Plaça de Víctor Siurana, 1, 25003 Lleida, Spain; rbatalla@macs.udl.cat (R.J.B.); ja.lopez@uib.cat (J.A.L.-T.); dvericat@macs.udl.cat (D.V.)
[3] Catalan Institute for Water Research (ICRA), 17003 Girona, Spain
[4] Faculty of Forest Sciences and Natural Resources, Universidad Austral de Chile, Independencia 631, Valdivia, Región de los Ríos, Chile
[5] Department of Biological and Agricultural Engineering, North Carolina State University, Raleigh, NC 27695, USA; birgand@ncsu.edu
[6] Institute for Geosciences and Environmental Research (IGE)-University Grenoble Alpes/IRD, 621 Avenue Centrale, 38400 Saint-Martin-d'Hères, France; michel.esteves@ird.fr
[7] Department of Agricultural and Environmental Science, University of Bari Aldo Moro, 70126 Bari, Italy; francesco.gentile@uniba.it
[8] School of Building & Civil Engineering, Cork Institute of Technology, Rossa Ave, Bishopstown, T12 P928 Cork, Ireland; joe.harrington@cit.ie
[9] School of Building and Society, University of Lyon, 92 Rue Pasteur, 69007 Lyon, France; oldrich.navratil@univ-lyon2.fr
[10] Mediterranean Ecogeomorphological and Hydrological Connectivity Research Team, Department of Geography, University of the Balearic Islands, Carretera de Valldemossa, km 7.5, 07122 Palma, Spain
[11] Institute of Agro-Environmental and Water Economy Research, INAGEA, University of the Balearic Islands Carretera de Valldemossa, km 7.5, 07122 Palma, Spain
[12] Institute of Earth and Environmental Science, University of Potsdam, Am Neuen Palais 10, 14469 Potsdam, Germany
[13] Forest Sciences and Technology Centre of Catalonia, Carrer de Sant Llorenç, 0, 25280 Solsona, Spain
* Correspondence: pjs7731@kwater.or.kr; Tel.: +82-42-629-4732

Received: 14 March 2019; Accepted: 24 April 2019; Published: 26 April 2019

Abstract: Fine particles or sediments are one of the important variables that should be considered for the proper management of water quality and aquatic ecosystems. In the present study, the effect of catchment characteristics on the performance of an already developed model for the estimation of fine sediments dynamics between the water column and sediment bed was tested, using 13 catchments distributed worldwide. The model was calibrated to determine two optimal model parameters. The first is the filtration parameter, which represents the filtration of fine sediments through pores of the stream bed during the recession period of a flood event. The second parameter is the bed erosion parameter that represents the active layer, directly related to the re-suspension of fine sediments during a flood event. A dependency of the filtration parameter with the catchment area was observed in catchments smaller than ~100 km^2, whereas no particular relationship was observed for larger catchments (>100 km^2). In contrast, the bed erosion parameter does not show a noticeable dependency with the area or other environmental characteristics. The model estimated the mass of fine sediments released from the sediment bed to the water column during flood events in the 13 catchments within ~23% bias.

Keywords: bed erosion; catchment area; filtration; sediment accumulation; sediment bed fluidization; sediment re-suspension

---

1. Introduction

Fine particles or sediments are considered one of the most important factors affecting the quality and functioning of fluvial environments. For instance, fine sediments are long-lasting sources of toxic substances in catchments; that is, contaminants, such as pathogens, heavy metals as well as nutrients, are transported attached to fine sediments [1–5]. Fine sediment dynamics also have various effects on the health of benthic communities and the overall aquatic ecosystems [6–9].

The suspended sediment yield is affected by various catchment characteristics such as climate, geology, soils, catchment area, and land cover [10–12]. Because of the spatial variability of these catchment characteristics, developing a universally applicable fine sediment transport model remains an important research challenge. Rainfall intensity, erodibility and runoff processes mainly govern the fine sediment dynamics at the basin outlet. Not only these factors but also the catchment sizes cause variations in the sediment supply and transport processes. For example,

Gao et al. [13] suggested that the suspended sediment load (Q_s) is dominated by short-time-interval processes in smaller catchments with less-developed drainage density and small capacity to store fine sediment (i.e., drainage area <0.1 km^2). As the drainage area increases, the homogeneity of the catchment decreases and drainage density gradually increases, leading to a greater contribution of remobilization of fine riverbed sediments and bank erosion to the overall sediment budget [13].

Suspended sediment concentration (hereafter C) is closely related to flow discharge (Q), but this relationship varies over time, from the flood scale to the annual scale. The C–Q relationship often shows orders of magnitude of scatter [14]. Such variability is explained by the fact that the rising limb of the flood generally shows a different C–Q relationship compared with the falling limb, leading to a hysteresis pattern in the relationship [13,15–19]. The supply of sediment from the channel system is often considered to be a significant source of sediment [18,20–22]. For instance, Klein [17] observed clockwise hysteresis, being mainly driven by the supply of sediments from the channel bed or from highly eroded hillslopes close to the outlet. In contrast, anticlockwise hysteresis can be observed when sediment is supplied from distant upstream sources. Recently, Yang et al. [23] derived a flow and sediment travel time model, verifying that clockwise hysteresis is observed when flow travel time is more extended than the sediment travel time, whereas anticlockwise hysteresis is observed in the opposite. More recently, Juez et al. [24], based in a series of laboratory tests, observed clockwise hysteresis driven by the supply of sediment from the channel bed, whereas anticlockwise hysteresis is observed when upstream supply of sediment has more contribution. These hysteresis patterns are one of the main reasons why single power-law models are generally insufficient to explain the scatter in the relationship between C and Q [14,25]. Seasonality of precipitation and land cover also causes scatter in C for a given Q [15,26–28]. For example, Alexandrov et al. [15] carried out a study in a semi-arid region and observed that autumn–spring convective storms with higher-intensity rainfall often produce higher C than winter frontal storms with lower intensity; much earlier,

Negev [26] and more recently Cantalice et al. [21] have suggested that the first flood in a given water year could have a higher C than subsequent floods of similar magnitude. The reason of these differences was attributed to the re-suspension of deposited sediment from bed during the first flood of the year. Seasonal variations of the flow due to snowmelt may also induce additional sediment supply from the channel bed and cause variations in the functional relationship between C and Q. Stubblefield et al. [29] observed an increase of the sediment supply from the channel bed when the flow rate was increased by snowmelt in a field study of Lake Tahoe. Interannual variations of the suspended sediment load with water discharge are also caused by larger-scale variations of the environment, such as climatic changes and related variability of discharge [30], or extreme events such as large floods [14].

A conceptual model coupling fine sediment dynamics with bedload transport was presented by Park and Hunt [31] based on systematic analysis of fine sediment and stream bed movement. This study led to the development of a model for the estimation of fine sediment accumulation and re-suspension from the bed [32]. It is worth mentioning that fine sediments are defined as "particles that are transported in suspension in surface waters and can also be accumulated in the sediment beds" [32]. Within this context, the objective of this study is (i) to analyze the applicability and robustness of the model developed by Park et al. [32] and (ii) to study the effect of catchment characteristics (e.g., catchment area, climate) on the performance of the model. The study was carried out based on data of 13 catchments with different drainage area and located in various hydro-climatic environments.

2. Materials and Methods

2.1. Research Sites and Data Sources

C and Q data from 13 catchments (ranging from 2.2 to 21,000 km^2) were used to test the robustness of the model under different environmental characteristics (Figure 1; Table 1; Table S1 in Supplementary Materials). Note that we consider environmental characteristics, as mainly, the catchment area and climatic condition of catchments.

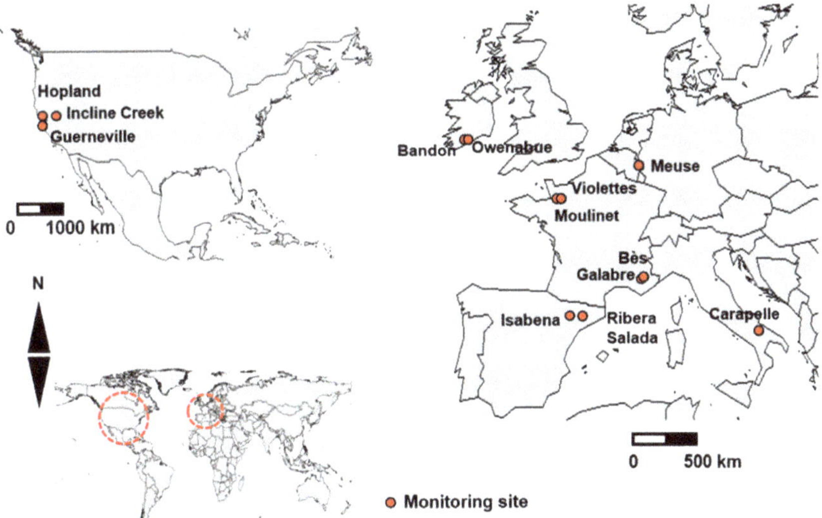

Figure 1. Location of the monitoring sites (see Table 1 and Table S1 for details of the catchments).

Table 1. Characteristics of the studied catchments and model parameters.

No	Sites		Catchment Area (km^2)	Climate	Mean Annual Precipitation (mm)	Dominant Bed Surface Materials	Observation		Q_c (m^3/s)	Q_{max} (m^3/s)	Background C (C_b)	M_{max} (Mg)	References
							Period	Frequency (min)					
1	Violettes	France	2.2	Temperate oceanic	900	Silty loess	1 June 2002–31 May 2003	10	0.05	0.15	4000Q	6.7	[33]
2	Moulinet	France	4.5	Temperate oceanic			1 June 2002–31 May 2003	10	0.1	0.41	300Q	9.8	
3	Incline Creek	Nevada, USA	7.4	Snow melt	890–1270	Sandy decomposed granite	4 April–24 May 2000	15	0.2	0.34	100(Q-0.2)	0.5	[34]
4	Galabre	France	20	Mediterranean mountainous	600–1200	Limestone and marls	3 October 2007–23December 2009	10	1	22	300Q	2200	[35]
5	Owenabue	Ireland	103	Temperate oceanic	~1200	Mudstone and sandstone	15 September 2009–15 September 2010	15	5	17	1.5Q	170	[16]
6	Bandon	Ireland	424	Temperate oceanic			10 February 2010–9 February 2011	15	10	110	0.1Q	480	
7	Bès	France	165	Mediterranean mountainous	600–1200	Limestone and marls	1 April 2008–31 December 2009	10	10	143	100Q	26,600	[35]
8	Ribera Salada	Spain	114	Mediterranean mountainous	760	Limestone and conglomerates	1 November 2005–30 October 2008	5	1	5.85	$10Q^{0.5} + 0.1Q^4$	53.4	[36]
9	Isabena	Spain	445	Mediterranean mountainous	770	Limestone, marls and clay-rocks	1 November 2007–30 September 2012	15	10	68	$0.02Q^3 + 100$	100,100	[37]
10	Carapelle	Southern Italy	506	Mediterranean	450–800	clayey-loamy-sandy	1 January 2007–31 December 2011	30	8	37	$5Q^2$	23,000	[38]
11	Hopland [†]	California, USA	938	Mediterranean	1000–1200	Sand-gravel and silty materials	1 October 2010–31 December 2014	15	10	425	2Q	26,560	[31,32,39]
12	Guerneville [†]	California, USA	3465	Mediterranean	1000–1200	Sand-gravel and silty materials	1 October 2009–31 September 2010/ 1 October 2012–31 December 2014	15	20	890	0.5Q	83,960	
13	Meuse	Belgian-Dutch border	21,000	European-continental	800– 1000	Limestone, shales and sandstone	1 October 1995–30 November 2010	1440	280	1700	0.03Q	187,800	[40]

[†] Data were obtained from a previous study [32].

These catchments are classified into 5 categories according to their climatic characteristics or geographical regions: (i) temperate oceanic, (ii) snowmelt, (iii) Mediterranean mountainous, (iv) Mediterranean, and (v) European-continental (Table 1).

The temperate oceanic climate is represented by two catchments located in Northwestern France, Violettes and Moulinet, where the mean longitudinal channel slope is about 1.8% in both sites [33,41]. The land in this region is used extensively for dairy cattle farming, including pastures. Cattle disturbance has been associated with bank erosion and increased suspended sediment concentration in the stream [42]. Turbidity sensors were installed at the outlet of each catchment and C was estimated from the relationship between turbidity and C [41]. The Q and C parameters were measured every 30 s, and 10-min average values were reported. Similar climatic and land use conditions are found in the Owenabue and Bandon catchments in Southern Ireland, where 90% of the land is used for pasture and tillage [16]. The Q and C values of these two catchments were provided by Ireland's National Office of Public Works for Q and the Cork Institute of Technology for C.

Snowmelt- and glacial melt-dominated streams have periodic flow rates and corresponding fine sediment concentration fluctuations [29]. One of such streams was the Incline Creek in Nevada, which drains into Lake Tahoe, CA, USA. For this stream, Q and C were measured at 15-min intervals between 4 April and 24 May 2000. This snowmelt-dominated catchment provides an extreme test of the model, given that there are only 24 h between flood events. The gauge elevation is 2100 m a.s.l., therefore, winter precipitation falls mainly in the form of snow. The flow regime of the United States Geological Survey (USGS) station (site number 103366993) shows daily cycles reflecting snowmelt conditions in spring. The C was continuously estimated from turbidity measurements, based on the turbidity-C relationship developed by Langlois et al. [34]. The earlier arrival of suspended sediment concentration peaks compared with peak water discharge shows clockwise hysteresis loops for almost all flood events.

The Mediterranean mountainous climate was assessed in four different catchments located in two different regions. Firstly, the two tributaries of the Bléone catchment, the Galabre and Bes rivers, located in the subalpine region of southeastern France, were evaluated. The climate of the catchment is characterized by a pronounced seasonality with the occurrence of frost in winter and high-intensity rainfall in summer. The peak water discharge during the spring season is affected by snowmelt in the Bès catchment where the median grain size of bed surface materials is 70 mm [35,43]. The main types of land cover found are forests, scrubland, sparse vegetation, and grassland. Continuous Q and C values were monitored at two gauging stations located at the outlet. Depending on its magnitude, the Q was regularly gauged with the salt (NaCl) dilution method and a current flow meter. The concentration C was estimated from turbidity data based on the method developed by Navratil et al. [44].

The Ribera Salada stream, located in the southern Pyrenees, and the river Isábena (which presents frequent flooding that causes relatively high sediment transport rates) located in the southern central Pyrenees, are others representative of the Mediterranean mountainous climate. Mean annual precipitation at both catchments is around 800 mm, being the monitoring period selected for the present study representative of the long-term hydrological regime. Predominant land uses are forest in headwaters and forest mixed with agriculture at the lowlands. Q and C (estimated from turbidity sensors by establishing rating curves between turbidity and C) were continuously measured at 15-min intervals, at the Inglabaga monitoring station (channel slope at around 1%) in the case of the former [36,45], and at the Capella gauging station (channel slope at around 0.4%) for the latter, and the median grain sizes of bed surface materials are 49.0 mm and 69.5 mm in the Ribera Salada and Isábena catchments, respectively [37,46].

Another different Mediterranean environment is that of the Carapelle catchment, located in the Puglia region of Southern Italy. It presents yearly precipitation that ranges from 450 to 800 mm, and land use is mostly agricultural where the mean slope of main channel is 1.8% [38,47]. Continuous Q and C values have been measured from 1 January 2007 to 31 December 2011 in this catchment, which is characterized by long periods with low flows and a prevalence of counter-clockwise hysteresis [48].

For this reason, a shorter period, between 3 March and 31 April 2009, was utilized for calibration in this study where 2009 was quite humid year (annual rainfall 786 mm) [38,47].

The Hopland and Guerneville catchments, both located in the Russian River, California, CA, USA, analyzed in a previous study [32] were included in the catchment list; the climate in this region is also Mediterranean-type, with warmer, drier summers and cooler, wetter winters, where the median grain sizes of bed surface materials are 7.9 mm and 7.1 mm in the Hopland and Guerneville catchments, respectively.

Finally, the wet European-continental climate was considered with the inclusion of the Meuse River catchment, located at the Belgian–Dutch border, with a length of 935 km and a catchment area of 36,000 km^2, where wet season is between October and April, and dry season is between May and September [40]. The main channel has steep slopes and land is dominantly used for both agriculture and forest [40]. Daily Q and C data reported over 15 years by the Dutch Institute for Inland Water Management and Waste Water Treatment (RIZA) are available from the upstream of the Eijsden gauging station, where the river length is approximately 700 km and the catchment area is ~21,000 km^2. Because the identification of model parameters requires sufficient resolution to discern the hysteresis of C to Q in rising and falling limbs during a flood event, daily data were applicable for model simulation in this relatively large catchment.

2.2. Model Description

The model applied in this study was developed by Park et al. [32]. The model estimates in-channel storage and re-suspension of fine sediment through three phases (Phase 1–3). In the model, it is considered that the effect of other catchment characteristics, such as channel morphology, on sediment dynamics is included through two parameters. This simplicity is one of the benefits for the application of this model. The model development process will not be described in detail in this study, but the main concepts are summarized.

2.2.1. Fine Sediment Accumulation

In Phase 1, when Q is less than the critical flow rate (Q_c) to initiate sediment bed material mobilization, the fine sediments in the water column are accumulated in the sediment bed through hyporheic flow. The change of accumulated fine sediments mass during the time period from t to t + Δt is represented by Equation (1), which was derived by considering that the mass of fine sediments accumulated in the sediment bed is proportional to the fine sediment concentration in surface water, C(t).

$$\Delta M(t) = \alpha C(t)\left[1 - \frac{M(t)}{M_{max}}\right]\Delta t \qquad \text{for } Q < Q_c, \tag{1}$$

where M is the mass of fine sediments accumulated within the pore space of the sediment bed. The maximum value of M, M_{max}, represents the sediment particle accumulation capacity of the sediment bed. The sediment particle removal parameter, α (L^3/T), represents the filtration and settling of fine sediment particles within the sediment bed.

2.2.2. Fine Sediment Re-Suspension

In Phase 2, during the rising flood with dQ/dt > 0 and Q > Q_c, fine sediments are released from the sediment bed into the water column, because the bed materials become fluidized when Q exceeds Q_c, initiating mobilization of the bed material.

An analysis of the data from Haschenburger [49] leads to the assumption that the erosion depth of the sediment bed during a flood event is an exponential function of the peak flow rate (Q_{peak}). From this approach, it is assumed that the maximum bed erosion occurs at the maximum flow rate (Q_{max}) during the observation period, when the release of all fine sediment particles within the sediment bed is expected. Thus M_{max} is observed at Q_{max} (Appendix A.1).

In the model, the ratio of fine sediments mass released from the sediment bed, $M_{f,model}$, to the maximum possible mass of fine sediments in storage, M_{max}, is expressed as an exponential function of the ratio of Q_{peak} to Q_{max}. Thus, the mass of fine sediments released from the sediment bed by flood event i is

$$M_{fi,model} = M_{max} \exp\left[-\beta\left(1 - \frac{Q(t_{p,i})}{Q_{max}}\right)\right],$$ (2)

where $t_{p,i}$ is time at Q_{peak} of flood event i and β a dimensionless sediment bed erosion parameter.

The mass of fine sediments remaining in the pore space of the sediment bed immediately after flood event i is estimated from the difference between the accumulated fine sediment mass in the sediment bed before flood event i and the mass of fine sediments released from the sediment bed by flood event i, as

$$M(t_{p,i}) = M(t_{s,i}) - M_{max} \exp\left[-\beta\left(1 - \frac{Q(t_{p,i})}{Q_{max}}\right)\right],$$ (3)

where $M(t)$ is the fine sediment mass accumulated in the sediment bed at time t, and $t_{s,i}$ is the time at the beginning of flood event i. In the model simulation, $M(t_{p,i})$ is restricted to be non- negative.

2.2.3. Fine Sediment Accumulation during Flood Recession

Finally, in Phase 3, in the falling limb with $dQ/dt < 0$ and $Q > Q_c$, fine sediments can be removed by filtration through hyporheic flow and stored within the pore space of the sediment bed. During flood recession, the available capacity for sediment storage is limited by partial fluidization of the sediment bed, which reduces the volume of porous media available for sediment accumulation (Appendix A.2). In the model, the available capacity for sediment storage in the sediment bed during the flood recession of a flood event, M_{cap}, is represented as

$$M_{cap}[Q(t)] = M_{max}\left\{1 - \exp\left[-\beta\left(1 - \frac{Q(t)}{Q_{max}}\right)\right]\right\},$$ (4)

The change of accumulated fine sediments mass during the flood recession period is estimated by substituting $M_{cap}[Q(t)]$ for M_{max} in Equation (1) as

$$\Delta M(t) = \alpha C(t)\left\{1 - \frac{M(t)}{M_{cap}[Q(t)]}\right\}\Delta t \qquad \text{for } Q > Q_c \text{ and } dQ/dt < 0,$$ (5)

2.2.4. Model Simulation

In the model, it is assumed that M is set to M_{max} as an initial condition of model simulation, and remains fixed at M_{max} before the start of the first flood event. Then, flood event i with a flow rate exceeding Q_c erodes the mass $M_{fi,model}$, which is estimated by Equation (2), into the water column. During the flow recession, M_{cap} is limited below M_{max}, and as the flow rate recedes, M_{cap} increases. In the case of another flood event occurring before the flow rate recedes below Q_c, additional sediment re-suspension occurs, which also can be estimated by Equation (2). In the case that the flow rate recedes below Q_c, sediment particles can accumulate in the pore space of the sediment bed up to the maximum capacity M_{max}, where the accumulation rate is limited by the available mass of fine sediments in the water column. The model estimates the mass of fine sediment storage and re-suspension during the repeated cycles of flood events.

2.3. Determination of the Model Parameters

The input parameters for the model simulation are Q_c, M_{max}, Q_{max}, and background suspended sediment concentration (C_b). These parameters were determined for the 13 catchments using Equations (6)–(8) following Park et al. [32] as described below.

Park et al. [31] observed the transition in the relationship between suspended sediment load (Q_s) and Q, which occurs when the flow initiates mobilization of bed materials. The value of Q at the transition corresponds to the Q that initiates bed mobilization, defined as Q_c. The suspended sediments in the water column above the sediment bed are separated according to the source of the sediments as "background suspended sediments from the catchment (C_b)" and "fine sediments released from the sediment bed during the rising limb of a flood event", where C_b is site-specific and represented by Equation (6):

$$C_b(t) = \gamma Q(t), \tag{6}$$

in which C_b is expressed in (mg/L), Q is expressed in (m^3/s), and γ is expressed in (mg·s/L/m^3).

The observed mass released from the sediment bed into the water column by a flood event ($M_{f,\,obs}$) is calculated from observed data using Equation (7), whereas $M_{f,\,model}$, the modeled mass of fine sediments released from the sediment bed into the water column by a flood event, is estimated by Equation (2):

$$M_{fi,obs} = \int_{t_{s,i}}^{t_{e,i}} Q(t)[C(t) - C_b(t)]dt, \tag{7}$$

The lower limit of integration $t_{s,i}$ is the beginning of flood event i, which is either the first occurrence when $Q(t_s) > Q_c$ or when $dQ(t)/dt$ transitions from negative to positive, while $Q > Q_c$ during multiple high-flow events. The upper limit of integration $t_{e,i}$ represents the time of the end of flood i, either when $Q(t_{e,i}) < Q_c$ or when dQ/dt transitions from negative to positive. The maximum value of $M_{f,\,obs}$ during the model calibration period is defined as M_{max} in the sediment bed, assuming that all fine sediments stored in the sediment bed are re-suspended during the maximum flood event when the highest peak flow rate (Q_{max}) is observed.

2.4. Model Evaluation

Two parameters, α and β, were utilized for model calibration. The fine sediment filtration parameter, α, represents the removal of fine sediments from the water column by filtration and settling of fine sediments within the pore space of the sediment bed of the catchment and the bed erosion parameter, β, represents the erosion of the sediment bed during a flood event. The optimal values of model parameters (α and β) were determined by a reiterative trial and error process to minimize the root-mean-square error to data standard deviation ratio (RSR), using Equations (8)–(10).

The model performance was also evaluated based on values of the term R, which is defined as the ratio of the modeled sediment mass released to the observed sediment mass released. The model performance is considered satisfactory when RSR < 0.70 [50] and when R approaches 1 when there is a total agreement between observed and predicted values.

$$R = \frac{\sum_{j=1}^{n} M_{fj,model}}{\sum_{j=1}^{n} M_{fj,obs}}, \tag{8}$$

where n is the number of flood events in the entire simulation period.

$$RSR = \frac{\sqrt{\sum_{i=1}^{n}(M_{fi,obs} - M_{fi,model})^2}}{\sqrt{\sum_{i=1}^{n}(M_{fi,obs} - \overline{M_{f,obs}})^2}}, \tag{9}$$

where n is the number of flood events and the mean observed mass released for all flood events is

$$\overline{M_{f,obs}} = \frac{1}{n} \sum_{i=1}^{n} M_{fi,obs}, \tag{10}$$

The period of data available for model simulation varies for each catchment. For catchments with more than three years of observations, about two-thirds of the data were used for model calibration and the remaining data were used for validation, except for the Meuse River where 15 years of data were available. Due to the longer period of observations in the Meuse River, five years were used for model calibration, whereas data from the remaining 10 years were used for model validation, in order to test the applicability of the model in a longer period.

3. Results

3.1. Model Parameters

The model parameters (i.e., Q_c, C_b, M_{max}, and Q_{max}) were determined from analysis of observed data in each of the 13 catchments (Figures S1 and S2 in Supplementary Materials) and are summarized in Table 1; a complete set of figures for all catchments is presented in Supplementary Materials, and in the next we present some observations on the data.

Slope breaks in the relationship between Q_s to Q were observed at flow rates of 5 m^3/s for Owenabue and 10 m^3/s for Bandon (Figure S1). The falling limb flow recessions asymptotically approached a linear relationship between C and Q, which defined the assumed background suspended sediment concentration dependence on Q as $C_b(Q) = 1.5Q$ for Owenabue and $C_b(Q) = 0.1Q$ for Bandon (Figure 2).

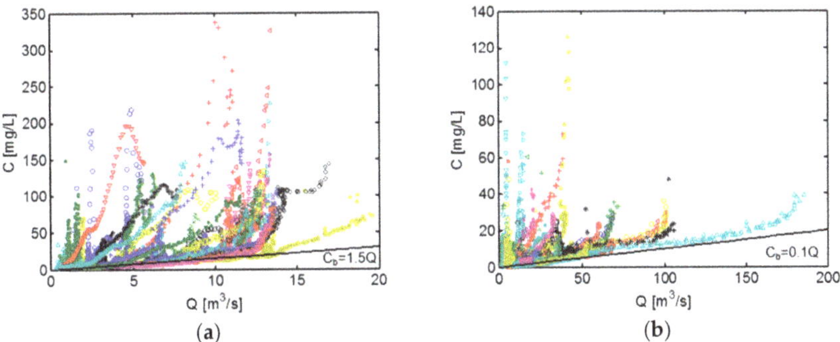

Figure 2. Fine sediment concentration data during falling limb recession of each flood event for Owenabue from 15 September 2009, to 15 September 2010 (**a**), and Bandon from 10 February 2010, to 9 February 2011 (**b**). The symbols with different colors and shapes represent different flood events.

Unlike in the other catchments, the range in Q was rather small for Incline Creek; that is, 0.15 to 0.45 m^3/s. In addition, the flow rates above 0.28 m^3/s are reported only within a resolution of 1 cfs (cubic feet per second), thus 0.028 m^3/s, which was the minimum resolution of the flow rate measurement for this USGS gauging site, leading to limited resolution of the recession curve and vertically aligned data beyond flow rates of 0.28 m^3/s (Figure S1).

The relationship of Q_s to Q for water discharge above 1 m^3/s for the Ribera Salada showed noticeable expansion of vertical scatter. Thus, Q_c was determined to be 1 m^3/s, which corresponds to previous observations [51]. In three catchments, Ribera Salada, Isabena, and Carapelle, considerable scatter was observed in the relationship between Q_s and Q, and power laws show a reasonably good fit with the falling limb recession of flood events (Figure S2). Thus, power laws were utilized for the estimation of the background fine sediment concentrations (Table 1). The Q_c values in two subalpine regions, Galabre and Bès, were determined to be 1 m^3/s and 10 m^3/s, respectively (Figure S1).

The hysteresis analysis of sequential flood events provides better insight into model parameter determination. For example, much less hysteresis is observed during the second flood event of the two sequential events with similar peak flow rates between 9 and 10 August 2002, in Moulinet, France

(Figure 3). This represents the depletion of stored suspended sediments by the preceding flood event and suggests that the suspended sediments were mostly supplied from the catchment during the second flood event. The observed mass of the fine sediments released from the sediment bed was 2.0 Mg for the first flood and 0.6 Mg for the second flood. The background suspended sediment concentration relationship $C_b = 300$ Q for Moulinet shows consistency with the suspended sediment concentration during the recession of the two flood events (Figure 3). A similar pattern of fine sediment dynamics in sequential flood events was also observed in the examples of hysteresis analysis between Q and C in other catchments. For example, much less hysteresis was observed during the second flood event of the two sequential flood events at Owenabue between 5 and 10 December 2009, at Incline Creek between 4 and 6 May 2000 and at the largest catchment Meuse between 1 December 1999 and 26 January 2000 (Figure S3). The background suspended sediment concentration relationship also shows consistency with the suspended sediment concentration during the recession of the two flood events at each site.

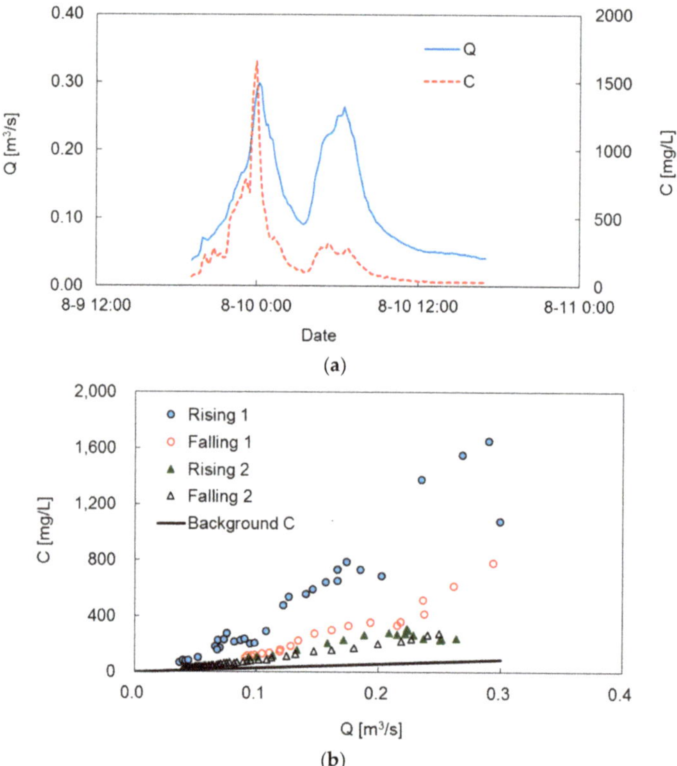

Figure 3. (a) Hydrograph (date format: month-day hour:min) and (b) hysteresis of sediment concentration (C) to suspended sediment load (Q) for sequential floods in Moulinet between 9 and 10 August 2002.

3.2. Model Calibration

Model calibration results are summarized in Table 2. The model was calibrated for each catchment to determine optimal model parameters (α and β) that minimize RSR (Figure S4 in Supplementary Materials). Figure 4 shows examples of model calibrations for two catchments, Owenabue and Bandon in Ireland, where the RSR was 0.49 and 0.36, respectively. The number of flood events in each catchment ranged from 22 to 79 during the calibration period. The filtration parameter, α, ranged from 0.022 to 1650 m^3/s, whereas relative consistency was observed for the bed erosion parameter, β, ranging from

2.4 to 5.3 (Figure S4 in Supplementary Materials). The model parameter sensitivity analysis shows that the model calibration stably converges to optimal values, and is more sensitive to change in α rather than β, in the two small catchments in France (Violettes and Moulinet), the snowmelt-dominated Incline Creek, and Carapelle (Figure S4). In the relatively larger catchments of Isabena, Guerneville, and Meuse, the model calibration also converges to optimal values, although it is more sensitive to change in β rather than α (Figure S4 and Park et al. [32]). In the other six catchments, Galabre, Owenabue, Bandon, Bès, Ribera Salada, and Hopland, the model calibrations are less sensitive to the change of α than to that of β when α is larger than the optimal value (Figure S4 and Park et al. [32]).

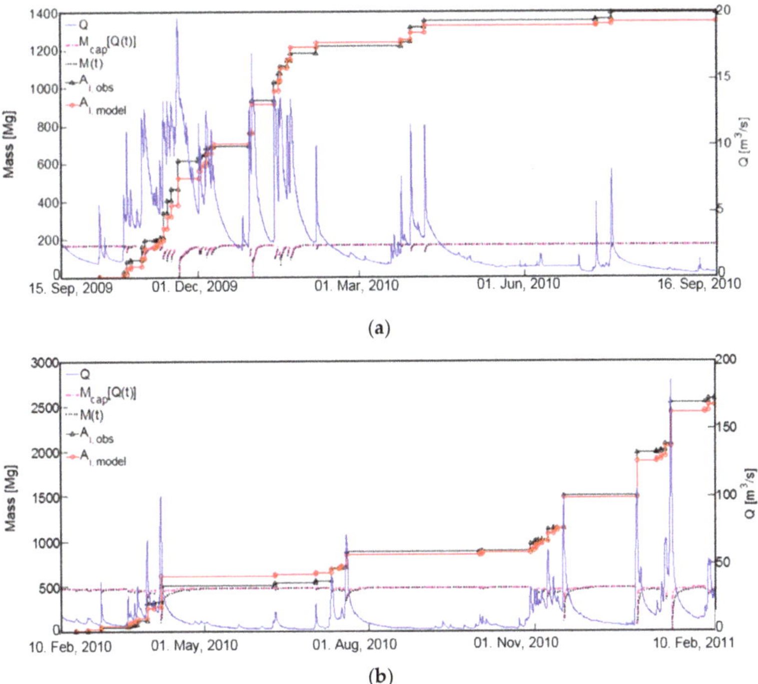

(a)

(b)

Figure 4. Model calibration output for (**a**) Owenabue from 15 September 2009 to 15 September 2010 and (**b**) Bandon from 10 February 2010 to 9 February 2011. The dashed magenta line represents the storage capacity of the sediment bed for fine sediments, M_{cap} [Q(t)], and the dotted black line represents the mass of the fine sediments stored in the sediment bed, M(t). The black line with triangles represents the observed cumulative mass of fine sediments released from the sediment bed by the first i flood events ($A_{i,obs}$) and the red line with circles represents the modeled cumulative mass of sediments released for the first i flood events ($A_{i,model}$).

The model-estimated cumulative mass of fine sediments released from the sediment bed shows a good fit to the observations for the 13 catchments (Table 2 and Figure S5). The average RSR value in these 13 catchments is 0.54, ranging from 0.33 to 0.97, where the largest RSR of 0.97 is observed in the Isabena catchment. The model also estimated an observation bias of less than 20% in 12 catchments, except for Violettes, where R was 1.23 (Table 2).

The mass of fine sediments released from channel beds, that is, the cumulative sum of $M_{f,\,obs}$, ranged from 18% to 65% of the total suspended sediment load in the 13 catchments during the model calibration periods (Table 2), which is consistent with previous studies [52–54].

Table 2. Model calibration results.

No	Sites	Optimal Model Parameter				Calibration	Number of Flood Events	Fine Sediment Mass		
		α (m^3/s)	β	RSR	R	Period		Released from Sediment Bed (Mg)	Total Mass Transported during Observation Period (Mg)	Released from Sediment Bed as Percent of Total (%)
1	Violettes	0.022	5.3	0.68	1.23	1 June 2002–31 May 2003	50	33	144	23
2	Moulinet	0.160	4.5	0.72	1.08	1 July 2002–30 June 2003	61	61	116	52
3	Incline Creek	0.25	5	0.73	0.94	4 April–24 May 2000	36	3.8	11.9	32
4	Galabre	20	2.4	0.64	1.05	3 October 2007–23 December 2009	39	14,000	26,000	54
5	Owenabue	300	4.7	0.49	0.97	15 September 2009–15 September 2010	32	1400	2500	56
6	Bandon	800	4.0	0.36	0.97	10 February 2010–9 February 2011	34	2580	3990	65
7	Bès	10	4.2	0.33	1.07	1 April 2008–31 December 2009	39	46,600	258,000	18
8	Ribera Salada	10	4.4	0.48	1.10	1 November 2005–30 October 2007	47	150	510	29
9	Isabena	10	3.9	0.97	0.93	1 November 2007–30 September 2010	79	623,000	1,063,000	58
10	Carapelle	20	3.6	0.48	0.92	3 March–23 April 2009	22	109,400	360,300	30
11	Hopland [†]	1000	4.4	0.36	1.02	1 October 2010–30 September 2013	69	137,000	330,000	42
12	Guerneville [†]	1050	4.3	0.35	0.83	1 October 2009–31 September 2010/ 1 October 2012–30 September 2013	25 (18 in 2010 water year, 8 in 2013 water year)	422,000	908,000	46
13	Meuse	1650	4.3	0.47	1.06	1 October 1995–30 September 2000	52	1,053,900	1,675,600	63

[†] Data were obtained from a previous study [32].

3.3. Model Validation

The model was applied to five catchments (Ribera Salada, Isabena, Hopland, Guerneville, and Meuse) where data for validation were available. For model validation, it was assumed that there were no observed suspended sediment data during the validation period. Thus, $C_b(t)$ was substituted for $C(t)$ as model input for phases 1 and 3. The number of flood events in the five catchments ranged from 10 to 101 during the validation period. The proportion of the cumulative mass of fine sediments released from the sediment bed during a flood event to total suspended load ranged from 44% to 66%. The mass was not compared in Hopland, where C data were not continuous during the validation period [32]. The validation results for each catchment are summarized in Table 3. The average RSR and R of the five catchments were 0.65 and 1.11, respectively. The model estimated the cumulative released mass of fine sediments well, with only 2% bias, for the Ribera Salada catchment (Figure 5a). The largest RSR of 1.04 was observed in the Isabena catchment, where R was 1.61 (Figure 5b). For the Meuse River, the model showed a good fit with the observations for the 10-year validation period with 20% bias (Figure 5c). Overall, the model bias in the five catchments ranged from 2% to 61% (Table 3).

3.4. Model Parameter Dependence on Catchment Characteristics

Figure 6a plots log α against the log of the catchment area, showing that there is an increase in α with area for smaller catchments. For catchments with areas of approximately 100 km^2 and larger, a limited dependence on the area is observed. Although no clear dependency of α on climatic condition was observed, it is notable that similar values of α, from 10 to 20, were observed in five catchments, four (Galabre, Bès, Ribera Salada, Isabena) in Mediterranean mountainous and one (Carapelle) in Mediterranean climate, regardless of catchment area. These five catchments are located around the Mediterranean Sea with Mediterranean-type climate, where climate and bedrock show similarities (for instance, limestone is present in many areas), except for Carapelle (a clayey-loamy dominated watershed). Relatively larger values of α (>100) were observed for the five largest catchments (i.e., Owenabue, Bandon, Hopland, Guerneville, and Meuse). The catchment data that could be analyzed thus far are limited, but it is encouraging that the filtration parameter is reasonably consistent, with area dependency and with similarity between sites of similar catchment area, among a wide range of catchments. Unlike the filtration parameter (α), the bed erosion parameter (β) values varied within a narrow range of 2.4 to 5.3 in the 13 catchments without notable dependency on catchment area or other environmental characteristics (Figure 6b).

Table 3. Model validation results.

Sites	Validation			Number of Flood Events	Fine Sediment Mass		
	RSR	R	Period		Released from Sediment Bed (Mg)	Total Mass Transported during the Observation Period (Mg)	Released from Sediment Bed as Percent of Total (%)
Ribera Salada	0.71	0.98	1 November 2007–30 October 2008	52	410	930	44
Isabena	1.04	1.61	1 October 2010–30 September 2012	42	188,500	285,000	66
Hopland [†]	0.35	1.22	Ten flood events between 1 October 2013, and 31 December 2014	10	-	-	-
Guerneville [†]	0.54	0.56	1 October 2013–31 December 2014	14	224,000	366,000	61
Meuse	0.62	1.20	1 October 2000–30 November 2010	101	1,668,000	2,884,000	58

[†] Data were obtained from the previous study [32].

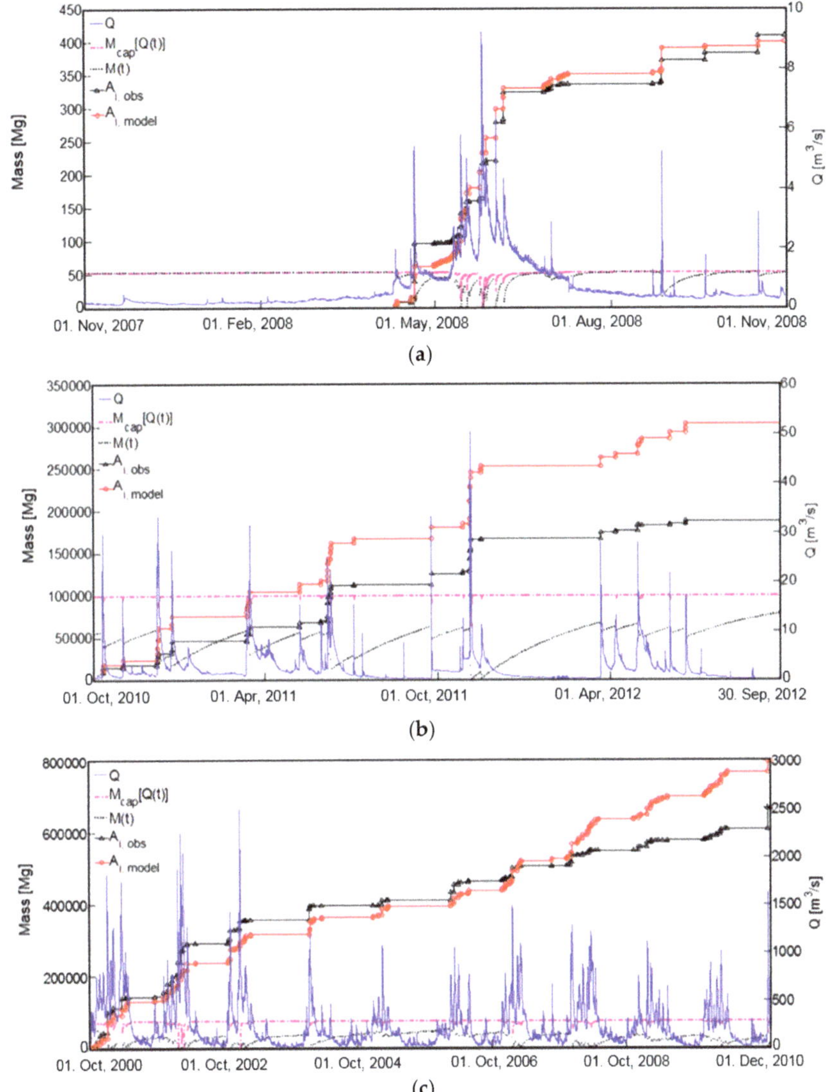

Figure 5. Model validation output for (**a**) the Ribera Salada from 1 November 2007 to 30 October 2008, (**b**) the Isabena from 1 October 2010 to 30 September 2012, and (**c**) the Meuse River from 1 October 2000 to 30 November 2010. The dashed magenta line represents the storage capacity of the sediment bed for fine sediments, $M_{cap}[Q(t)]$, and the dotted black line represents the mass of fine sediments stored in the sediment bed, $M(t)$. The black line with triangles represents the observed cumulative mass of fine sediments released from the sediment bed by the first i flood events ($A_{i,obs}$) and the red line with circles represents the modeled cumulative mass of sediments released for the first i flood events ($A_{i,model}$).

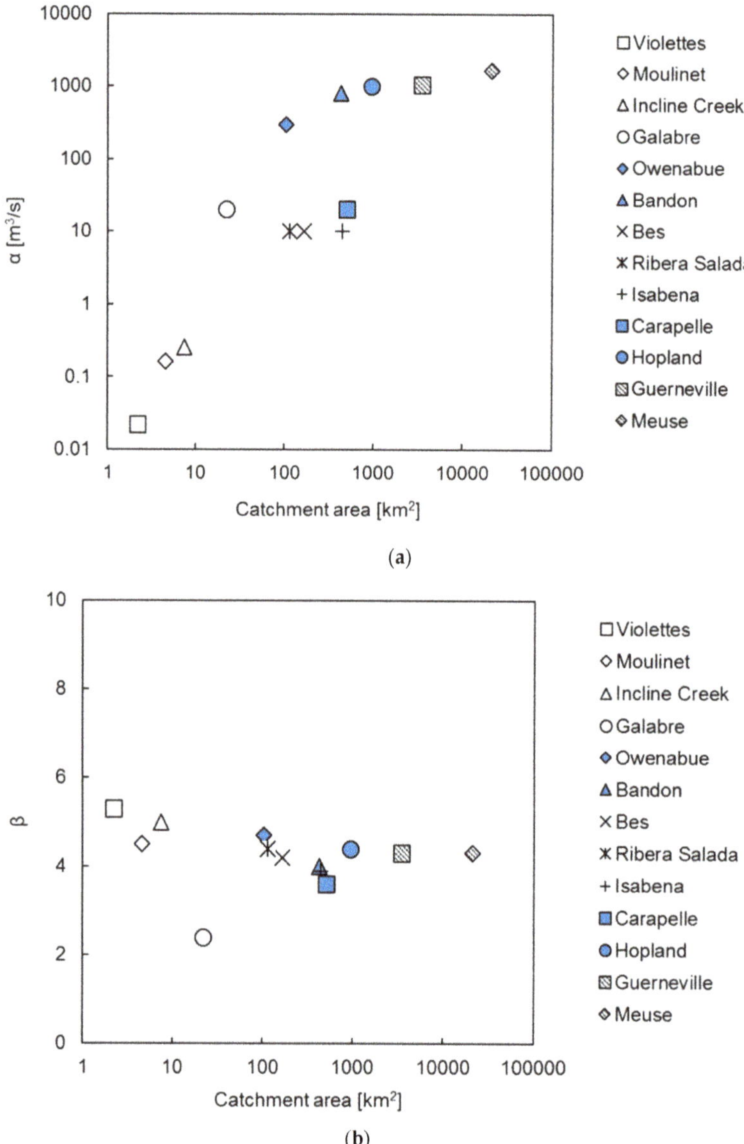

Figure 6. Model parameter dependency on the catchment area. (**a**) the relationship between α and the catchment area and (**b**) the relationship between β and the catchment area.

4. Discussion

The storage and re-suspension model developed by Park et al. [32] was applied to multiple catchments in contrasted climatic conditions with different catchment areas and other catchment characteristics (e.g., soil properties, and land use). Area dependency of the filtration parameter was observed, and it is notable that five catchments with Mediterranean-type climates show consistency in α, with ranges from 10 to 20 regardless of catchment scale. Relatively larger values of α (>100) were observed in larger catchments (i.e., Owenabue, Bandon, Hopland, Guerneville, and Meuse) where climatic conditions and bed material compositions were more variable (Table 1). In the case of the bed erosion parameter, the values were within a narrow range. It is clear that both α and β were affected by various environmental characteristics in each catchment. Understanding the effects of these various environmental characteristics on the model parameters, including the reason for the possible area dependency of α and relative consistency of α in the five catchments of Mediterranean-type climate, is suggested as a subject for future study.

The purpose of this study was to verify the general applicability of the storage and re-suspension model for catchments with various environmental characteristics and to understand the effect of environmental characteristics on the model performance. Therefore, providing specific values of model parameters for each catchment was out of the scope of this paper. Model calibration in 13 catchments shows a good fit with the real observations and thus verifies the possible applicability of the model, whereas there are certain areas where the applicability of the model can be improved in future studies, as described below.

(i) The first is to minimize uncertainties in determining the model input parameters Q_c, Q_{max}, C_b, and M_{max} for each catchment. Uncertainties in the determination of these model parameters can have various causes, such as limited periods of observation in the catchment and variability of sediment dynamics in the natural river system, which induce noticeable scatter of Q_s, even at the same water discharge. For example, as shown in the previous study [32], a larger peak flow rate (1050 m^3/s) than Q_{max} (890 m^3/s) was observed in Guerneville on 12 December 2014, during the validation period, which was not observed during the calibration period. The model-estimated mass of fine sediments released from the sediment bed was 72,600 Mg, which is only ~50% of the amount during the flood event [32].

In Carapelle, a Q_{max} of 37 m^3/s and M_{max} of 23,000 Mg were observed on 5 March 2009, where the maximum flow rate during the model calibration period was 120 m^3/s, registered on 6 March 2009. It is interesting that the flood with the largest mass of fine sediment release in Carapelle had a flow rate of only one third of the highest flow rates on record. As in Cantalice et al. [21], this could be explained by the fact that C in the first flood is related to the re-suspension of deposited sediments, whereas it decreases in the subsequent events as considerable amount of sediment was re-suspended in the previous flood event. This suggests that high rates of fine sediment erosion are possible also at moderate flow rates.

An example of model parameter uncertainty can also be drawn from the model simulation result for the Isabena. There is relatively greater disagreement between the model estimation and observations of the fine sediment mass released from the sediment bed in the Isabena, where the RSR of the model calibration is 0.97. This disagreement may be attributed to the wide range in observed suspended sediment concentrations during flood recession periods and baseflows, leading to considerable uncertainty in the assumed dependence of the background suspended sediment concentration on Q. In the Isabena, C_b was hardly discernable in the linear scale plot; thus, C_b was determined from the log–log scale (Figure S2). The wide ranges of scatter in the relationship between C and Q in the two subalpine regions, Galabre and Bès, also cause uncertainty in the determination of Q_c and C_b in the two catchments. Unlike the Isabena, even with these constraints with respect to the application of the model, the calibrated model provides good representations of the fine sediment release during flood events, with 5% and 7% bias of the observation in the two catchments, Galabre and Bès, respectively.

Overall, model calibration and validation results in this study provide a good estimation of the observed sediment dynamics. However, consistent and longer-term observations will reduce possible uncertainties, including model parameter determination, and further improve model performance.

(ii) Secondly, there are possible issues that can cause considerable model bias, such as episodic events that may produce bank erosion. For example, the model reasonably estimated the observed mass of fine sediments released from the sediment bed in Violettes (Table 2). The observed mass of fine sediments released from the sediment bed ($M_{f,\,obs}$) was less than 1 Mg when Q was less than 0.1 m^3/s, except for the two flood events on 16 and 27 October 2002. The model-estimated mass of fine sediments released from the sediment bed ($M_{f, model}$) was less than 1 Mg for all flood events including the two events specified above, while $M_{f, obs}$ was 3.5 Mg and 2.1 Mg, where the Q of the two flood events on 16 and 27 October was 0.09 m^3/s and 0.06 m^3/s, respectively. Thus, a considerable underestimation of $M_{f, obs}$ was observed for these two flood events. This underestimation by the model may be related to the additional supply of sediment by episodic events, such as bank erosion, associated with cattle trampling in riparian pastures from March to October [33,42].

The two model parameters in this study, α and β, successfully accounted for all variability despite contrasting environmental conditions, whereas episodic events may be considered little and may increase the uncertainty of the numerical model as it was assumed that the mass of fine sediment re-suspended from the sediment bed is proportional to the bed erosion depth which is considered as an exponential function of bed shear stress. Thus, better insights into the characteristics and episodic events of each catchment would provide a better understanding of the possible reasons for the model bias and thus clues for improving the model in future studies.

Various natural characteristics can affect fine sediment dynamics, and thus should be considered for the development of sediment dynamics models for rivers. For example, in catchments with low geomorphic activity, it is possible to obtain statistically significant multivariate models to predict suspended sediment concentrations [36]. However, in catchments with greater sedimentary activity, the results of these models fail to be significant through all the time, which indicates that sediment supply and the role of the riverbed acting as a sediment source or sink play fundamental roles [55]. Riverbed sediment clogging by cohesive sediment (<63 μm) is also one of factors that have possible effect on sediment transport processes and it can be considered that in sites with sandy or sand-gravel such as the Incline Creek, the Hopland and the Guerneville, the bed material is not dominantly cohesive while the sediment is cohesive in sites with clay or silty bed material dominated such as the Violettes, the Moulinet, and the Isábena. The limitation of available information for the detailed characteristics and conditions of field sites limits the practical applicability of sediment model in many cases. Including more parameters would improve model performance but would also increase model complexity, and would require more effort for data observation and thus reduce the practical applicability of the model [32,56]. The model developed in this study includes only two model parameters, α and β, but shows good ability for estimating fine sediment storage mass in 13 catchments with various environmental characteristics, which is an obvious benefit of this model.

5. Conclusions

The general applicability of a storage and re-suspension model was tested in this study. The model was applied to 13 catchments with different climatic conditions (e.g., precipitation and hydrological conditions) and catchment area. The initial model parameters, Q_c, Q_{max}, C_b, and M_{max}, were determined from the observed data. The observed cumulative mass of fine sediments released from the bed in relation to the total suspended load during the model calibration period ranges from 18 to 65%.

The model performance was evaluated using the statistical parameters RSR and R. The optimal model simulation parameters, α and β, were determined to be values that minimize the RSR based on trial and error. The RSR of the model calibration ranges from 0.33 to 0.97, with an average value of 0.54, and the R value ranges from 0.83 to 1.23, with an average of 1.01. The value of the filtration parameter, α, ranges from 0.022 to 1650 m^3/s; a clear area dependency was observed up to an approximate catchment area of less than 100 km^2. The bed erosion parameter, β, was set within a narrower range than α, between 2.4 and 5.3.

It is also noticeable that relatively small values of α from 10 to 20 were observed in five catchments located around the Mediterranean Sea with similar climate, while larger values of α (>100) were observed in five catchments with largest area.

Overall, the model estimated the mass of fine sediments released from the sediment bed in the 13 catchments within ~23% bias.

Supplementary Materials: The following are available online at http://www.mdpi.com/2073-4441/11/5/878/s1, Figure S1: Q vs Q$_s$, Figure S2: Background line, Figure S3: Hysteresis analysis, Figure S4: Sensitivity analysis, Figure S5: Model calibration, Table S1: Study sites.

Author Contributions: Conceptualization: J.P.; Methodology: J.P.; Software: J.P.; Validation: J.P., R.J.B., F.B., M.E., F.G., J.R.H., O.N., J.A.L.-T., D.V.; Formal analysis: J.P., R.J.B., F.B., M.E., F.G., J.R.H., O.N., J.A.L.-T., D.V.; Investigation: J.P., R.J.B., F.B., M.E., F.G., J.R.H., O.N., J.A.L.-T., D.V.; Resources: J.P.; R.J.B., F.B., M.E., F.G., J.R.H., O.N., J.A.L.-T., D.V.; Data curation: J.P., R.J.B., F.B., M.E., F.G., J.R.H., O.N., J.A.L.-T., D.V.; Writing—original draft preparation: J.P.; Writing—review and editing: J.P., R.J.B., F.B., M.E., F.G., J.R.H., O.N., J.A.L.-T., D.V.; Visualization: J.P.; Supervision: J.P.; Project administration: J.P.; Funding acquisition: R.J.B., F.B., M.E., F.G., J.R.H., O.N., J.A.L.-T., D.V.

Funding: J. A. López-Tarazón is in receipt of a Vicenç Mut postdoctoral fellowship (CAIB PD/038/2016), while Damià Vericat is a a Serra Húnter Fellow at the University of Lleida. Authors acknowledge the support from the Economy and Knowledge Department of the Catalan Government through the Consolidated Research Group 2017 SGR 459 Fluvial Dynamics Research Group (RIUS). The data in Galabre and Bès were collected during the STREAMS (Sediment TRansport and Erosion Across Mountains) project, funded by the French National Research Agency (ANR/ BLAN06-1_13915).

Acknowledgments: This study applied a filtration and bed erosion model developed from the analysis of the United States Geological Survey field monitoring data, which was easily accessed from "California datacube" developed by the UC Berkeley Water Center in collaboration with the Lawrence Berkeley National Laboratory and Microsoft Research. Great thanks to Dr. Jacques L. Langlois for kindly provided data in Incline Creek, also thanks to Professor Hans Middelkoop at Utrecht University, the Netherlands, for kindly provided link to the flow and suspended sediment data for Meuse River.

Conflicts of Interest: The authors declare no conflict of interest. The funders had no role in the design of the study; in the collection, analyses, or interpretation of data; in the writing of the manuscript, or in the decision to publish the results.

Abbreviations

M	mass
L	length
T	time
Mg	megagram (equal to 1000 km)
Q (L^3/T)	water discharge
Q$_s$ (M/T)	fine sediment loading rate
Q$_c$ (L^3/T)	critical flow rate required to initiate the mobilization of sediment bed material
Q$_{peak}$ (L^3/T)	peak flow rate of a flood event
Q$_{max}$ (L^3/T)	maximum recorded flow rate during the observation period
C (M/L^3)	concentration of fine sediments within the water column
C$_b$ (M/L^3)	background suspended sediment concentration from the catchment
M (M)	mass of fine sediments accumulated within the pore space of the sediment bed
M$_{max}$ (M)	maximum mass of fine sediments accumulated within the pore space of the sediment bed, representing the capacity of the sediment bed for fine sediments accumulation
M$_{f,obs}$ (M)	Observed mass of fine sediments released from the sediment bed into the water column
M$_{f,model}$ (M)	model-estimated mass of fine sediments released from the sediment bed into the water column
M$_{cap}$ (M)	capacity for fine sediment storage in the sediment bed
A$_{i,obs}$ (M)	observed cumulative mass of fine sediments released in the first i flood events of the season
A$_{i,model}$ (M)	model-estimated cumulative mass of fine sediments released in the first i flood events of the season
α (L^3/T)	sediment removal parameter representing filtration and settling of fine sediments within the sediment bed of the catchment
β	dimensionless sediment bed erosion parameter
t$_{s,i}$ (T)	time at the start of a flood event i
t$_{p,i}$ (T)	time at the peak flow rate of a flood event i

Appendix A. Details of Model Description

Appendix A.1. Fine Sediment Re-Suspension

An analysis of the data from Haschenburger [49] leads to an assumption that the average depth of bed erosion was an exponential function of bed shear stress as

$$\text{bed erosion depth at } Q_{\text{peak}} \propto \exp\left(\beta' Q_{\text{peak}}\right), \tag{A1}$$

where β' is a bed erosion parameter.

From this approach, it is assumed that the maximum bed erosion depth occurs at Q_{max} during the observation period as

$$\text{maximum bed erosion depth at } Q_{\text{max}} \propto \exp(\beta' Q_{\text{max}}), \tag{A2}$$

The mass of fine sediments released from the sediment bed by a flood with Q_{peak} is assumed to be proportional to the bed erosion depth and thus, M_{max} would be expected at Q_{max}. The ratio of $M_{\text{f,model}}$, to M_{max} is

$$\frac{M_{\text{f,model}}}{M_{\text{max}}} = \exp\left[\beta' Q_{\text{peak}} - \beta' Q_{\text{max}}\right] = \exp\left[-\beta\left(1 - \frac{Q_{\text{peak}}}{Q_{\text{max}}}\right)\right], \tag{A3}$$

where β, a dimensionless sediment bed erosion parameter, is defined as $\beta' Q_{\text{max}}$.

Thus, the mass of fine sediments released form the sediment bed by flood event i is

$$M_{\text{fi,model}} = M_{\text{max}} \exp\left[-\beta\left(1 - \frac{Q(t_{p,i})}{Q_{\text{max}}}\right)\right], \tag{A4}$$

where $t_{p,i}$ is time at Q_{peak} of flood event i.

Appendix A.2. Fine Sediment Accumulation during Flood Recession

In this study, it is assumed that partial bed fluidization or erosion during falling limb of a flood event reduces the volume of porous media available for particle accumulation. For flood recession period with flow rate $Q(t)$, Equations (A1) and (A2) are modified using the substitution $\beta = \beta' Q_{\text{max}}$ and applied to the model as

$$\text{bed erosion depth at } Q(t) \propto \exp\left[\beta \frac{Q(t)}{Q_{\text{max}}}\right], \tag{A5}$$

$$\text{maximum bed erosion depth at } Q_{\text{max}} \propto \exp[\beta], \tag{A6}$$

The available capacity for fine sediment storage is estimated by subtracting the bed erosion depth during flow recession at a flow rate $Q(t)$ from the maximum bed erosion depth. Thus in the model, the available capacity for fine sediment storage in the sediment bed is represented during the falling limb of a flood event by

$$\frac{M_{\text{cap}}[Q(t)]}{M_{\text{max}}} = \frac{\exp[\beta] - \exp\left[\beta \frac{Q(t)}{Q_{\text{max}}}\right]}{\exp[\beta]} = 1 - \exp\left[-\beta\left(1 - \frac{Q(t)}{Q_{\text{max}}}\right)\right], \tag{A7}$$

$$M_{\text{cap}}[Q(t)] = M_{\text{max}}\left\{1 - \exp\left[-\beta\left(1 - \frac{Q(t)}{Q_{\text{max}}}\right)\right]\right\}, \tag{A8}$$

References

1. Droppo, I.G.; Liss, S.N.; Williams, D.; Nelson, T.; Jaskot, C.; Trapp, B. Dynamic existence of waterborne pathogens within river sediment compartments. Implications for water quality regulatory affairs. *Environ. Sci. Technol.* **2009**, *43*, 1737–1743. [CrossRef] [PubMed]
2. Søndergaard, M.; Jensen, J.P.; Jeppesen, E. Role of sediment and internal loading of phosphorus in shallow lakes. *Hydrobiologia* **2003**, *506*, 135–145. [CrossRef]
3. Singer, M.B.; Aalto, R.; James, L.A.; Kilham, N.E.; Higson, J.L.; Ghoshal, S. Enduring legacy of a toxic fan via episodic redistribution of California gold mining debris. *Proc. Natl. Acad. Sci. USA* **2013**, *110*, 18436–18441. [CrossRef]
4. Herrero, A.; Vila, J.; Eljarrat, E.; Ginebreda, A.; Sabater, S.; Batalla, R.J.; Barceló, D. Transport of sediment borne contaminants in a Mediterranean river during a high flow event. *Sci. Total Environ.* **2018**, *633*, 1392–1402. [CrossRef] [PubMed]
5. Quesada, S.; Tena, A.; Guillén, D.; Ginebreda, A.; Vericat, D.; Martínez, E.; Navarro-Ortega, A.; Batalla, R.J.; Barceló, D. Dynamics of suspended sediment borne persistent organic pollutants in a large regulated Mediterranean river (Ebro, NE Spain). *Sci. Total Environ.* **2014**, *473*, 381–390. [CrossRef] [PubMed]
6. Suttle, K.B.; Power, M.E.; Levine, J.M.; McNeely, C. How fine sediment in riverbeds impairs growth and survival of juvenile salmonids. *Ecol. Appl.* **2004**, *14*, 969–974. [CrossRef]
7. Béjar, M.; Gibbins, C.; Vericat, D.; Batalla, R.J. Effects of suspended sediment transport on invertebrate drift. *River Res. Appl.* **2017**, *33*, 1655–1666. [CrossRef]
8. Buendia, C.; Gibbins, C.N.; Vericat, D.; Batalla, R.J.; Douglas, A. Detecting the structural and functional impacts of fine sediment on stream invertebrates. *Ecol. Indic.* **2013**, *25*, 184–196. [CrossRef]
9. Buendia, C.; Gibbins, C.N.; Vericat, D.; Batalla, R.J. Effects of flow and fine sediment dynamics on the turnover of stream invertebrate assemblages. *Ecohydrology* **2014**, *7*, 1105–1123. [CrossRef]
10. Pelletier, J.D. A spatially distributed model for the long-term suspended sediment discharge and delivery ratio of drainage basins. *J. Geophys. Res.* **2012**, *117*, F2. [CrossRef]
11. Warrick, J.A.; Mertes, L.A.K. Sediment yield from the tectonically active semiarid Western Transverse Ranges of California. *Geol. Soc. Am. Bull.* **2009**, *121*, 1054–1070. [CrossRef]
12. Syvitski, J.P.; Morehead, M.D.; Bahr, D.B.; Mulder, T. Estimating fluvial sediment transport: The rating parameters. *Water Resour. Res.* **2000**, *36*, 2747–2760. [CrossRef]
13. Gao, P.; Nearing, M.A.; Commons, M. Suspended sediment transport at the instantaneous and event time scales in semiarid watersheds of southeastern Arizona, USA. *Water Resour. Res.* **2013**, *49*, 6857–6870. [CrossRef]
14. Warrick, J.A.; Madej, M.A.; Goni, M.A.; Wheatcroft, R.A. Trends in the suspended-sediment yields of coastal rivers of northern California, 1955-2010. *J. Hydrol.* **2013**, *489*, 108–123. [CrossRef]
15. Alexandrov, Y.; Laronne, J.B.; Reid, I. Intra-event and inter-seasonal behaviour of suspended sediment in flash floods of the semi-arid northern Negev, Israel. *Geomorphology* **2007**, *85*, 85–97. [CrossRef]
16. Harrington, S.T.; Harrington, J.R. An assessment of the suspended sediment rating curve approach for load estimation on the Rivers Bandon and Owenabue, Ireland. *Geomorphology* **2013**, *185*, 27–38. [CrossRef]
17. Klein, M. Anti clockwise hysteresis in suspended sediment concentration during individual storms: Holbeck catchment, Yorkshire England. *Catena* **1984**, *11*, 251–257. [CrossRef]
18. Carson, M.A.; Taylor, C.H.; Grey, B.J. Sediment production in a small Appalachian watershed during spring runoff: the Eaton Basin, 1970-1972. *Can. J. Earth Sci.* **1973**, *10*, 1707–1734. [CrossRef]
19. Williams, G.P. Sediment concentration versus water discharge during single hydrologic events in rivers. *J. Hydrol.* **1989**, *111*, 89–106. [CrossRef]
20. Harvey, J.W.; Drummond, J.D.; Martin, R.L.; McPhillips, L.E.; Packman, A.I.; Jerolmack, D.J.; Stonedahl, S.H.; Aubeneau, A.F.; Sawyer, A.H.; Larsen, L.G.; et al. Hydrogeomorphology of the hyporheic zone: Stream solute and fine particle interactions with a dynamic streambed. *J. Geophys. Res.* **2012**, *117*, G00N11. [CrossRef]
21. Cantalice, J.R.B.; Cunha, M.; Stosic, B.D.; Piscoya, V.C.; Guerra, S.M.S.; Singh, V.P. Relationship between bedload and suspended sediment in the sand-bed Exu River, in the semi-arid region of Brazil. *Hydrol. Sci. J.* **2013**, *58*, 1789–1802. [CrossRef]
22. Piqué, G.; López-Tarazón, J.A.; Batalla, R.J. Variability of in-channel sediment storage in a river draining highly erodible areas (the Isábena, Ebro Basin). *J. Soil. Sediment.* **2014**, *14*, 2031–2044. [CrossRef]

23. Yang, C.C.; Lee, K.T. Analysis of flow-sediment rating curve hysteresis based on flow and sediment travel time estimations. *Int. J. Sediment Res.* **2018**, *33*, 171–182. [CrossRef]

24. Juez, C.; Hassan, M.A.; Franca, M.J. The origin of fine sediment determines the observations of suspended sediment fluxes under unsteady flow conditions. *Water Resour. Res.* **2018**, *54*, 5654–5669. [CrossRef]

25. Walling, D.E. The sediment delivery problem. *J. Hydrol.* **1983**, *65*, 209–237. [CrossRef]

26. Negev, M. *Analysis of Data on Suspended Sediment Discharge in Several Streams in Israel*; Hydrological Paper No. 12; Israel Hydrological Service: Jerusalem, Israel, 1969; pp. 8–17.

27. Alexandrov, Y.; Cohen, H.; Laronne, J.B.; Reid, I. Suspended sediment load, bed load, and dissolved load yields from a semiarid drainage basin: A 15-year study. *Water Resour. Res.* **2009**, *45*, W08408. [CrossRef]

28. Brasington, J.; Richards, K. Turbidity and suspended sediment dynamics in small catchments in the Nepal Middle Hills. *Hydrol. Process.* **2000**, *14*, 2559–2574. [CrossRef]

29. Stubblefield, A.P.; Reuter, J.E.; Goldman, C.R. Sediment budget for subalpine watersheds, Lake Tahoe, California, USA. *Catena* **2009**, *76*, 163–172. [CrossRef]

30. Achite, M.; Ouillon, S. Suspended sediment transport in a semiarid watershed, Wadi Abd, Algeria (1973–1995). *J. Hydrol.* **2007**, *343*, 187–202. [CrossRef]

31. Hunt, J.R. Coupling fine particle and bedload transport in gravel-bedded streams. *J. Hydrol.* **2017**, *552*, 532–543.

32. Park, J.; Hunt, J.R. Modeling fine particle dynamics in gravel-bedded streams: Storage and re-suspension of fine particles. *Sci. Total Environ.* **2018**, *634*, 1042–1053. [CrossRef]

33. Lefrançois, J.; Grimaldi, C.; Gascuel-Odoux, C.; Gilliet, N. Suspended sediment and discharge relationships to identify bank degradation as a main sediment source on small agricultural catchments. *Hydrol. Process.* **2007**, *21*, 2923–2933. [CrossRef]

34. Langlois, J.L.; Johnson, D.W.; Mehuys, G.R. Suspended sediment dynamics associated with snowmelt runoff in a small mountain stream of Lake Tahoe (Nevada). *Hydrol. Process.* **2005**, *19*, 3569–3580. [CrossRef]

35. Navratil, O.; Evrard, O.; Esteves, M.; Legout, C.; Ayrault, S.; Némery, J.; Mate-Marin, A.; Ahmadi, M.; Lefévre, I.; Poirel, A.; Bonté, P. Temporal variability of suspended sediment sources in an alpine catchment combining river/rainfall monitoring and sediment fingerprinting. *Earth Surf. Proc. Land.* **2012**, *37*, 828–846. [CrossRef]

36. Tuset, J.; Vericat, D.; Batalla, R. Rainfall, runoff and sediment transport in a Mediterranean mountainous catchment. *Sci. Total Environ.* **2016**, *540*, 114–132. [CrossRef] [PubMed]

37. López-Tarazón, J.A.; Batalla, R.J. Dominant discharges for suspended sediment transport in a highly active Pyrenean river. *J. Soil. Sediment.* **2014**, *14*, 2019–2030. [CrossRef]

38. Gentile, F.; Bisantino, T.; Corbino, R.; Milillo, F.; Romano, G.; Trisorio Liuzzi, G. Monitoring and analysis of suspended sediment transport dynamics in the Carapelle torrent (southern Italy). *Catena* **2010**, *80*, 1–8. [CrossRef]

39. Ritter, J.R.; Brown, W.M., III. *Turbidity and Suspended-Sediment Transport in the Russian River Basin*; Open-File Report 72-316; United States Geological Survey: Menlo Park, CA, USA, 1971; pp. 15–29.

40. Doomen, A.; Wijma, E.; Zwolsman, J.J.; Middelkoop, H. Predicting suspended sediment concentrations in the Meuse River using a supply-based rating curve. *Hydrol. Process.* **2008**, *22*, 1846–1856. [CrossRef]

41. Birgand, F.; Lefrancois, T.; Grimaldi, C.; Novince, E.; Gilliet, N.; Odoux, C.G. Mesure des flux et échantillonnage des matières en suspension sur de petits cours déau. *Ingénieries* **2004**, *40*, 21–35.

42. Vongvixay, A.; Grimaldi, C.; Dupas, R.; Fovet, O.; Birgand, F.; Gilliet, N.; Gascuel-Odoux, C. Contrasting suspended sediment export in two small agricultural catchments: Cross-influence of hydrological behaviour and landscape degradation or stream bank management. *Land Degrad. Dev.* **2018**, *29*, 1385–1396. [CrossRef]

43. Evrard, O.; Navratil, O.; Ayrault, S.; Ahmadi, M.; Némery, J.; Legout, C.; Lefévre, I.; Poirel, A.; Bonté, P.; Esteves, M. Combining suspended sediment monitoring and fingerprinting to determine the spatial origin of fine sediment in a mountainous river catchment. *Earth Surf. Proc. Land.* **2011**, *36*, 1072–1089. [CrossRef]

44. Navratil, O.; Esteves, M.; Legout, C.; Gratiot, N.; Nemery, J.; Willmore, S.; Grangeon, T. Global uncertainty analysis of suspended sediment monitoring using turbidimeter in a small mountainous river catchment. *J. Hydrol.* **2011**, *398*, 246–259. [CrossRef]

45. Vericat, D.; Batalla, R.J.; Gibbins, C.N. Sediment entrainment and depletion from patches of fine material in a gravel-bed river. *Water Resour. Res.* **2008**, *44*, W11415. [CrossRef]

46.	López-Tarazón, J.A.; Batalla, R.; Vericat, D.; Francke, T. Suspended sediment transport in a highly erodible catchment: the River Isábena (Southern Pyrenees). *Geomorphology* **2009**, *109*, 210–221. [CrossRef]

47.	Bisantino, T.; Bingner, R.; Chouaib, W.; Gentile, F.; Trisorio Liuzzi, G. Estimation of runoff, peak discharge and sediment load at the event scale in a medium-size Mediterranean watershed using the AnnAGNPS model. *Land Degrad. Dev.* **2015**, *26*, 340–355. [CrossRef]

48.	García-Rama, A.; Pagano, S.G.; Gentile, F.; Lenzi, M.A. Suspended sediment transport analysis in two Italian instrumented catchments. *J. Mt. Sci.* **2016**, *13*, 957–970. [CrossRef]

49.	Haschenburger, J.K. A probability model of scour and fill depths in gravel-bed channels. *Water Resour. Res.* **1999**, *35*, 2857–2869. [CrossRef]

50.	Moriasi, D.N.; Arnold, J.G.; Van Liew, M.W.; Bingner, R.L.; Harmel, R.D.; Veith, T.L. Model evaluation guidelines for systematic quantification of accuracy in watershed simulations. *Trans. ASABE* **2007**, *50*, 885–900. [CrossRef]

51.	Vericat, D.; Batalla, R.J. Sediment transport from continuous monitoring in a perennial Mediterranean stream. *Catena* **2010**, *82*, 77–86. [CrossRef]

52.	Collins, A.L.; Walling, D.E. Fine-grained bed sediment storage within the main channel systems of the Frome and Piddle catchments, Dorset, UK. *Hydrol. Process.* **2007**, *21*, 1448–1459. [CrossRef]

53.	Walling, D.E.; Owens, P.N.; Leeks, G.J. The role of channel and floodplain storage in the suspended sediment budget of the River Ouse, Yorkshire, UK. *Geomorphology* **1998**, *22*, 225–242. [CrossRef]

54.	Walling, D.E.; Collins, A.L.; Jones, P.A.; Leeks, G.J.L.; Old, G. Establishing fine-grained sediment budgets for the Pang and Lambourn LOCAR catchments, UK. *J. Hydrol.* **2006**, *330*, 126–141. [CrossRef]

55.	López-Tarazón, J.A.; Batalla, R.J.; Vericat, D.; Balasch, J. Rainfall, runoff and sediment transport relations in a mesoscale mountainous catchment: The River Isábena (Ebro basin). *Catena* **2010**, *82*, 23–34. [CrossRef]

56.	Bhosekar, A.; Ierapetritou, M. Advances in surrogate based modeling, feasibility analysis and optimization: A review. *Comput. Chem. Eng.* **2017**, *108*, 250–267. [CrossRef]

 *water*

Article

A New Mass-Conservative, Two-Dimensional, Semi-Implicit Numerical Scheme for the Solution of the Navier-Stokes Equations in Gravel Bed Rivers with Erodible Fine Sediments

Maurizio Tavelli [*,†], **Sebastiano Piccolroaz** [†], **Giulia Stradiotti** [†], **Giuseppe Roberto Pisaturo** and **Maurizio Righetti**

Faculty of Science and Technology, Free University of Bozen-Bolzano, Universitätsplatz - Piazza Università 5, 39100 Bozen-Bolzano, Italy; spiccolroaz@unibz.it (S.P.); gstradiotti@unibz.it (G.S.); gpisaturo@unibz.it (G.R.P.); mrighetti@unibz.it (M.R.)
* Correspondence: mtavelli@unibz.it
† These authors contributed equally to this work.

Received: 14 December 2019; Accepted: 18 February 2020; Published: 3 March 2020

Abstract: The selective trapping and erosion of fine particles that occur in a gravel bed river have important consequences for its stream ecology, water quality, and overall sediment budgeting. This is particularly relevant in water bodies that experience periodic alternation between sediment supply-limited conditions and high sediment loads, such as downstream from a dam. While experimental efforts have been spent to investigate fine sediment erosion and transport in gravel bed rivers, a comprehensive overview of the leading processes is hampered by the difficulties in performing flow field measurements below the gravel crest level. In this work, a new two-dimensional, semi-implicit numerical scheme for the solution of the Navier-Stokes equations in the presence of deposited and erodible sediment is presented, and tested against analytical solutions and performing numerical tests. The scheme is mass-conservative, computationally efficient, and allows for a fine discretization of the computational domain. Overall, this makes the model suitable to appreciate small-scales phenomena such as inter-grain circulation cells, thus offering a valid alternative to evaluate the shear stress distribution, on which erosion and transport processes depend, compared to traditional experimental approaches. In this work, we present proof-of-concept of the proposed model, while future research will focus on its extension to a three-dimensional and parallelized version, and on its application to real case studies.

Keywords: sediment transport; sediment entrainment; clogging; colmation; numerical modeling

1. Introduction

River damming produces alterations on the natural river functionality, both in the water discharge, as well as in the sediment transport and connectivity [1,2]. In particular, large dams [3] are estimated to trap more than 99% of the sediments entering the reservoir [4]. This causes the progressive silting of the reservoir and inhibits the sediment load in the river flow downstream of the dam, thus altering the river morphology [5] and the aggradation/degradation dynamics that are closely linked to the balance between upstream sediment supply and local transport capacity conditions [6]. Sediment supply-limited conditions from upstream cause the (selective) erosion of finer particles from the granular bed, until the flow is unable to move the coarser grains and new equilibrium conditions are reached [7]. During this degradation process, the median size of the bed material progressively coarsens and the sediment transport rate decreases, leading to a process known as bed armoring [8–10].

Occasionally, armored stream beds may be subject to high sediment loads, for example during dam flushing or removal operations, or during flood events associated to large sediment input from lateral inflows. Under these conditions, finer particles infiltrate into the void spaces of the immobile coarse bed grains, according to a selective trapping mechanism that is reciprocal to the selective erosion that caused bed armoring [11]. If the infiltration of fine particles into the coarse bed interstices is extensive, the volume of voids drops increasing the compactness of the stream bed texture, thus decreasing its hydraulic conductivity and increasing its resistance to flow (e.g., [12,13]). This process is known as colmation or clogging (see, for example, [14] for a review), and has significant impacts on stream ecology (e.g., [15–18]), exchanges of water, dissolved substances and heat with the underlying hyporheic zone and groundwater [19], and flow and turbulence structure [20–22]. Under high flow conditions, the armour layer can break up and the entire river bed becomes mobilized, hence resetting the bed morphology and grain size distribution. However, if the erosion capacity of the flow is not sufficient to remove the coarse grains, as soon as the sediment load from upstream declines, the reestablishment of sediment supply-limited conditions reactivates selective erosion of finer particles, sustaining the armoring of the stream bed.

The cleaning dynamics controlling the erosion of finer particles from coarse granular beds is inherently different from that typical of uniform sized beds. In fact, in armored beds, the presence of macro-roughness due to the coarser particles alters the flow structure and, consequently, the distribution of the stress components below the gravel crest level. In these beds, besides turbulent stresses, form-induced stresses and form drag also contribute to the total shear stress distribution [23]. In addition, the vertical component of the stress responsible for lifting and transporting fine material was found to be decreasing below gravel crest level [22,24,25]. This alteration is mirrored in the reduced sediment entrainment and transport capacity of the flow, which is affected also by the smaller fine sediment-water active interface with respect to that of a uniform bed.

It is therefore clear that the traditional formulae derived for uniform bed cases fail to describe erosion and sediment transport processes over immobile gravel beds. In fact, these formulae do not account for the reduction in the effective part of the shear stress, nor the reduction in the fine sediment-water active interface. In this respect, performing laboratory experiments is a common methodological practice for investigating selective transport dynamics in armored beds and gaining useful elements to derive empirical formulas of fine sediment transport (e.g., [26–29]). Experimental research is typically carried out in laboratory flumes using laser-scanner (e.g., [29]) or digital photogrammetry (e.g., [30]) to measure the changes in the topography. The track of these changes in the fine sediment level inside the gravel matrix is usually coupled with measures of transport rate (trough sieves or density cells (e.g., [27,29]) or concentration (e.g., [26,31]) to quantify the fine sediment transport and/or erosion rate between the gravel. Based on these experimental approaches, useful fine sediment transport formulae has been proposed in previous literature, as for instance in [26–29].

Despite the above cited empirical formulae and despite the examples of direct measurements of the flow field in the roughness layer [23] (e.g., [22,24,25,32]), a comprehensive framework on the inter-grain flow and sediments dynamics in gravel bed rivers still poses some scientific challenges. These challenges are primarily due to the operational difficulties to perform velocity measurements far below the gravel crest level and to quantify the relative contribution of the form drag to the total shear stress [25]. In addition, a fair comparison among existing studies is not obvious due to the differences in the bed topographies, which chiefly controls the distribution of the shear stress components [25], thus hampering the derivation of general considerations. In this context, fine-scale numerical models can offer a valid alternative to overcome the inherent difficulties of fine-resolution, inter-grain experimental measurements. At the same time, they can easy the investigation of the role of the geometry in affecting the stresses distribution, provided that the setup and repetition of laboratory experiments with different configurations is not a minor matter. While examples of Direct Numerical

Simulation (DNS) over rough bed configurations do exist (e.g., [33–35]), the inclusion of sediment active layers in fine-resolution hydrodynamics model is a relatively unexplored area of research.

In this study, we present and test a new semi-implicit numerical scheme for the solution of the two-dimensional Navier-Stokes equations, in which we included the possibility to easily simulate sediment entertainment and transport processes. The scheme, based on the method proposed by [36–38], is mass-conservative, computationally efficient, and able to solve the small-scale structures that characterize inter-grain flow field. In this study, we present proof-of-concept and preliminary results of this model as a first step towards its extension to a complete three-dimensional model coupled with a turbulence closure scheme. To this end, we focus on the validation of the proposed model against numerical tests and on showing its potential for applications in the context of fine sediment transport dynamics in gravel bed rivers.

2. Materials and Methods

2.1. Numerical Model

2.1.1. Governing Equations

The governing equations are the two-dimensional incompressible Navier-Stokes, that are composed by the momentum conservation:

$$
\begin{aligned}
\frac{\partial u}{\partial t} + \frac{\partial uu}{\partial x} + \frac{\partial uw}{\partial z} &= -\frac{\partial p}{\partial x} + \nu \left(\frac{\partial^2 u}{\partial x^2} + \frac{\partial^2 u}{\partial z^2} \right), \\
\frac{\partial w}{\partial t} + \frac{\partial wu}{\partial x} + \frac{\partial ww}{\partial z} &= -\frac{\partial p}{\partial z} + \nu \left(\frac{\partial^2 w}{\partial x^2} + \frac{\partial^2 w}{\partial z^2} \right) - g,
\end{aligned}
\tag{1}
$$

and the incompressibility condition:

$$
\frac{\partial u}{\partial x} + \frac{\partial w}{\partial z} = 0,
\tag{2}
$$

where u, w are the components of the velocity field in the $x-$ and $z-$ direction, $\nu = \mu/\rho$ is the kinematic viscosity coefficient, μ is the dynamic viscosity coefficient, ρ is the fluid density, p is the pressure term normalized by the fluid density, and g is the gravity acceleration. If the pressure is assumed to be locally hydrostatic, it can be expressed in terms of the atmospheric pressure p_a and the hydraulic head η as:

$$
p = p_a + g(\eta - z).
\tag{3}
$$

By means of Equation (3) it is possible to rewrite the momentum Equations (1) in terms of the hydraulic head η:

$$
\begin{aligned}
\frac{\partial u}{\partial t} + \frac{\partial uu}{\partial x} + \frac{\partial uw}{\partial z} &= -g\frac{\partial \eta}{\partial x} + \nu \left(\frac{\partial^2 u}{\partial x^2} + \frac{\partial^2 u}{\partial z^2} \right), \\
\frac{\partial w}{\partial t} + \frac{\partial wu}{\partial x} + \frac{\partial ww}{\partial z} &= -g\frac{\partial \eta}{\partial z} + \nu \left(\frac{\partial^2 w}{\partial x^2} + \frac{\partial^2 w}{\partial z^2} \right).
\end{aligned}
\tag{4}
$$

Further introducing the mass conservation law for the suspended sediment concentration c (here defined as volume concentration, hence simplifying the mass conservation law assuming constant sediment density), and defining its relation with the sediment level $S = S(x, z, t)$, two more equations need to be added to the system:

$$
\frac{\partial c}{\partial t} + \frac{\partial uc}{\partial x} + \frac{\partial (w - w_c)c}{\partial z} = f_{cS}(S, c)
\tag{5}
$$

$$
\frac{\partial S}{\partial t} = f_{Sc}(S, c),
\tag{6}
$$

where w_c is the falling velocity of the suspended sediment; f_{Sc} and f_{cS} are two functions that regulate the mass transfer between the passive concentration c and the sediment level S. Note that in Equations (5) and (6) the sediment dynamics are intrinsically assumed to be driven only by the sedimentation/erosion process through the functions f_{Sc} and f_{cS}. The suspended concentration is expressed as volume of suspended sediment per unit of control volume (i.e., volume concentration relative to each computational cell), thus it can range in $c \in [0, 1 - \phi]$, where ϕ indicates the sediment porosity. Similarly, the sediment level S is defined per unit of volume, thus it can vary in the range $S \in [0, 1]$, where $S = 1$ corresponds to a computational cell completely filled by the sediment. Note that the suspended sediment is assumed to be passively transported by the fluid, while the deposited sediment level actively interacts with the fluid dynamics as in general it may vary in time, thus changing the boundaries of the fluid domain. The exchange between suspended and deposited sediment must satisfy the mass conservation law:

$$\frac{d}{dt}[c + S(1 - \phi)] = 0. \tag{7}$$

2.1.2. Staggered Mesh

The Partial Differential Equations (PDE) system composed by Equations (2), (4) and (5) is numerically solved in a domain $\Omega(t) \subseteq \Omega \in \Re^2$ discretized by a regular mesh $\Omega = \bigcup_{i,k} \Omega_{i,k}$, where $\Omega_{i,k} = [x_{i-\frac{1}{2}}, x_{i+\frac{1}{2}}] \times [z_{k-\frac{1}{2}}, z_{k+\frac{1}{2}}]$ are rectangular control volumes centered in (x_i, z_k). The mesh is assumed to be homogeneous with $\Delta x = x_{i+\frac{1}{2}} - x_{i-\frac{1}{2}}$ and $\Delta z = z_{k+\frac{1}{2}} - z_{k-\frac{1}{2}}$, for $i = 1, \ldots, I_{max}$ and $k = 1, \ldots, K_{max}$. The main quantities are defined following a staggered approach. In particular, the hydraulic head η, the suspended concentration c and the sediment level S are defined in the cell centers, namely $\eta = \eta_{ik}$, $c = c_{ik}$, $S = S_{ik}$. The horizontal velocity is defined in the center of the vertical edges, while the vertical velocity in the center of horizontal edges, namely, $u = u_{i\pm\frac{1}{2},k}$ and $w = w_{i,k\pm\frac{1}{2}}$.

The system is solved for discrete times $t = t^0, t^1, \ldots, t^n, \ldots$, hence the solution at a certain time t^n is marked with the upper index n, namely $\xi(t^n) := \xi^n$, while the time interval between two consecutive steps is $\Delta t^n = t^{n+1} - t^n$. Note that the effective volume in each control cell can have different values due to the presence of the sediment, and can be computed as $V_{i,k} = \Delta x \Delta z(1 - S_{ik})$. Since $S_{i,k} \in [0, 1]$, the volume computed in this way satisfies $0 \leq V_{i,k} \leq \Delta x \Delta z$. The effective edge length can change as well, being defined above the sediment top: for any time t^n the effective vertical and horizontal edges lengths are defined as $\Delta z^n_{i\pm\frac{1}{2},k}$ and $\Delta x^n_{i,k\pm\frac{1}{2}}$, see Figure 1. For non-saturated cells (i.e., $S < 1$), the effective edges lengths are computed using an upwind procedure:

$$\Delta z^n_{i+\frac{1}{2},k} = \begin{cases} \dfrac{V^n_{i,k}}{\Delta x} & u^n_{i+\frac{1}{2},k} > 0 \\[2mm] \dfrac{V^n_{i+1,k}}{\Delta x} & u^n_{i+\frac{1}{2},k} < 0 \\[2mm] \dfrac{\max(V^n_{i,k}, V^n_{i+1,k})}{\Delta x} & u^n_{i+\frac{1}{2},k} = 0 \end{cases} \qquad \Delta x^n_{i,k+\frac{1}{2}} = \begin{cases} 0 & S_{i,k} = 1 \\[4mm] \Delta x & S_{i,k} \neq 1 \end{cases} \tag{8}$$

We further assume that $\Delta z^n_{i\pm\frac{1}{2},k} = \Delta x^n_{i,k\pm\frac{1}{2}} = 0$ for any (i, k) where $S_{i,k} = 1$.

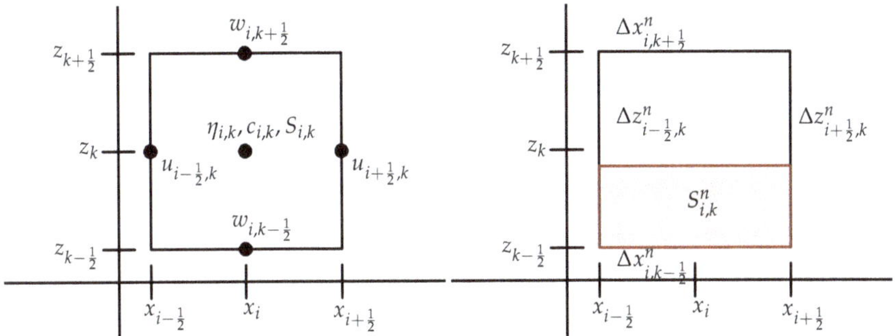

Figure 1. Schematic of the used staggered mesh.

2.1.3. Numerical Method

In order to avoid the development of a fully nonlinear system, the convective and the viscous terms in Equation (4) are discretized explicitly, while the velocity field and the hydraulic head in Equations (2) and (4) are discretized implicitly. A finite volume approximation of the continuity Equation (2) reads:

$$\Delta z^n_{i+\frac{1}{2},k} u^{n+1}_{i+\frac{1}{2},k} - \Delta z^n_{i-\frac{1}{2},k} u^{n+1}_{i-\frac{1}{2},k} + \Delta x^n_{i,k+\frac{1}{2}} w^{n+1}_{i,k+\frac{1}{2}} - \Delta x^n_{i,k-\frac{1}{2}} w^{n+1}_{i,k-\frac{1}{2}} = 0. \tag{9}$$

Consistently, a finite difference approximation for the momentum Equations (4) reads:

$$u^{n+1}_{i+\frac{1}{2},k} = Fu^n_{i+\frac{1}{2},k} - \Delta t\, g\, \frac{\eta^{n+1}_{i+1,k} - \eta^{n+1}_{i,k}}{\Delta x} \tag{10}$$

$$w^{n+1}_{i,k+\frac{1}{2}} = Fw^n_{i,k+\frac{1}{2}} - \Delta t\, g\, \frac{\eta^{n+1}_{i,k+1} - \eta^{n+1}_{i,k}}{\Delta z}, \tag{11}$$

where $Fu^n_{i+\frac{1}{2},k}, Fw^n_{i,k+\frac{1}{2}}$ contain all the explicit contributions of the convective and viscous terms, that in this paper are expressed through the explicit conservative formulation proposed in [39]. Substituting the discrete velocities $u^{n+1}_{i\pm\frac{1}{2},k}, w^{n+1}_{i,k\pm\frac{1}{2}}$ as expressed in Equations (10) and (11) into Equation (9), we obtain a linear system for the unknown hydraulic head:

$$\frac{\Delta t\, g}{\Delta x}\left[\Delta z^n_{i+\frac{1}{2},k}\left(\eta^{n+1}_{i+1,k} - \eta^{n+1}_{i,k}\right) - \Delta z^n_{i-\frac{1}{2},k}\left(\eta^{n+1}_{i,k} - \eta^{n+1}_{i-1,k}\right)\right]$$
$$+\frac{\Delta t\, g}{\Delta z}\left[\Delta x^n_{i,k+\frac{1}{2}}\left(\eta^{n+1}_{i,k+1} - \eta^{n+1}_{i,k}\right) - \Delta x^n_{i,k-\frac{1}{2}}\left(\eta^{n+1}_{i,k} - \eta^{n+1}_{i,k-1}\right)\right] = b^n_{i,k}, \tag{12}$$

where the term $b^n_{i,k}$ contains all the known terms evaluated at the time t^n:

$$b^n_{i,k} = \Delta z^n_{i+\frac{1}{2},k} Fu^n_{i+\frac{1}{2},k} - \Delta z^n_{i-\frac{1}{2},k} Fu^n_{i-\frac{1}{2},k} + \Delta x^n_{i,k+\frac{1}{2}} Fw^n_{i,k+\frac{1}{2}} - \Delta x^n_{i,k-\frac{1}{2}} Fw^n_{i,k-\frac{1}{2}} \tag{13}$$

Equation (12) is a five-points diagonal system, symmetric and at least semi-positive definite (e.g., [36]). Therefore, it can be solved using a matrix-free implementation of the conjugate gradient method. Once the hydraulic head η^{n+1} is known, the velocity field can be easily updated through the explicit formulae in Equations (10) and (11).

As for the scalar transport, similarly to [38], we refer to a semi-implicit finite volume approximation based on upwind fluxes:

$$c^*_{i,k} = c^n + \frac{\Delta t}{\Delta x \Delta z} \left[\Delta z^n_{i+\frac{1}{2},k} c^{n,up}_{i+\frac{1}{2},k} \left(u^{n+1}_{i+\frac{1}{2},k} \right) - \Delta z^n_{i-\frac{1}{2},k} c^{n,up}_{i-\frac{1}{2},k} \left(u^{n+1}_{i-\frac{1}{2},k} \right) \right.$$
$$\left. \Delta x^n_{i,k+\frac{1}{2}} c^{n,up}_{i,k+\frac{1}{2}} \left(w^{n+1}_{i,k+\frac{1}{2}} - w_s \right) - \Delta x^n_{i,k-\frac{1}{2}} c^{n,up}_{i,k-\frac{1}{2}} \left(w^{n+1}_{i,k-\frac{1}{2}} - w_s \right) \right], \tag{14}$$

where $c^{n,up}_.$ is the upwind contribution defined as:

$$c^{n,up}_{i+\frac{1}{2},k}(V) = \frac{1}{2} \left[c^n_{i,k}(|V| + V) + c^n_{i+1,k}(|V| - V) \right] \tag{15}$$

$$c^{n,up}_{i,k+\frac{1}{2}}(V) = \frac{1}{2} \left[c^n_{i,k}(|V| + V) + c^n_{i,k+1}(|V| - V) \right] \tag{16}$$

If there is no sedimentation/erosion, then $c^{n+1}_{i,k} = c^*_{i,k}$, otherwise the mass transfer is performed assuming that the fluid is in a local-static equilibrium. First, the dynamic part of the suspended sediment is computed through Equation (14), then the result is used to update explicitly the new sediment level through the discrete version of Equation (6):

$$S^{n+1}_{i,k} = S^n_{i,k} + \frac{\Delta t}{\Delta x \Delta z} Q_{SC}(S^n_{i,k}, c^*_{i,k}), \qquad \text{where} \qquad Q_{SC}(S^n_{i,k}, c^*_{i,k}) = \int_{\Omega_{i,k}} f_{SC}(S^n_{i,k}, c^*_{i,k}). \tag{17}$$

The sediment level at the time step $n + 1$ as resulting from Equation (17) allows for updating the volumes/edges lengths specified in Equation (8). Finally, the concentration at the time step $n + 1$ is updated according to the discrete version of the mass conservation law (7) that reads:

$$c^{n+1}_{i,k} = c^*_{i,k} + \left(S^n_{i,k} - S^{n+1}_{i,k} \right) (1 - \phi). \tag{18}$$

We note that by substituting Equation (14) into (18) and integrating $c^{n+1}_{i,k}$ over the water column, one obtains the discrete mass conservation law for the suspended transport, where the source term is the discrete version of $\dfrac{\partial z_b}{\partial t}(1 - \phi)$, with z_b total sediment level. However, the form in (14)–(18) is more general since it does not require a single value function $z_b = z_b(x, t)$, but it can possibly be applied to cases with multiple sediment/water interfaces.

For the chosen explicit discretization of the nonlinear convective terms, the method is stable under a Courant–Friedrichs–Lewy (CFL) type restriction based on the fluid velocity (e.g., [40]),

$$\Delta t \leq \frac{1}{\dfrac{|u|}{\Delta x} + \dfrac{|w|}{\Delta z} + \dfrac{2\nu}{\Delta x^2} + \dfrac{2\nu}{\Delta z^2}}. \tag{19}$$

The method becomes unconditionally stable if an Eulerian-Lagrangian scheme is adopted, in combination with a sub-step approach for the evolution of the concentration and sediment level [37,41].

2.1.4. Crank-Nicholson Time Discretization

For non-stationary problem we can improve the temporal accuracy by means of the so called θ-method. In order to do that, u^{n+1} and w^{n+1} in Equation (9) need to be substituted by $u^{n+\theta}$ and $w^{n+\theta}$, while η^{n+1} in (10), (11) and (14) needs to be substituted by $\eta^{n+\theta}$. $(u, v, \eta)^{n+\theta}$ are defined as:

$$(u, v, \eta)^{n+\theta} = \theta \cdot (u, v, \eta)^{n+1} + (1 - \theta) \cdot (u, v, \eta)^n, \tag{20}$$

where θ is an implicit factor to be taken in the interval $(0.5, 1]$ (e.g., [42]). The final system for the hydraulic head reads:

$$\frac{\Delta t\, g\, \theta^2}{\Delta x}\left[\Delta z^n_{i+\frac{1}{2},k}\left(\eta^{n+1}_{i+1,k}-\eta^{n+1}_{i,k}\right)-\Delta z^n_{i-\frac{1}{2},k}\left(\eta^{n+1}_{i,k}-\eta^{n+1}_{i-1,k}\right)\right]$$

$$+\frac{\Delta t\, g\, \theta^2}{\Delta z}\left[\Delta x^n_{i,k+\frac{1}{2}}\left(\eta^{n+1}_{i,k+1}-\eta^{n+1}_{i,k}\right)-\Delta x^n_{i,k-\frac{1}{2}}\left(\eta^{n+1}_{i,k}-\eta^{n+1}_{i,k-1}\right)\right]=b^n_{i,k} \tag{21}$$

where $b^n_{i,k}$ becomes:

$$b^n_{i,k}=\theta\left(\Delta z^n_{i+\frac{1}{2},k}Fu^n_{i+\frac{1}{2},k}-\Delta z^n_{i-\frac{1}{2},k}Fu^n_{i-\frac{1}{2},k}+\Delta x^n_{i,k+\frac{1}{2}}Fw^n_{i,k+\frac{1}{2}}-\Delta z^n_{i-\frac{1}{2},k}Fw^n_{i,k-\frac{1}{2}}\right)$$

$$+\theta\left(1-\theta\right)\frac{\Delta t\, g}{\Delta x}\left[\Delta z^n_{i+\frac{1}{2},k}\left(\eta^n_{i+1,k}-\eta^n_{i,k}\right)-\Delta z^n_{i-\frac{1}{2},k}\left(\eta^n_{i,k}-\eta^n_{i-1,k}\right)\right]$$

$$+\theta\left(1-\theta\right)\frac{\Delta t\, g}{\Delta z}\left[\Delta x^n_{i,k+\frac{1}{2}}\left(\eta^n_{i,k+1}-\eta^n_{i,k}\right)-\Delta x^n_{i,k-\frac{1}{2}}\left(\eta^n_{i,k}-\eta^n_{i,k-1}\right)\right]$$

$$-\left(1-\theta\right)\left(\Delta z^n_{i+\frac{1}{2},k}u^n_{i+\frac{1}{2},k}-\Delta z^n_{i-\frac{1}{2},k}u^n_{i-\frac{1}{2},k}+\Delta x^n_{i,k+\frac{1}{2}}w^n_{i,k+\frac{1}{2}}-\Delta z^n_{i,k-\frac{1}{2}}w^n_{i,k-\frac{1}{2}}\right). \tag{22}$$

We note that these modifications do not affect the structure of the linear system for the hydraulic head, since they are just rescaling it through a factor θ^2.

2.2. Validation Tests

The numerical model was first validated against some standard numerical tests, for which reference solutions are available. In the first validation test, we considered the Blasius analytical solution for the laminar velocity profile in the boundary layer above a semi-infinite plate [43]. This validation test was obtained by simulating a plate 0.2 m long, assuming a uniform upstream velocity parallel to it and equal to $u_\infty = 0.3$ m/s. The two-dimensional x-z domain $\Omega = [-0.01, 0.2] \times [0.0, 0.031]$ m was discretized according to a 600×300 grid, for a total of $180,000$ elements having horizontal and vertical dimension of $\Delta x = 3.5 \times 10^{-4}$ m and $\Delta z = 1.03 \times 10^{-4}$ m, respectively. We set $\theta = 0.51$, $g = 1$ m/s^2, and $\nu = 10^{-6}$ m^2/s. As for the boundary conditions (BCs), we assumed the velocity u_∞ as the left BC, transmissive BCs at the right and at the top edges, and no-slip BC at the bottom plate, beginning from $x = 0$ in order to trigger the boundary layer.

The second validation test was performed considering the so called lid-driven cavity problem, which is another classical benchmark test [44]. The problem consists in a cavity $\Omega = [-0.5, 0.5]^2$ m where the initial velocity field is $(u, w) = 0.0$ m. We set $\theta = 1$, and $g = 1$ m/s^2. We imposed $(u, w) = (1, 0)$ m/s as the top BC and no-slip BCs at the other three boundaries. We repeated the test for two different values of the Reynolds number, $Re = 400$ and $Re = 1000$, assuming the sediment level $S \equiv 0$. For both tests we discretized the domain according to a square grid 400×400, for a total of $160,000$ elements having horizontal and vertical dimensions of $\Delta x = 2.5 \times 10^{-3}$ m and $\Delta z = 2.5 \times 10^{-3}$ m, respectively. This benchmark test was chosen due to its analogy to the inter-grain regions, where circulation cells develop bounded laterally and at the bottom by the grains and by the fine sediment, respectively.

The same lid-driven cavity test was performed introducing an erodible sediment at the bottom of the cavity, characterized by density $\rho_s = 1553$ kg/m^3 and porosity $\phi = 0.46$. This resulted in considering mobile boundaries of the fluid domain due to the variation in time of the bed level z_b, according to the following equation:

$$\frac{\partial z_b}{\partial t}=\frac{D-E}{(1-\phi)}, \tag{23}$$

where E and D are erosion and deposition rates, respectively. In Equation (23) we assumed $D = 0$, which corresponds to simulating a non-equilibrium erosive process. We considered the same domain Ω, grid discretization, BCs, and g as in the original lid-driven cavity test, but we filled the cavity with sediment up to $1/3$ of the cavity height, set $\theta = 0.51$, and performed the test for just one value of the Reynolds number, i.e., $Re = 1000$. We defined the lowering of the fine sediment bed through the van Rijn erosion rate formula [45], which, expressed in $[m/s]$, reads as follows:

$$E = 0.00033\sqrt{g\Delta d_s}d_*^{0.3}T^{1.5},\tag{24}$$

where Δ is the fine sediment relative density $\Delta = \dfrac{\rho_s}{\rho} - 1$, d_s is the fine sediment median diameter, d_* is the dimensionless grain size $d_* = d_s\left(\dfrac{\Delta g}{\nu^2}\right)^{1/3}$, and T is the dimensionless excess of shear stress $T = \dfrac{\Theta - \Theta_{cr}}{\Theta_{cr}}$. In the above formulas, ρ_s is the fine sediment density, Θ is the Shields parameter, and Θ_{cr} its critical value. The Shields parameter Θ is defined as:

$$\Theta = \left(\frac{\tau_b}{\Delta\rho g d_s}\right),\tag{25}$$

where τ_b is the shear stress at the bottom, here defined as:

$$\tau_b = \mu\left.\frac{\partial u(z)}{\partial z}\right|_{z=z_b}.\tag{26}$$

To fasten the simulation, the erosion rate E from Equation (24) was multiplied by a factor 100. This test was introduced to validate the water and sediment mass conservation properties of the model, and to verify its robustness when dealing with time-varying changes in the boundaries of the fluid domain.

Numerical Experiments

Gravel bed rivers are often represented in laboratory experiments by covering the flume bed with spheres or hemispheres (e.g., [26,28,46,47]). We therefore assumed two simplified topographic configurations with the gravel represented by homogeneous spheres with different spacing, according to the most common simplified configurations used in laboratory flumes: in-line arrangement (Figure 2a) and closest-packing arrangement (Figure 2b). In particular, in the first case, we examined the section where two consecutive spheres touch each other (Figure 2d), while in the second case we examined the section with the maximum inter-sphere spacing (Figure 2e).

To investigate the differences between a simplified and a real bed topography, we considered also a real gravel topography obtained from point-laser-scanning a longitudinal section in a laboratory flume covered by gravel (Figures 2c). In this case, we selected a 20 cm long section (Figure 2f), whose granulometric distribution was characterized by $D_{50} = 26.5$ mm and $D_{90} = 29.5$ mm. In order to set up equivalent geometry configurations, the two simplified cases described above were drawn considering spheres of 30.0 mm diameter.

In addition to the bed topography of the gravel matrix, in the numerical experiments we considered also the presence of inter-grain inerodible fine sediments at fixed level, to see the variations in the flow field depending on the depth. The fine sediment interface was schematized as a horizontal line, whose level was assigned according to four different filling rates: $Z = 0$, 0.25, 0.50, and 0.75, where $Z = \dfrac{D/2 + z_b}{D/2}$ and D is either the diameter of the spheres in the simplified case, or D_{90} in the real topography. The vertical coordinate z was defined positive upwards, with $z = 0$ at the gravel crest.

All the simulations were performed considering a 0.2 m long domain with a rigid top boundary at 0.03 m above the gravel crest. The domain was discretized according to a 400×600 grid for a total of 240,000 elements having horizontal and vertical dimension of $dx = 5 \times 10^{-4}$ m and $dz = 10^{-4}$ m, respectively. We set $\theta = 1$, and $g = 9.81$ m/s^2, and $\nu = 10^{-6}$ m^2/s. The upstream and downstream boundary conditions were as follows: a uniform horizontal inflow velocity equal to 0.3 m/s and transmissive conditions downstream, with an assigned upstream/downstream pressure gradient equal to 2.5×10^{-3} m/m. No slip boundary conditions were assigned at the bottom, and free slip conditions were assigned at the top boundary. We notice that solving the incompressible Navier-Stokes equations has the undeniable advantage of reducing the computational burden, while the simulation setup is equivalent to considering a free-surface flow with a bed slope equal to the pressure gradient.

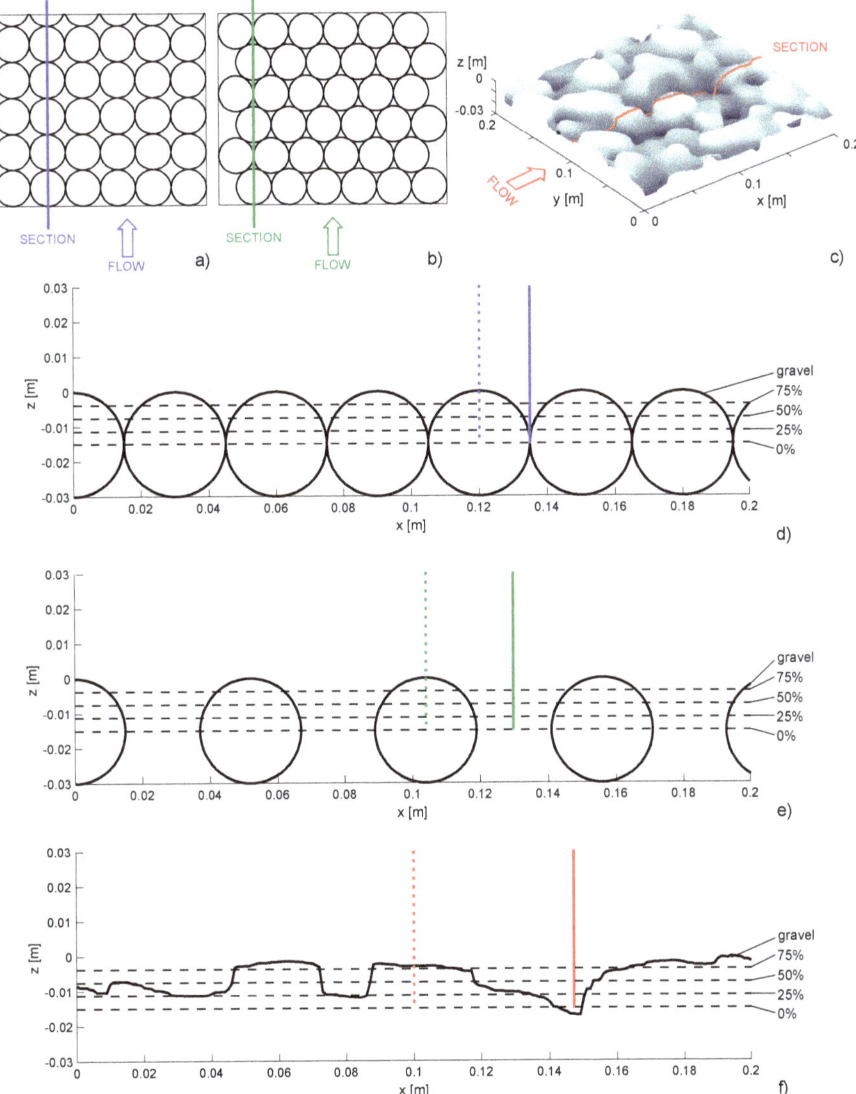

Figure 2. Schematic of the simulated gravel bed topographies: (**a**) in-line arrangement, view from above; (**b**) closest-packing arrangement, view from above; (**c**) real gravel bed, 3D view; (**d**) in-line arrangement, longitudinal section; (**e**) closest-packing arrangement, longitudinal section; (**f**) real gravel bed, longitudinal section. In panels (**d**–**f**), the horizontal dashed lines indicate the fine sediment filling rate, while the vertical lines indicate the gravel crest (dotted) and inter-grain cavity (continuous) sections.

3. Results and Discussion

3.1. Validation Tests

The Blasius solution is characterized by the following non linear third-order Ordinary Differential Equation (ODE):

$$f''' + f \cdot f'' = 0 \qquad f(0) = 0, \quad f'(0) = 0, \quad \lim_{\xi \to \infty} f'(\xi) = 1, \tag{27}$$

where $f' = u/u_\infty$ and $\xi = z\sqrt{\dfrac{u_\infty}{2\nu x}}$ is the so called Blasius coordinate. f is the primitive function of f', f'' is the second order derivative, f''' is the third order derivative. The reference solution is then obtained using a tenth-order Discontinuous Galerkin (DG) ODE solver [48]. In Figure 3a, the simulated velocity field is presented at time $t = 10\,s$, and clearly shows the formation of the Blasius boundary layer.

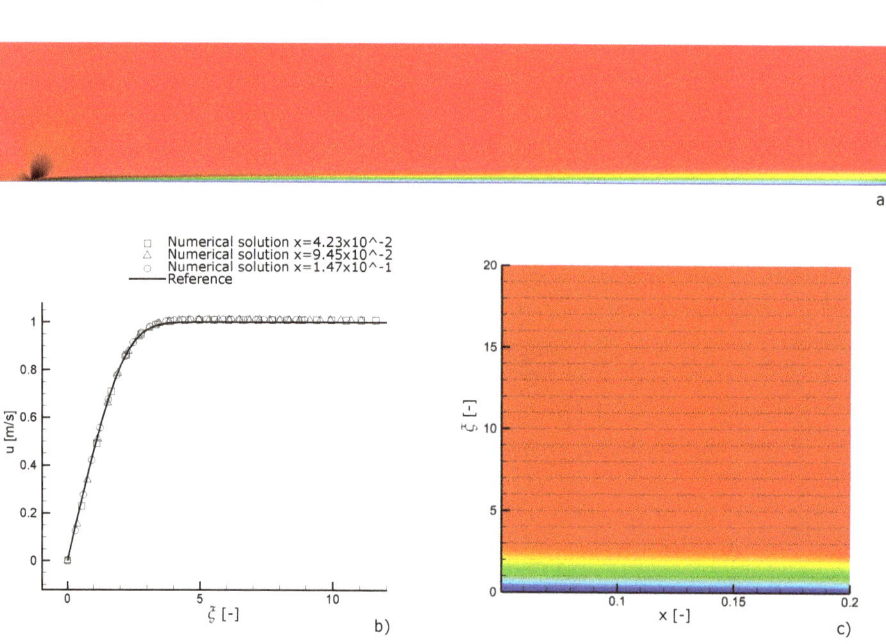

Figure 3. (**a**) Field map of the horizontal velocity component u, showing the formation of the Blasius boundary layer; (**b**) comparison between the numerical solution and the analytical Blasius solution at three different locations of the domain; and (**c**) magnitude of the horizontal velocity component u in the (x, ξ) plane.

The horizontal velocity profile in the plane (x, ξ) and the velocity profile taken from the 25%, 50%, and 75% of the domain are compared with the exact Blasius solution in Figure 3b. A good agreement between the analytical solution and the numerical results can be observed, despite the small value of ν (i.e., $10^{-6}\,m^2/s$). Furthermore, since the Blasius solution depends only on ξ, we expect the velocity profile to remain constant in the (x, ξ) plane, as is confirmed in Figure 3c.

As for the classical (i.e., fix bed) lid-driven cavity test, we calculated the velocity field in the cavity and compared it to the available reference solution given by [44]. Figures 4a and 5a show the velocity streamlines in the cavity at the final time $t_{end} = 100$ s, while the comparison with the reference solution is shown in Figures 4b and 5b. In particular, these latter plots show the normal velocities passing through the lines $\{x = 0\}$ and $\{z = 0\}$. For both the values of the Reynolds number, the figures show good agreement between the analytical solution and the numerical results, as it is also confirmed by the correct formation of secondary circulation cells at the two lower corners of the domain, as found also by [44].

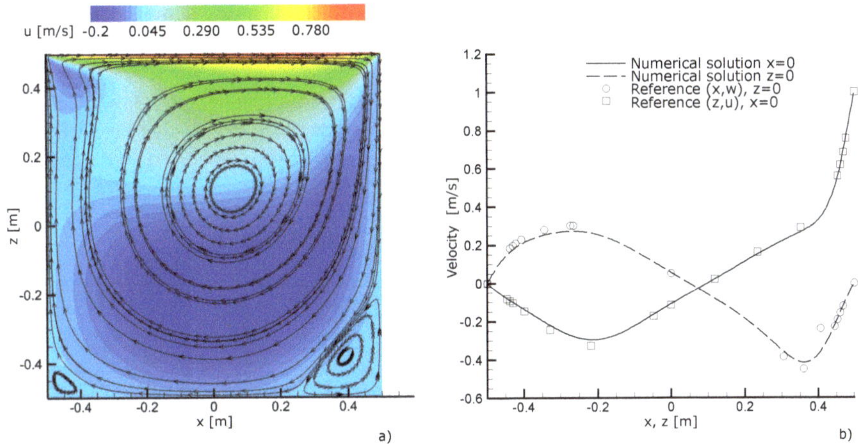

Figure 4. (**a**) Streamlines in the cavity at the final time $t = 100$ s; and (**b**) comparison with the reference solution [44] for $Re = 400$.

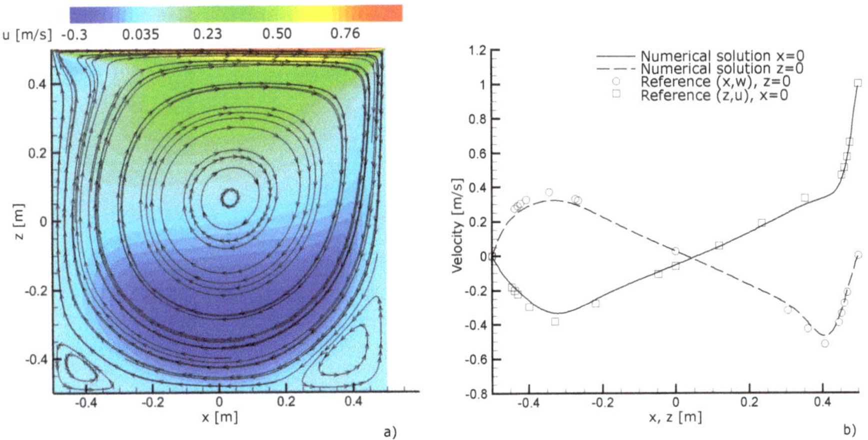

Figure 5. (**a**) Streamlines in the cavity at the final time $t = 100$ s; and (**b**) comparison with the reference solution [44] for $Re = 1\,000$.

The results of the same lid-driven cavity test, but performed on an erodible bed, are summarized in Figure 6. The figure shows the velocity field and the suspended sediment concentration field at times $t = 20, 50, 100, 300$ s, from which it is possible to appreciate that the erosion process is controlled by the main circulation cell, while the secondary circulation cells change in time according to the evolution of the water-sediment interface. The change in the suspended sediment concentration is due to the increase in the amount of sediments entrained from the bottom, as reflected by the progressive lowering of the bottom level in panels e-h. Since the cavity is a closed system, the progressive erosion of the bottom increases the total amount of suspended sediments in the domain.

In Figure 7 we report the measured mass exchange. The left side shows the time evolution of the total mass of suspended sediment and the total mass of deposited sediment, whereas the right side shows the balance between the free water mass and the mass of water constrained in the deposited sediment voids. Both sediment and water mass conservation are satisfied during the entire simulation with a precision determined by the tolerance used in the conjugate gradient algorithm (10^{-10} in all cases shown here), hence possibly going down to the machine epsilon. Overall, this last validation test indicates the robustness of the model when dealing with sediment erosion and transport processes and with the associated time-varying evolution of the fluid domain boundaries. Model robustness is also indicated by the results obtained considering a 100×100 computational grid (10,000 elements having horizontal and vertical dimension of $\Delta x = 10^{-2}$ m and $\Delta z = 10^{-2}$ m; see Figure S1 in the Supplementary Materials), which are coherent with those shown in Figure 6, although in this latter case we clearly see a higher level of detail and lower numerical diffusion.

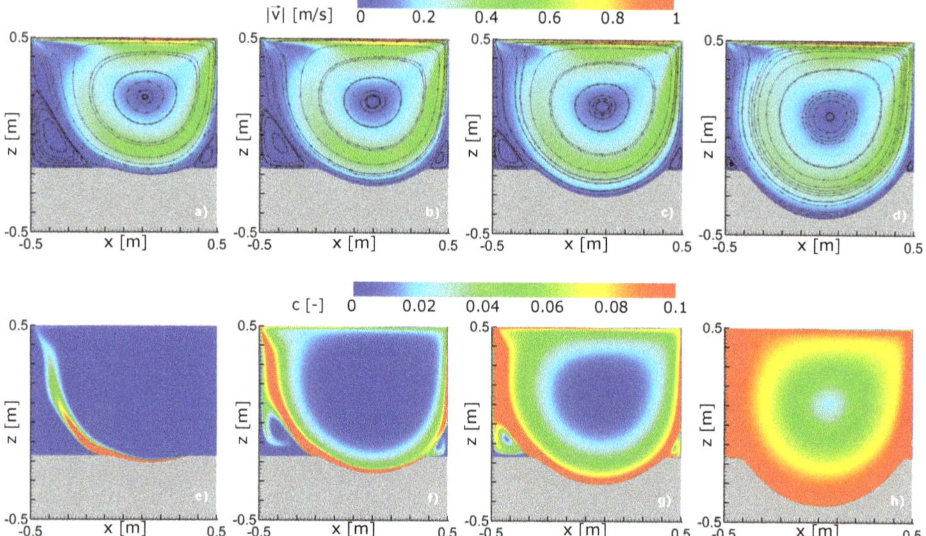

Figure 6. (**a–d**) Evolution of the streamlines in the cavity with the erodible bed at times $t = 20, 50, 100, 300$ s and coloured velocity magnitude $|\vec{v}|$, where $\vec{v}$ denotes the velocity vector; and (**e–h**) volume concentration of suspended sediment.

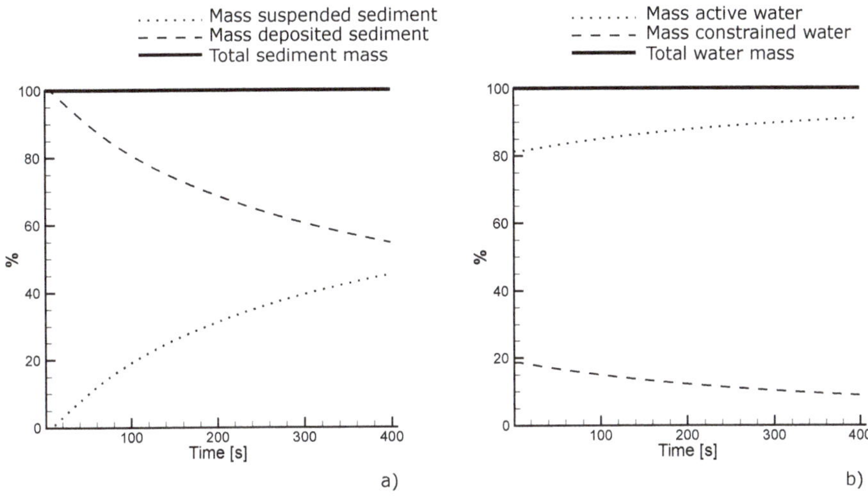

Figure 7. Time evolution of (**a**) sediment and (**b**) water mass, rescaled with respect to the initial total mass.

3.2. Numerical Experiments

In this section, we present the numerical results obtained for the simplified and real gravel bed topographies, that were simulated as representative cases of typical laboratory flume configurations. The simulated flow fields are shown in Figures 8–10 for all fine sediment filling rates-topography combinations. Results are presented in terms of horizontal velocity component and streamlines (left panels), and vorticity ω (right panels), where in a two-dimensional flow ω is defined as follows:

$$\vec{\omega} = \left(\frac{\partial w}{\partial x} - \frac{\partial u}{\partial z} \right) \vec{y}, \tag{28}$$

$\vec{y}$ being the unit vector along the third dimension y.

The fine grid resolution used to discretize the computational domain allowed to well capture the inter-grain flow structures, including the development of secondary circulations and stagnation points. In all cases, the flow field shown in Figures 8–10 refers to the end of the simulation (15 s), when the influence of the initial conditions was substantially lost and the flow field reached statistical steady state conditions with the development of periodic eddies generated by the variable topography (Figure 11).

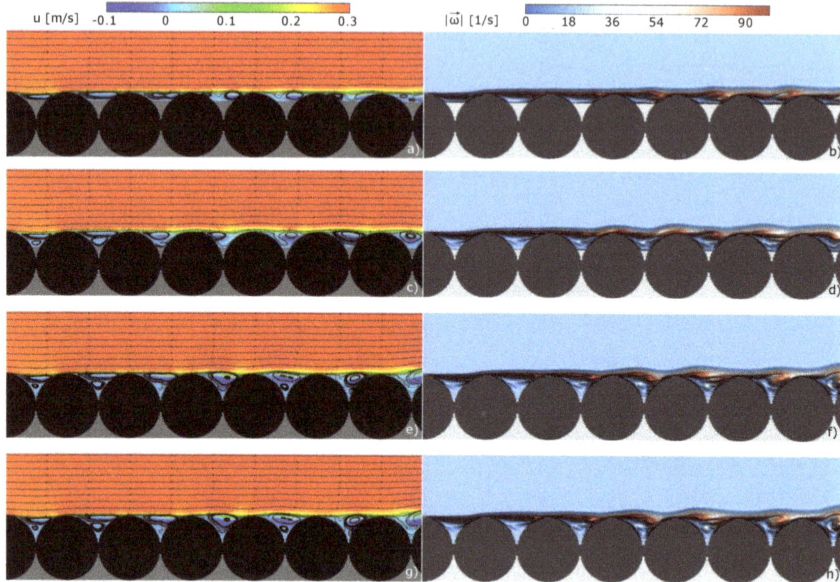

Figure 8. Horizontal velocity component u and streamlines (**left panels**), and magnitude of the vorticity $|\vec{\omega}|$ (**right panels**) for the spheres in-line arrangement.

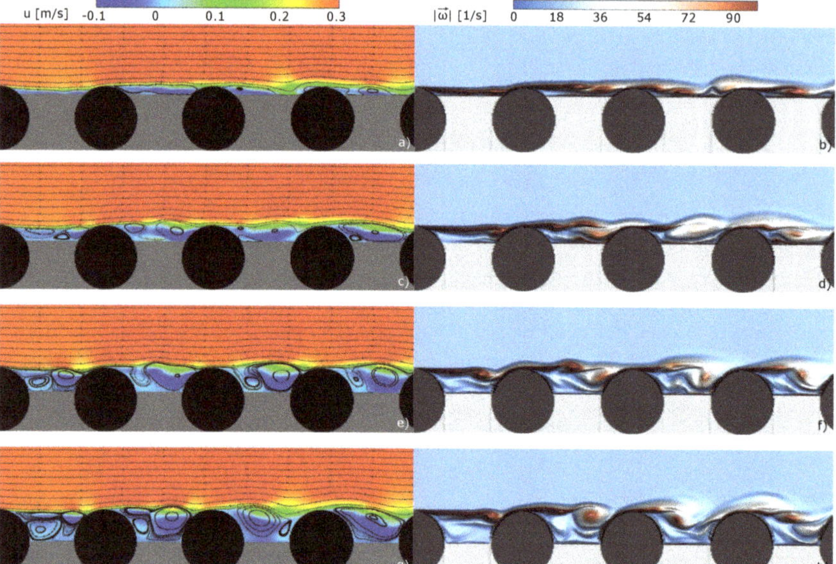

Figure 9. Horizontal velocity component u and streamlines (**left panels**), and magnitude of the vorticity $|\vec{\omega}|$ (**right panels**) for the spheres closest-packing arrangement.

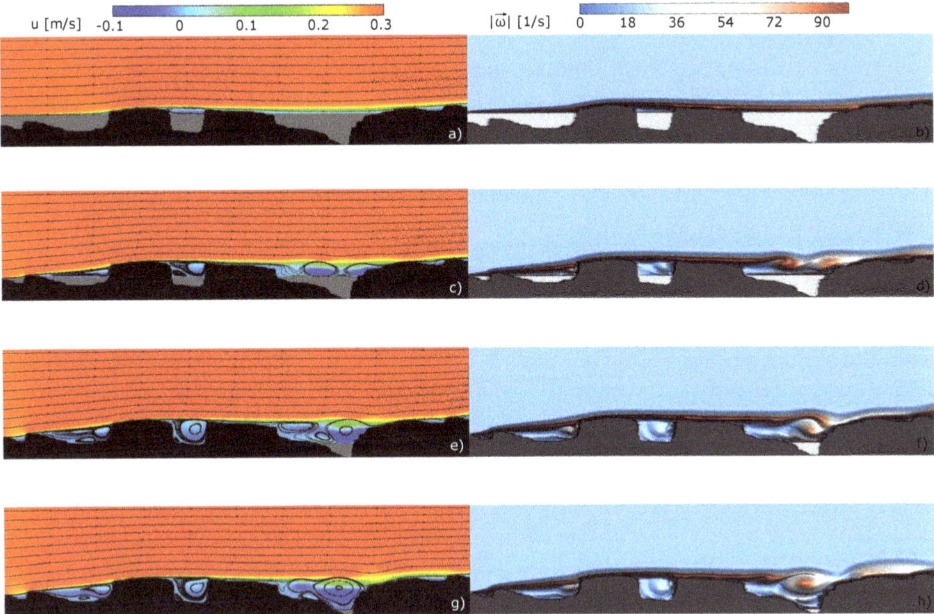

Figure 10. Horizontal velocity component u and streamlines (**left panels**), and magnitude of the vorticity $|\vec{\omega}|$ (**right panels**) for the real gravel topography configuration.

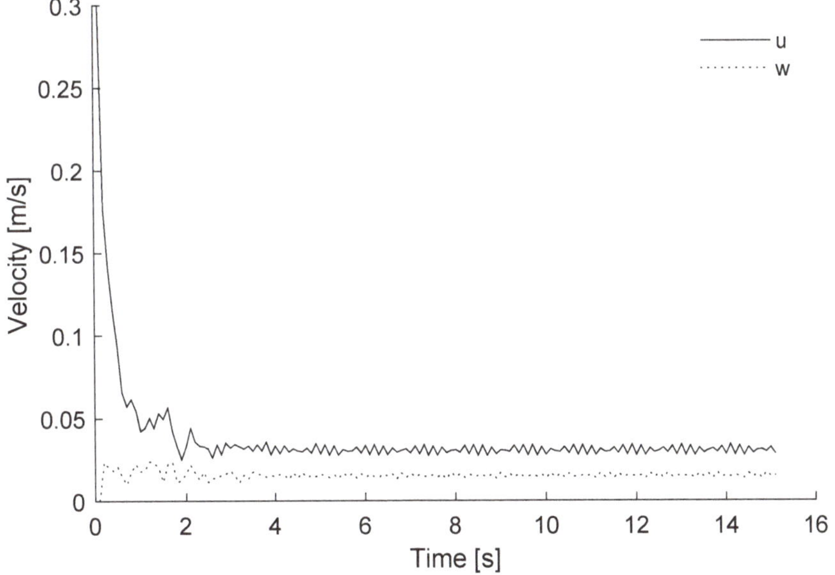

Figure 11. Time evolution of the horizontal (u) and vertical (w) velocity component intensities, averaged over the region below the gravel crest level, and relative to the real gravel bed configuration with fine sediment filling rate $Z = 0.5$.

Despite advancements in experimental technologies exploited in recent studies (e.g., [22,24,25,32]), a detailed evaluation of the inter-grain flow field is still hampered by the small spatial scales of the geometries and processes at hand. However, being able to simulate the flow field at the inter-grain scale is certainly a desired goal, in that it allows for deriving the total shear stress distribution, which is the key quantity controlling fine sediment dynamics. According to the double-averaging approach, in gravel bed topographies, the total shear stress is expressed as the combination of viscous, turbulent, form-induced stresses, and form drag [23]. Quantifying the relative influence of the effective components (i.e., turbulent and form-induced stresses) and dispersive component (i.e., the form drag) is crucial to accurately estimate the entrainment and transport of the inter-grain fine sediment. In this regard, the above results suggest that the proposed numerical model has a great potential for applications in the context of fine sediment transport dynamics in gravel bed rivers. Among the major advantages is the possibility to acquire the fine resolution flow field in continuous, whereas experimental measurements are typically performed at discrete positions that get rarer going further in depth below the gravel crest. This means loosing information that a continuous numerical measurement can ensure, such as the existence of the inter-grain secondary circulations observed in Figures 8–10. Such structures determine a double inflection in the velocity profile in all topographic configurations for decreasing fine sediment filling rates. This clearly emerges from the vertical profiles of the horizontal component of the velocity shown in Figure 12, at chosen sections representative of the gravel crest and of the inter-grain cavity (see Figure 2). The presented profiles are averaged along 5 s of the simulation (from 10 to 15 s, with an output resolution of 0.1 s), for the same fine sediment filling rates-topography combinations of Figures 8–10. The instantaneous velocity profiles used to compute the averaged profiles are also shown for the inter-grain cavity section (thin continuous lines), which indicate the high non-stationary behavior of the flow field in the roughness layer [23]. The same figure but for the vertical component of the velocity is shown in Figure 13. As expected, both Figures 12 and 13 show clear differences depending on fine sediment filling rates and gravel bed topography. The bed topography influence on the flow field confirms that special attention should be paid when using simplified gravel bed geometries in laboratory flume experiments. In fact, despite they offer undeniable advantages in terms of simplification of the experimental setup (e.g., by allowing for an easier definition of porosity, granulometric distribution, and bed topography), a spheres-covered bed may not be entirely representative of a natural water-worked gravel bed due to different macro-roughness structures. Strong differences are particularly evident for the vertical component of the velocity (Figure 13), which is responsible for lifting (i.e., eroding) fine material at the bottom [22,24,25]. In this regard, we note that the instantaneous profiles of vertical velocity undergo large changes from negative (i.e., downward velocity) to positive (i.e., upward velocity) values in all topographic configurations. Time variability in the velocity field is also evident from the profiles of the horizontal component (Figure 12), indicating the existence of highly non-stationary inter-grain recirculation cells. This is confirmed also in the vorticity field and particle tracking videos available in the Supplementary Materials.

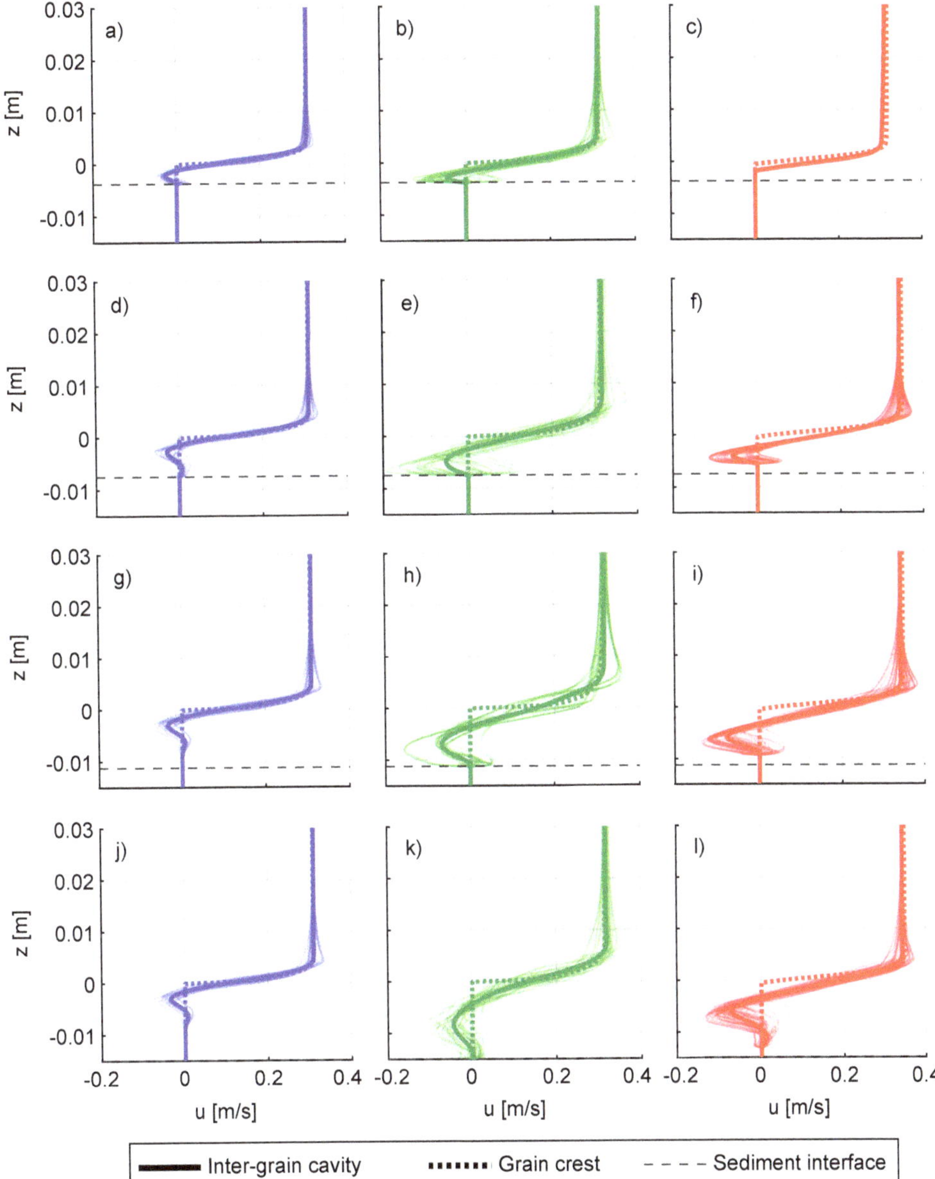

Figure 12. Vertical profiles of the horizontal velocity component *u* at chosen sections representative of the gravel crest and of the inter-grain cavity (see Figure 2), for the in-line arrangement (**left panels**), closest-packing arrangement (**central panels**), and real gravel bed configuration (**right panels**). Thick lines indicate profiles averaged from 10 to 15 s of simulation, while thin lines indicate instantaneous profiles every 0.1 s.

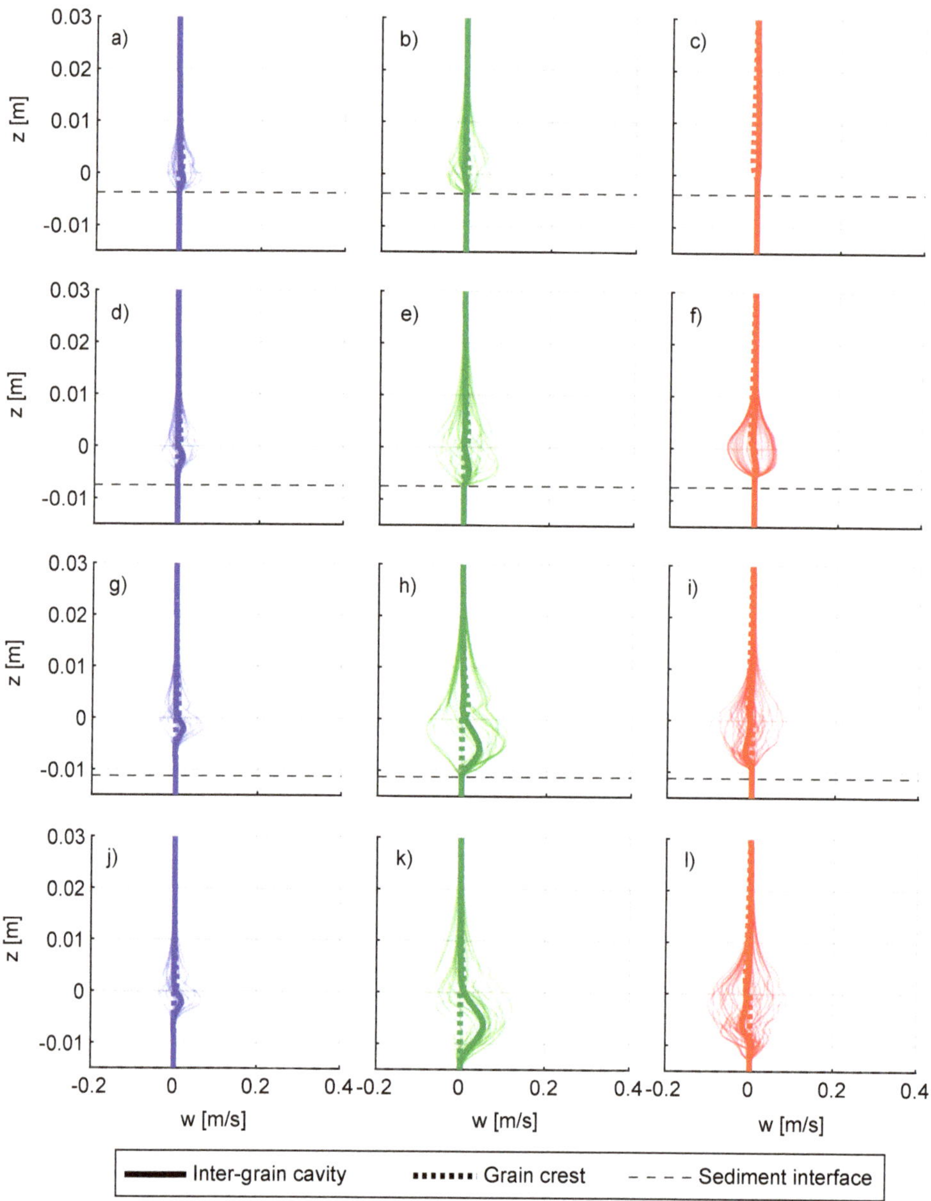

Figure 13. Vertical profiles of the vertical velocity component w at chosen sections representative of the gravel crest and of the inter-grain cavity (see Figure 2), for the in-line arrangement (**left panels**), closest-packing arrangement (**central panels**), and real gravel bed configuration (**right panels**). Thick lines indicate profiles averaged from 10 to 15 s of simulation, while thin lines indicate instantaneous profiles every 0.1 s.

4. Conclusions

A second order semi-implicit numerical scheme on staggered Cartesian meshes for the incompressible Navier-Stokes equations, based on the method proposed by [36–38] in presence of a time dependent sedimentation/erosion process, was derived. In the scheme, we defined the hydraulic head in the cells centers and the velocity at the cells interfaces. By formally substituting the discrete momentum equations into the discrete continuity equation, we obtained a symmetric semi-positive definite linear system where the only unknown is the hydraulic head at the new time step. The system is then solved using a fast iterative linear solver such as the conjugate gradient algorithm [49]. We note that the method is built in such a way that the computation in each cell involves only its direct neighbors. This makes the algorithm particularly suitable to parallelization, since the data that need to be synchronized are limited to the single layer of cells surrounding each parallel region.

For the entrainment and deposition of the sediment we used an explicit finite volume scheme in combination with a general mass flux between suspended and deposited sediment. The deposition of the suspended sediment changes the effective domain sizes in terms of volume and edges length in the cells. The method is mass-conservative and limited in the time discretization by a classical CFL-type time restriction based on the local fluid velocity. However, if the convective-viscous terms are solved by an Eulerian-Lagrangian method combined with a local time stepping/subcycling approach for the sediment dynamics, the method becomes unconditionally stable. Furthermore, compared to [36], the pressurized system allows for avoiding the solution of the mildly nonlinear contribution through the Nested-Newton approach [50], which ultimately reduces the computational time thus allowing for a fine resolution in the mesh.

The method was validated against some classical benchmarks, i.e., the Blasius boundary layer and the lid-driven cavity test. Moreover, a modified version of the lid-driven cavity test was run, to verify the conservation of sediment and water mass in presence of erodible sediment, and the robustness of the model in presence of time-varying boundaries of the fluid domain.

Once validated, the model was used to simulate three simple cases representative of typical experiments for gravel bed rivers at different filling rates. The numerical results for the inter-grain flow field show the formation of main and secondary circulation cells, forced by the presence of the macro-roughness elements, which generates a double inflection in the time-averaged velocity profile for the lower filling rates. This information would probably be lost if performing only experimental measurements of the flow dynamics, which are possible just at discrete points, especially below the gravel crest. Furthermore, the use of a numerical model simplifies the repetition of the experiments considering different topographies, which contributes at improving the understanding of the geometry role in the stresses distribution.

We note that the model is two-dimensional, therefore it does not account for the three-dimensional effects and its results are not immediately representative of the real case. While a full description of the inter-grain flow field and turbulence structure would require the use of a three-dimensional model coupled with a proper turbulence closure scheme, even in this form the proposed model provides useful clues on the approximation effects introduced when using simplified geometries to represent real topographies.

Future work will concern the extension of the present approach to the complete VOF (Volume Of Fluid) method such as proposed in [36], and its extension to the three-dimensions in the presence of erodible sediment, together with the inclusion of a proper turbulence closure and high-performance parallelization standards.

Supplementary Materials: The following are available online at http://www.mdpi.com/2073-4441/12/3/690/s1, Figure S1: the numerical results for the lid-driven cavity test with erodible bed obtained using a coarser domain discretization; video S1: the vorticity field for the closest-packing arrangement; video S2: particle tracking for the closest-packing arrangement.

Author Contributions: Conceptualization, M.T., S.P. and G.S.; methodology, M.T., S.P., and G.S.; validation, M.T.; formal analysis, M.T., S.P. and G.S.; data curation M.T. and S.P.; writing–original draft preparation, G.S. and S.P.; writing–review and editing, all authors; supervision, M.R. All authors have read and agreed to the published version of the manuscript.

Funding: Part of this work was supported by the projects Sediplan-r (FESR1002) and LTFD Laboratory of Thermo Fluid Dynamics (FESR1029), financed by the European Regional Development Fund (ERDF) Investment for Growth and Jobs Programme 2014-2020, and CRC project HM: Hydropeaking mitigation, financed by the Free University of Bozen-Bolzano.

Acknowledgments: We are grateful to E. Spilone for relevant discussion during the preparation of this work.

Conflicts of Interest: The authors declare no conflict of interest.

References

1. Schleiss, A.J.; Franca, M.J.; Juez, C.; Cesare, G.D. Reservoir sedimentation. *J. Hydraul. Res.* **2016**, *54*, 595–614. [CrossRef]

2. Kaffas, K.; Hrissanthou, V.; Sevastas, S. Modeling hydromorphological processes in a mountainous basin using a composite mathematical model and ArcSWAT. *CATENA* **2018**, *162*, 108–129. [CrossRef]

3. ICOLD. *Sedimentation and Sustainable Use of Reservoirs and River Systems*; Technical Report; International Commission on Large Dams: Paris, France, 2009.

4. Williams, G.P.; Wolman Gordon, M. *Downstream Effects of Dams on Alluvial Rivers*; Geological Survey Professional Paper 1286; U.S. Government Printing Office: Washington, DC, USA, 1984. [CrossRef]

5. Brandt, S. Classification of geomorphological effects downstream of dams. *CATENA* **2000**, *40*, 375–401. [CrossRef]

6. Juez, C.; Hassan, M.A.; Franca, M.J. The Origin of Fine Sediment Determines the Observations of Suspended Sediment Fluxes Under Unsteady Flow Conditions. *Water Resour. Res.* **2018**, *54*, 5654–5669. [CrossRef]

7. Kondolf, G.M. Hungry Water: Effects of Dams and Gravel Mining on River Channels. *Environ. Manag.* **1997**, *21*, 533–551. [CrossRef]

8. Shen, H.W.; Lu, J. Development and Prediction of Bed Armoring. *J. Hydraul. Eng.* **1983**, *109*, 611–629. [CrossRef]

9. Dietrich, W.; Kirchner, J.; Ikeda, H.; Iseya, F. Sediment Supply and Development of Coarse Surface Layer in Gravel Bedded Rivers. *Nature* **1989**, *340*. [CrossRef]

10. Wilcock, P.R.; DeTemple, B.T. Persistence of armor layers in gravel-bed streams. *Geophys. Res. Lett.* **2005**, *32*. [CrossRef]

11. Bui, V.; Bui, M.; Rutschmann, P. Advanced Numerical Modeling of Sediment Transport in Gravel-Bed Rivers. *Water* **2019**, *11*, 550. [CrossRef]

12. Schälchli, U. The clogging of coarse gravel river beds by fine sediment. *Hydrobiologia* **1992**, *235–236*, 189–197. [CrossRef]

13. Wu, F.C.; Huang, H.T. Hydraulic Resistance Induced by Deposition of Sediment in Porous Medium. *J. Hydraul. Eng.* **2000**, *126*. [CrossRef]

14. Wharton, G.; Mohajeri, S.H.; Righetti, M. The pernicious problem of streambed colmation: A multi-disciplinary reflection on the mechanisms, causes, impacts, and management challenges. *Wiley Interdiscip. Rev. Water* **2017**, *4*, e1231. [CrossRef]

15. Brunke, M.; Gonser, T. The ecological significance of exchange processes between rivers and groundwater. *Freshw. Biol.* **1997**, *37*, 1–33. [CrossRef]

16. Kemp, P.; Sear, D.; Collins, A.; Naden, P.; Jones, I. The impacts of fine sediment on riverine fish. *Hydrol. Process.* **2011**, *25*, 1800–1821. [CrossRef]

17. Jones, J.; Collins, A.; Naden, P.; Sear, D. The relationship between fine sediment and macrophytes in rivers. *River Res. Appl.* **2012**, *28*, 1006–1018. [CrossRef]

18. Jones, J.I.; Murphy, J.F.; Collins, A.L.; Sear, D.A.; Naden, P.S.; Armitage, P.D. The impact of fine sediment on macro-invertebrates. *River Res. Appl.* **2012**, *28*, 1055–1071. [CrossRef]

19. Krause, S.; Hannah, D.M.; Fleckenstein, J.H. Hyporheic hydrology: Interactions at the groundwater-surface water interface. *Hydrol. Process.* **2009**, *23*, 2103–2107. [CrossRef]

20. Sambrook Smith, G.H.; Nicholas, A.P. Effect on flow structure of sand deposition on a gravel bed: Results from a two-dimensional flume experiment. *Water Resour. Res.* **2005**, *41*. [CrossRef]

21. Wren, D.; Langendoen, E.; Kuhnle, R. Effects of sand addition on turbulent flow over an immobile gravel bed. *J. Geophys. Res. Earth Surf.* **2011**, *116*, 1–12. [CrossRef]

22. Mohajeri, S.H.; Righetti, M.; Wharton, G.; Romano, G.P. On the structure of gravel-bed flow with intermediate submergence: Implications for sediment transport. *Adv. Water Resour.* **2016**, *92*, 90–104. [CrossRef]

23. Nikora, V.; Goring, D.; McEwan, I.; Griffiths, G. Spatially averaged open-channel flow over rough bed. *J. Hydraul. Eng.* **2001**, *127*, 123–133. [CrossRef]

24. Mignot, E.; Barthelemy, E.; Hurter, D. Double-averaging analysis and local flow chracterization of near-bed turbulence in gravel-bed channel flow. *J. Fluid Mech.* **2009**, *618*, 279–303. [CrossRef]

25. Dey, S.; Das, R. Gravel-bed hydrodynamics: Double-averaging approach. *J. Hydraul. Eng.* **2012**, *138*, 707–725. [CrossRef]

26. Grams, P.E.; Wilcock, P.R. Equilibrium entrainment of fine sediment over a coarse immobile bed. *Water Resour. Res.* **2007**, *43*. [CrossRef]

27. Kuhnle, R.A.; Wren, D.G.; Langendoen, E.J.; Rigby, J.R. Sand transport over an immobile gravel bed substrate. *J. Hydraul. Eng.* **2013**, *139*, 167–176. [CrossRef]

28. Grams, P.E.; Wilcock, P.R. Transport of fine sediment over a coarse, immobile riverbed. *J. Geophys. Res. Earth Surf.* **2015**, *119*, 188–211. [CrossRef]

29. Kuhnle, R.A.; Langendoen, E.J.; Wren, D.G. Prediction of sand transport over immobile gravel from supply-limited to capacity conditions. *J. Hydraul. Eng.* **2017**, *143*. [CrossRef]

30. Bertin, S.; Friedrich, H. Effects of Sand Addition and Bed Flushing on Gravel Bed Surface Microtopography and Roughness. *Water Resour. Res.* **2019**. [CrossRef]

31. Kuhnle, R.A.; Wren, D.G.; Langendoen, E.J. Erosion of Sand from a Gravel Bed. *J. Hydraul. Eng.* **2016**, *142*, 04015052. [CrossRef]

32. Wren, D.G.; Kuhnle, R.A.; Langendoen, E.J.; Rigby, J.R. Turbulent Flow and Sand Transport over a Cobble Bed in a Laboratory Flume. *J. Hydraul. Eng.* **2014**, *140*, 04014001. [CrossRef]

33. Fornarelli, F.; Vittori, G. Oscillatory boundary layer close to a rough wall. *Eur. J. Mech. B/Fluids* **2009**, *28*, 283–295. [CrossRef]

34. Ghodke, C.D.; Apte, S.V. DNS study of particle-bed–turbulence interactions in an oscillatory wall-bounded flow. *J. Fluid Mech.* **2016**, *792*, 232–251. [CrossRef]

35. Mazzuoli, M.; Blondeaux, P.; Simeonov, J.; Calantoni, J. Direct numerical simulation of the oscillatory flow around a sphere resting on a rough bottom. *J. Fluid Mech.* **2017**, *822*, 235–266. [CrossRef]

36. Casulli, V. A semi-implicit numerical method for the free-surface Navier–Stokes equations. *Int. J. Numer. Methods Fluids* **2014**, *74*, 605–622. [CrossRef]

37. Casulli, V.; Zanolli, P. Semi-implicit numerical modeling of nonhydrostatic free-surface flows for environmental problems. *Math. Comput. Model.* **2002**, *36*, 1131–1149. [CrossRef]

38. Casulli, V.; Zanolli, P. High resolution methods for multidimensional advection–diffusion problems in free-surface hydrodynamics. *Ocean Model.* **2005**, *10*, 137–151. [CrossRef]

39. Stelling, G.S.; Duinmeijer, S.P.A. A staggered conservative scheme for every Froude number in rapidly varied shallow water flows. *Int. J. Numer. Methods Fluids* **2003**, *43*, 1329–1354. [CrossRef]

40. Gross, E.S.; Bonaventura, L.; Rosatti, G. Consistency with continuity in conservative advection schemes for free-surface models. *Int. J. Numer. Methods Fluids* **2002**, *38*, 307–327. [CrossRef]

41. Casulli, V. Eulerian-Lagrangian methods for the Navier-Stokes equations at high Reynolds number. *Int. J. Numer. Methods Fluids* **1988**, *8*, 1349–1360. [CrossRef]

42. Casulli, V.; Cattani, E. Stability, accuracy and efficiency of a semi-implicit method for three-dimensional shallow water flow. *Comput. Math. Appl.* **1994**, *27*, 99–112. [CrossRef]

43. Blasius, H. Grenzschichten in Flüssigkeiten mit kleiner Reibung. *Z. Math. Phys.* **1908**, *56*, 1–37.

44. Ghia, U.; Ghia, K.; Shin, C.T. High-resolutions for incompressible flow using the Navier-Stokes equations and a multigrid method. *J. Comput. Phys.* **1982**, *48*, 387–411. [CrossRef]

45. van Rijn, L.C. Sediment pick-up functions. *J. Hydraul. Eng.* **1984**, *110*, 1494–1502. [CrossRef]

46. Manes, C.; Pokrajac, D.; McEwan, I.; Nikora, V. Turbulence structure of open channel flows over permeable and impermeable beds: A comparative study. *Phys. Fluids* **2009**, *21*, 125109. [CrossRef]

47. Yager, E.M.; Kirchner, J.W.; Dietrich, W.E. Calculating bed load transport in steep boulder bed channels. *Water Resour. Res.* **2007**, *43*. [CrossRef]

48. Dumbser, M. Arbitrary high order PNPM schemes on unstructured meshes for the compressible Navier–Stokes equations. *Comput. Fluids* **2010**, *39*, 60–76. [CrossRef]
49. Hestenes, M.R.; Stiefel, E. Methods of conjugate gradients for solving linear systems. *J. Res. Natl. Bur. Stand.* **1952**, *49*, 409–436. [CrossRef]
50. Casulli, V.; Zanolli, P. Iterative solutions of mildly nonlinear systems. *J. Comput. Appl. Math.* **2012**, *236*, 3937–3947. [CrossRef]

Article

A Fuzzy Transformation of the Classic Stream Sediment Transport Formula of Yang

Konstantinos Kaffas [1], Matthaios Saridakis [2], Mike Spiliotis [2,*], Vlassios Hrissanthou [2] and Maurizio Righetti [1]

[1] Faculty of Science and Technology, Free University of Bozen–Bolzano, 39100 Bozen–Bolzano, Italy; Konstantinos.Kaffas@unibz.it (K.K.); Maurizio.Righetti@unibz.it (M.R.)
[2] Department of Civil Engineering, Democritus University of Thrace, 67100 Xanthi, Greece; msarida@civil.duth.gr (M.S.); vhrissan@civil.duth.gr (V.H.)
* Correspondence: mspiliot@civil.duth.gr

Received: 29 November 2019; Accepted: 11 January 2020; Published: 16 January 2020

Abstract: The objective of this study is to transform the arithmetic coefficients of the total sediment transport rate formula of Yang into fuzzy numbers, and thus create a fuzzy relationship that will provide a fuzzy band of in-stream sediment concentration. A very large set of experimental data, in flumes, was used for the fuzzy regression analysis. In a first stage, the arithmetic coefficients of the original equation were recalculated, by means of multiple regression, in an effort to verify the quality of data, by testing the closeness between the original and the calculated coefficients. Subsequently, the fuzzy relationship was built up, utilizing the fuzzy linear regression model of Tanaka. According to Tanaka's fuzzy regression model, all the data must be included within the produced fuzzy band and the non-linear regression can be concluded to a linear regression problem when auxiliary variables are used. The results were deemed satisfactory for both the classic and fuzzy regression-derived equations. In addition, the linear dependence between the logarithmized total sediment concentration and the logarithmized subtraction of the critical unit stream power from the exerted unit stream power is presented. Ultimately, a fuzzy counterpart of Yang's stream sediment transport formula is constructed and made available to the readership.

Keywords: stream sediment transport; total load; sediment concentration; Yang formula; fuzzy regression; fuzzy coefficients; fuzzy logic

---

1. Introduction

The need for knowledge of the amount of sediment reaching specific points of streams and river segments became evident from the early 20th century [1–3]. As a consequence of that, the investigation of the sediment transport processes and mechanisms emerged as a high significance research topic for hydrologists, physicists and engineers in the years that followed. Sediments constitute an integral part of river flows, relentlessly forming the shape of fluvial systems and variously affecting everything in their path [4,5]. Water-quality issues, changes in the wet cross-section, increased flooding risk and obstruction of navigation, as a result of excessive depositions, effects on the aquatic ecosystems, decline of macrophyte growth, clogging of spawning gravel, pressures inflicted on coastal zones, effective diminution of dams' storage volume, due to excessive sedimentation, and extreme erosion rates in the case of sediment-starved water (usually below storage dams—theory of hungry water) [6–11], are some of the effects of sediments, which constitute the driving force behind the investigation of sediment transport processes, as well as modeling and quantification efforts. Moreover, knowledge about the interrelated interactions among water-biota-sediment in natural rivers is one of the central issues in today's sustainable river management [12].

The total sediment load results as the sum of the suspended load and the bed load, with the suspended load being the largest part of it. According to the literature, bed and bank erosion, in rivers, can be considered as a percentage of 10–20% of the total load [13–15], although this largely depends on whether they are sandy-bed or gravel-bed rivers [16]. Naturally, the finer the bed material is, the more easily it is entrained and transported downstream. Hence, the bed load ratio—as a fraction of the total load—increases, as the bed material becomes finer.

The result of decades of intensive research on river sedimentology and sediment transport is an amplitude of formulas, models, and theoretical concepts, aiming at the estimation of sediment load in natural streams. Depending on their target, these models can be divided into three principal classes: (a) bed-load models [17,18], (b) suspended-load models [19,20], (c) total-load models [21,22]. Despite most of the above-cited models were developed half a century—or more—ago, their theoretical basis and fundamental equations are so powerful, that even today they dominate the stream sediment transport research. The models for total sediment load can be further categorized as follows [23]: (a) stochastic models and regression models [24–26], (b) energy models [22,27,28], (c) shear stress models [20,29,30].

Yang first introduced his unit stream power theory for the determination of total sediment concentration, in open channels, in 1972 [27]. This new theory questioned the assumption, made by conventional sediment transport equations, that sediment transport rate could be determined on the basis of water discharge, average flow velocity, energy slope, or shear stress [31]. Yang [22], primarily, implemented his unit stream power theory for sandy-bed open channels, and thus developed a formula applicable for bed material with particle size less than 2 mm. In 1984, Yang [32] extended his unit stream power equation from sand transport to gravel transport, for gravel beds with particle sizes between 2 mm and 10 mm. Yang's unit stream power theory has been extensively applied in the literature, and with more than 2000 citations, it constitutes one of the most esteemed formulas for the determination of total sediment yield.

Fuzzy logic has proved a particularly useful tool in the hands of engineers, and its use in recent decades has been widespread in hydrology, hydraulics and sediment transport [33–35]. Fuzzy linear regression provides a functional fuzzy relationship between dependent and independent variables [36], where uncertainty manifests itself in the coefficients of the independent variables.

Fuzzy logic has been utilized in a variety of cases to study the sediment transport processes, as well as to estimate the total sediment concentration. As a recent paradigm, Chachi et al. [36] introduced a fuzzy regression method based on the Multivariate Adaptive Regression Splines (MARS) technique, to estimate suspended load, based on discharge and bed-load transport data, using fuzzy triangular numbers. The comparison of the model's results with real data and two other fuzzy regression models (fuzzy least-absolutes and fuzzy least-squares regressions) showed that the fuzzy regression model performs well for predicting the fuzzy suspended load, by discharge, as well as the fuzzy bed load transport data. In 2018, Spiliotis et al. [37] transformed the threshold—expressed by a dimensionless critical shear stress—for incipient sediment motion into a fuzzy set, by means of Zanke's formula [38], for the computation of the dimensionless critical shear stress, by using fuzzy triangular numbers instead of crisp values. The fuzzy band produced included almost all the used experimental data with a functional spread. The same group of researchers carried out similar studies, with an adaptive fuzzy-based regression and data from several gravel-bed rivers from mountain basins of Idaho, USA [39], and with conventional fuzzy regression analysis and a goal programming-based fuzzy regression using experimental data [40]; the results were satisfactory in both cases. In 2015, Özger and Kabataş [41] successfully applied fuzzy logic and combined wavelet and fuzzy logic techniques (WFL) to predict suspended sediment load data which, then, were compared with monthly measured suspended sediment data from Corukhi River and miscellaneous East Black Sea basins. Kişi, in 2009 [42], and Kişi et al. [43] efficiently elaborated evolutionary fuzzy models (EFMs) and triangular fuzzy membership functions for suspended sediment concentration estimation using data from the US Geological Survey (USGS). Lohani et al. [44] applied Zadeh's [45] fuzzy rule-based approach to derive stage-discharge-sediment concentration relationships. Firat (2010) [46] used an Adaptive Neuro-Fuzzy Inference System (ANFIS) approach as a monthly total sediment forecasting system.

The present study aims to redefine the coefficients of the stream sediment transport formula of Yang [22] with a fuzzy regression, using the very same experimental data that Yang used for the original equation. Basically, it is intended to build a functional "fuzzy twin" of the original equation, which will provide a fuzzy band for the total sediment concentration for natural sandy-bed rivers. The study initiated with the collection, analysis and processing of the primary experimental data, which, by itself, was a painstaking process. Finally, the 93.3% of the original experimental data was possible to be collected. Based on this data, a fuzzy "duplicate" of Yang's equation was built, by means of the fuzzy regression model of Tanaka [47]. In addition, the original sediment transport equation was reconstructed, by means of classic multiple linear regression, in order to validate the quality of data by comparing the calculated coefficients with the original ones. Apart from the coefficients, an efficiency assessment was carried out on the basis of comparison between the measured crisp total sediment concentrations and the calculated concentrations with a fuzzy band. It was shown that all the elaborated methods produced successful results for both the classic and the fuzzy multiple regressions.

It is the authors' belief that fuzzy logic efficiently deals with the uncertainties that naturally envelop the complex sediment transport processes, by providing a fuzzy band for the final result—whichever this might be.

2. Unit Stream Power Theory of Yang for Sediment Transport in Natural Rivers

In 1972, Yang [27], with the introduction of the unit stream power theory, fundamentally questioned the applicability of most sediment transport models which until then argued that sediment transport rate could be determined on the basis of physical magnitudes, such as discharge, flow velocity, energy slope or shear stress.

Yang defines the unit stream power as the velocity-slope product. The rate of energy per unit weight of water available for transporting water and sediment in an open channel of reach length x and total drop Y is [31]:

$$\frac{dY}{dx} = \frac{dx}{dt}\frac{dY}{dx} = VS = \text{unit stream power} \tag{1}$$

where Y is the elevation above a datum which also equals the potential energy per unit weight of water above a datum; x is the longitudinal distance; V is the mean flow velocity; S is the energy slope; and VS is the unit stream power.

To determine total sediment concentration, Yang regarded a relation between several physical quantities of the following form:

$$\varphi(C_t, VS, V_*, v, \omega, d_{50}) = 0 \tag{2}$$

where C_t is the total sediment concentration (ppm), with wash load excluded; V_* is the shear velocity (m/s); v is the water kinematic viscosity (m^2/s); ω is the fall velocity (m/s); and d_{50} is the median particle diameter (m).

By means of Buckingham's π theorem, the total sediment concentration can be expressed as a function of dimensionless parameters, as follows:

$$C_t = \varphi'(VS/\omega, V_*/\omega, \omega \cdot d_{50}/v) = 0 \tag{3}$$

Yang added a critical unit stream power in the formula, to account for incipient motion of sediment, and after dimensional analysis, he derived the following equation for the total sediment concentration:

$$\begin{aligned} \log C_F = {} & 5.435 - 0.286\log\frac{\omega d_{50}}{v} - 0.457\log\frac{V_*}{\omega} \\ & + \left(1.799 - 0.409\log\frac{\omega d_{50}}{v} - 0.314\log\frac{V_*}{\omega}\right)\log\left(\frac{VS}{\omega} - \frac{V_{cr}S}{\omega}\right) \end{aligned} \tag{4}$$

$$\frac{V_{cr}}{\omega} = \frac{2.5}{\log(V_*d_{50}/v - 0.06)} + 0.66, \qquad \text{if} \qquad 1.2 < \frac{V_*d_{50}}{v} < 70 \tag{5}$$

$$\frac{V_{cr}}{\omega} = 2.05, \qquad \text{if} \qquad \frac{V_*d_{50}}{v} \geq 70 \tag{6}$$

where C_F is the calculated total sediment concentration (ppm); and $V_{cr}S$ is the critical unit stream power, derived as the product of mean critical flow velocity and energy slope.

Equation (4) is the dimensionless unit stream power equation that can be used to calculate the total sediment concentration, in ppm by weight, in both laboratory flumes and natural sandy-bed rivers, with median particle size less than 2 mm. Knowing the discharge and the geometric characteristics of the channel, and with simple calculations, the aforementioned sediment concentration can easily be transformed into any form of sediment load, sediment yield, or sediment discharge.

Yang's unit stream power theory has been applied in a plethora of cases in literature, both continuously [48,49] and event-based [50,51]. Quaintly, nonetheless successfully, it has also been applied for estimating overland flow erosion capacity [52,53]. Because of the fact that Yang's equations for total load [22,54] were built with data in the sand-size range, their application should be limited only in sandy rivers. However, Moore and Burch [52] proved that Equation (4) can be applied equally well to predict the sediment transport rate in sheet and rill flows, when soil particles are in ballistic dispersion. It should be mentioned, however, that Moore and Burch used a constant value, of 0.002 m/s, for the critical unit stream power [31].

3. "Fuzzy Twin"—The Physical Meaning

As mentioned above, the ultimate goal of this research is to build a functional "fuzzy twin" of the unit stream power formula of Yang. In an effort to explain the physical meaning of the term "fuzzy twin", it is considered meaningful to separately analyze "fuzzy" and "twin".

While a portion of the engaging parameters, such as the flow velocity, the flow depth, the bed slope and the water temperature, can be determined with a fairly high precision in natural streams, still the overall uncertainty that blankets the stream sediment transport processes, let alone the determination of the in-stream sediment concentration, is appreciably high. This is not only associated with the bed morphology and the grain size distribution, but also with the constantly altering flow conditions that prevail in natural rivers. Yet, any uncertainty due to measurement errors seems to be small compared to uncertainties in the computational part. This is due to simplifications and assumptions made by sediment transport formulas, in which reality is usually poorly reflected. Hence, apart from the measurement errors, the fuzzy band is even more meant to deal with uncertainties in the computational part, namely uncertainties that have to do with the representation of all the involved physical processes in a formula. Just to give an example, fall velocity, for instance, can be measured with much greater accuracy than it can be computed by any existing formula. Obvious reasons for this are that in all fall velocity formulas the particle is considered a sphere, and is usually represented by the median particle diameter, d_{50}, and not by its actual diameter, as well as the disregard of turbulence. Going further, the uncertainty raises by the subjectivity in the estimation of the incipient motion criterion [54] and the turbulence impact on sediment transport. To better identify the source of uncertainty in Yang's formula, the assumption of one-dimensional, uniform and steady flow (especially, in the case of natural rivers), as well as the regression analysis between sediment transport rate and stream discharge, which partly neglects the physical mechanisms of the sediment transport phenomenon, must be considered, as well. The uncertainties would significantly be reduced in the case of an analytical physically based model. The complex nature of sediment transport and the associated uncertainties have been very well documented in literature [54–59]. In terms of uncertainty, Kleinhans (2005) [57] compares the notoriety of the sediment transport problem with that of the roughness problem and he stresses the necessity of calibration. In such cases, fuzzy regression, contrarily to conventional solutions such as classic regression, offers an efficient and applicable solution, by producing a fuzzy band within which the measured values are most likely included. Indeed, Azamathulla et al. [60] state that classic regression does not efficiently cope with the uncertainties that dominate both input and output data and instead they use a Fuzzy Inference System (FIS) as a prediction model. Hence, "fuzzy" is justified by the fact that the computed sediment concentration is not a crisp value, as it would be if the classic formula of Yang Equation (4) had been used, but a range of values which is expected to contain the observed data.

As already mentioned, and as it is thoroughly presented in Section 4, the construction of the fuzzy total sediment concentration formula is based on the exact same datasets that Yang used 47 years ago to derive his unit stream power formula. Hence, both formulas were built upon the same foundation and this makes them "twins".

4. Materials and Methods

4.1. Experimental Data for the Derivation of Yang's Formula

Yang determined the coefficients of Equation (4) by considering the logarithmic total sediment concentration, $logC_F$, as the dependent variable and the $log(\omega \cdot d_{50}/v)$, $log(V_*/\omega)$, $log(VS/\omega - V_{cr}S/\omega)$, $log(\omega \cdot d_{50}/v) \cdot log(VS/\omega - V_{cr}S/\omega)$, $log(V_*/\omega) \cdot log(VS/\omega - V_{cr}S/\omega)$, as the independent variables, and applying a multiple regression analysis for 463 sets of data in laboratory flumes.

These data were obtained by the following hydraulic and sediment transport related surveys, in laboratory flumes:

- Nomicos (1956) [61]
- Vanoni and Brooks (1957) [62]
- Kennedy (1961) [63]
- Stein (1965) [64]
- Guy et al. (1966) [65]
- Williams (1967) [66]
- Schneider (1971) [67]

It should be mentioned that this research initiated with collecting and organizing the experimental data of the above surveys which, by itself, was a very laborious task. Moreover, an appreciable effort was put in dealing with inaccuracies and incorrect values found in bibliography, in order to record the exact and correct data, as they result from the initial surveys.

4.1.1. Nomicos' Data (1956)

Nomicos [61] investigated the friction characteristics of streams with sediment load. Velocity and sediment profiles were measured, and friction factor and von Karman's constant were calculated in a 40-foot long and 0.875-foot wide flume. Nomicos conducted seven sets of experiments with 43 runs under uniform flow conditions and with various bed configurations, using sands ranging from 0.1 mm to 0.16 mm. Yang [22] utilized a portion of 12 runs, with the same particle size (0.152 mm) and flow depth (0.24 foot), for his regression analysis.

4.1.2. Vanoni and Brooks' Data (1957)

Vanoni and Brooks [62] carried out a total of 94 experimental runs, in the context of four different experiments, in two laboratory flumes with fine sand of several size distributions under uniform flow conditions. Fine sand with median particle diameter of 0.137 mm was used for the channel bed. During these experiments, the relationship between sediment transport rates and the hydraulic variables was investigated. A number of 14 runs, from the experiments conducted in a 60-foot long and 2.79-foot wide flume, was used by Yang.

4.1.3. Kennedy's Data (1961)

In 1961, Kennedy [63] carried out a series of experiments with the objective of investigating the factors involved in the formation of antidunes, the characteristics of stationary waves, as well as the effect of these on the friction factor and sediment transport. Three experiments, in three flumes with different geometric characteristics, were executed for the needs of Kennedy's survey. More specifically, fine sand of 0.233 mm and 0.549 mm was used in a 40-foot long and 0.875-foot wide flume, and 0.233 mm sand was used in a 60-foot long and 2.79-foot wide flume. These are the very same flumes utilized by Nomicos [61], and Vanoni and Brooks [62], in their laboratory experiments. A corresponding number of 14, 13 and 14 sets of data were considered by Yang from each of Kennedy's experiments.

4.1.4. Stein's Data (1965)

Stein [64] executed experiments for the determination of total load and total bed materials by fractional sampling, static and dynamic dune properties, and head losses encountered by flow over an alluvial bed. Stein's results showed that in the presence of moving dunes, mean flow velocity appeared to be the decisive parameter for the determination of total load and total bed load. The experiments were conducted in a 100-foot long and 4-foot wide flume with a bed material of 0.4 mm. From the 73 runs of Stein, Yang [22] selected 42 sets of experimental data.

4.1.5. Guy, Simons and Richardson's Data (1966)

The primary purpose of Guy et al. [65] was to summarize and make available to the public, the results of the hydraulic and sediment data that were collected by Simons et al. [68] in a unique series of experiments at Colorado State University, between 1956 and 1961. During these experiments, 339 equilibrium runs were executed in order to determine the effects of bed material size, water temperature, and fine sediment in the flow on the hydraulic and transport variables.

More than half (286 sets of data) of the 463 sets of data Yang used for his unit stream power sediment transport formula, were derived from the Guy et al. [65] survey. A number of 10 sets of experiments with different conditions were conducted in two 150-foot long and 8-foot wide, and 60-foot long and 2-foot wide flumes for fine sand beds with a variety of median particle diameters for the bed material, ranging from 0.19 mm to 0.93 mm.

4.1.6. Williams' Data (1967)

Williams (1967) [66] used the coarser sand (1.35 mm), compare to the other studies, in a 52-foot long and 1-foot wide laboratory flume in order to study sediment transport in a series of 37 runs with bed forms ranging from an initial plane bed to antidunes. For the range of conditions examined, unique relationships were found between any two variables as long as depth was constant [66]. All the 37 sets of data from William's survey were imported to Yang's multiple regression.

4.1.7. Schneider's Data (1971)

Schneider's data [67] constitutes an exception, compare to the rest of the surveys, as it is the only data not coming from a published research. Indeed, it was observed, by the authors, that in several of Yang's relative publications from 1973 onwards, Schneider's data is cited as "personal communication". Yang [22] has published the values and ranges of the physical quantities of this data (i.e., 1.67–6.45 m/s, for mean flow velocity, 18–17,152 ppm, for total sediment concentration, etc.), yet the 31 sets of data of Schneider remain unknown. Despite the appreciable efforts put by the authors to obtain this data, this was not possible.

The data of the aforementioned surveys, which Yang used, in 1973, to construct his well-known and widely used formula, for the determination of total sediment concentration, are summarized in Table 1. Though all calculations in the mathematical part of this study were executed in System International (SI) units, the values in Table 1 are given in the original US customary units, so that they are more easily recognizable and correlated to previous literature. It must be highlighted that the values displayed in this table, whether they were obtained directly in the correct units or they were first converted (i.e., lbs/ft/s into ppm, for measured total sediment concentration, or °F into °C, for temperature), are the exact values with no rounding, whatsoever. This is said because there are some slight or, in some cases, major differences with earlier publishing.

Table 1. Experimental data for unit stream power equation.

Particle Size, d_{50} (mm)	Channel Length, L (ft)	Channel Width, W (ft)	Water Depth, h (ft)	Temperature, T (°C)	Average Velocity, V (ft/s)	Water Surface Slope, $S \times 10^3$	Total Sediment Concentration, C_t (ppm)	Number of Data, N
Nomicos' data (1956) [61]								
0.152	40	0.875	0.241	25.0–26.0	0.80–2.63	2.0–3.9	300–5767	12
Vanoni and Brooks' data (1957) [62]								
0.137	60	2.79	0.147–0.346	18.9–27.4	0.77–2.53	0.7–2.8	37–3000	14
Kennedy's data (1961) [63]								
0.233	40	0.875	0.147–0.346	24.5–30.1	1.57–3.42	2.6–16.0	730–34,700	14
0.549	40	0.875	0.074–0.346	24.3–27.0	1.65–4.65	5.5–27.2	1680–35,900	14
0.233	60	2.79	0.145–0.356	23.0–27.3	1.35–3.45	1.7–22.9	490–58,500	13
Stein's data (1965) [64]								
0.4	100	4.0	0.59–1.20	20.0–28.9	1.38–5.51	0.61–10.79	93–24,249	42
Guy et al. data (1966) [65]								
0.19	150	8.0	0.49–1.09	12.3–19.7	1.04–4.74	0.43–9.50	29–47,300	29
0.27	150	8.0	0.45–1.13	10.2–18.5	1.24–4.93	0.46–10.22	12–35,800	18
0.28	150	8.0	0.30–1.07	10.2–17.6	1.04–4.93	0.45–10.07	12–42,400	33
0.48(0.45)	150	8.0	0.19–1.00	9.0–20.0	0.75–6.18	0.39–10.10	10–15,100	34
1.2(0.93)	150	8.0	0.38–1.11	16.7–21.7	1.46–6.07	0.37–12.80	15–10,200	32
0.32	60	2.0	0.54–0.74	7.0–34.3	1.24–5.73	0.86–16.20	55–49,300	29
0.33 (uniform)	60	2.0	0.49–0.52	19.8–20.3	1.17–6.93	0.88–11.40	47–18,400	12
0.33 (graded)	60	2.0	0.48–0.53	19.6–24.1	1.06–6.34	0.47–9.80	12–22,500	14
0.50(0.47)	150	8.0	0.30–1.33	10.7–24.5	4.69–17.45	0.42–9.60	23–17,700	50
0.59(0.54)	60	2.0	0.59–0.89	16.9–25.1	1.37–6.27	0.38–19.28	17–50,000	35
Williams' data (1967) [66]								
1.35	52	1.0	0.094–0.517	11.9–30.8	1.27–3.49	1.10–22.18	10–9223	37
Schneider's data (1971) [67]								
0.25	–	8.0	1.012–2.822	20.4–22.4	1.67–6.45	0.10–4.97	18–17,152	31

Yang [22] used, in his analysis, only data in the sand size range 0.0625 mm < d < 2 mm. It is important to note that the particle size, d_{50}, is the median sieve diameter of the sediment, while Guy et al. [65] published their data in terms of fall diameter. According to Yang [22], the difference between these two measurements of particle size is insignificant when either one is smaller than 0.4 mm. The fall diameter was converted, by Yang, into sieve diameter by means of Figure 7 of Report 12 of the Inter-Agency Committee on Water Resources (1957) [69]. The numbers shown in parentheses, in Table 1, refer to the fall diameters for the coarse sand.

Missing the 31 sets of data of Schneider, out of a total of 463 sets of data, resulted in obtaining 432 of them, which correspond to the 93.3% of the total amount of data. As far as the authors are concerned, this is the closest one can get in collecting the dataset upon which the unit stream power sediment transport equation of Yang was based. All the work presented in this study, is based on this, nearly complete, dataset.

4.2. Fuzzy Regression

A fuzzy set can be seen as a mapping from a general set X to the closed interval [0, 1]. A fuzzy set can be expressed by a membership function, which shows to what degree an element lies in the examined fuzzy set. A membership function is confined in the interval [0, 1], with a membership degree of 0 indicating that the element does not belong to the set and a membership degree of 1 indicating that the element fully belongs to the set. Subsequently, an object with a membership degree between 0 and 1 will belong to the set to some degree [37].

A fuzzy number is a fuzzy set which, furthermore, satisfies the properties of convexity and normality. It is defined in the axis of real numbers and its membership function is a piecewise continuous function [70].

The (soft) α-cut set of the fuzzy number A, with $0 < \alpha \leq 1$, is defined as follows [71]:

$$[A]_a = \{x | \mu_A(x) \geq a, \quad x \in R\} \tag{7}$$

where $\mu_A(x)$ the membership function of the fuzzy number A; and R is the set of real numbers.

An interesting point is that the crisp set including all the elements with non-zero membership function is the 0-strongcut which can be defined as follows [72]:

$$A_{0^+} = \{x | \mu(x) > 0, \quad x \in R\} \tag{8}$$

More analytically, according to Equation (8), above the 0-cut is an open interval that does not contain the boundaries. For this reason, and in order to have a closed interval containing the boundaries, Hanss [73] suggested the phrase worst-case interval W, which is the union of the 0-strongcut and the boundaries [74].

Linear regression analysis is used to model the linear relationship between the independent variables and the dependent variable. Most collected data in the present study constitute independent variables and the derivative regression model should approximate the results of the dependent variable measurements according to the criteria specified by the analyst. In the fuzzy linear regression model, the difference between the computational data and the actual values (measurements) is assumed to be due to the structure of the system. The proposed model carries this uncertainty back to its coefficients or, in other words, our inability to construct a precise relationship, is directly introduced into the model, on the fuzzy parameters [75,76]. Based on the above reasoning, the coefficients for the independent variables are chosen to be fuzzy numbers. This study also deals with cases where both the input data (independent variables) and the derived output (dependent variable) are classic numbers. The problem of fuzzy linear regression is reduced to a linear programming problem according to the following steps [77]:

1. The model is as follows:

$$\widetilde{Y}_j = \widetilde{A}_0 + \widetilde{A}_1 x_{1j} + \widetilde{A}_2 x_{2j} + \ldots + \widetilde{A}_n x_{nj} \tag{9}$$

where $\widetilde{Y}$ is the fuzzy dependent variable; $j = 1, \ldots, m$; $i = 1, \ldots, n$; $\widetilde{A}_i = (a_i, c_i)$ are symmetric fuzzy triangular numbers selected as coefficients; and x is the independent variable (Figure 1). In addition, n is the number of independent variables; m is the number of data; a is the central value (where $\mu = 1$); and c is the semi-width.

2. Determination of the degree h at which the data $[(x_{1j}, x_{2j}, \ldots, x_{nj}), y_j]$ is aimed to be included in the estimated number Y_j:

$$\mu_{Yj}(y_j) \geq h, \quad j = 1, \ldots, m \tag{10}$$

The constraints express the concept of inclusion in case that the output data are crisp numbers. In the examined case of the widely used model of Tanaka [47], a more soft definition of the fuzzy subsethood is used compared to the Zadeh [42] definition. Hence, the inclusion of a fuzzy set A into the fuzzy set B with the associated degree $0 \leq h \leq 1$ is defined as follows:

$$[A]_h \subseteq [B]_h \tag{11}$$

In our case, since the data are crisp (for each individual data), the set A is only a crisp value (a point of data which must be included in the produced fuzzy band) and the fuzzy set B is a fuzzy triangular number. Hence, Equation (11) is equivalent to:

$$\sum_{i=0}^{n} a_i x_{ij} - (1-h)\sum_{i=0}^{n} c_i |x_{ij}| \leq y_j \leq \sum_{i=0}^{n} a_i x_{ij} + (1-h)\sum_{i=0}^{n} c_i |x_{ij}|, \quad j = 1, \ldots, m \tag{12}$$

It must be clarified that the above equations hold for a specified h-cut and not for every α-cut. Normally, the 0-strongcut is used since greater levels of h lead to a greater uncertainty.

3. Determination of the minimization function (objective function) J. In the conventional fuzzy linear regression model, the objective function, J, is the sum of the produced fuzzy semi-widths for the data:

$$J = \left\{ mc_0 + \sum_{j=1}^{m} \sum_{i=1}^{n} c_i |x_{ij}| \right\} \tag{13}$$

where c_0 is the semi-width of the constant term; and c_i semi-width of the other fuzzy coefficients.

Since fuzzy symmetric triangular numbers are selected as fuzzy coefficients, it can be proved that the objective function is the sum of the semi-widths of the produced fuzzy band regarding the available data:

$$J = \left\{ mc_0 + \sum_{j=1}^{m} \sum_{i=1}^{n} c_i |x_{ij}| \right\} = \frac{1}{2} \sum_{j=1}^{m} \left(Y_j^+ - Y_j^- \right) \tag{14}$$

where Y_j^+, Y_j^- the right and the left-hand side of the 0-strongcut, respectively.

4. The problem results in the following linear programming problem:

$$\left. \begin{aligned} & \min\left\{ mc_0 + \sum_{j=1}^{m} \sum_{i=1}^{n} c_i |x_{ij}| \right\} \\ & \sum_{i=0}^{n} a_i x_{ij} - (1-h)\sum_{i=0}^{n} c_i |x_{ij}| = y_h^L \leq y_j \\ & \sum_{i=0}^{n} a_i x_{ij} + (1-h)\sum_{i=0}^{n} c_i |x_{ij}| = y_h^R \geq y_j \end{aligned} \right\} \tag{15}$$

where $c_i \geq 0$, for $i = 0, 1, \ldots , n$.

In addition, many times, when data are classic numbers, we can easily approximate non-linear cases with the fuzzy linear regression model with the help of auxiliary variables. In this case, the total uncertainty (cumulative width) indicates incomplete complexity, whereas non-physical behavior is an indicator of overtraining [77], due to adoption of excessive complexity in non-linear models.

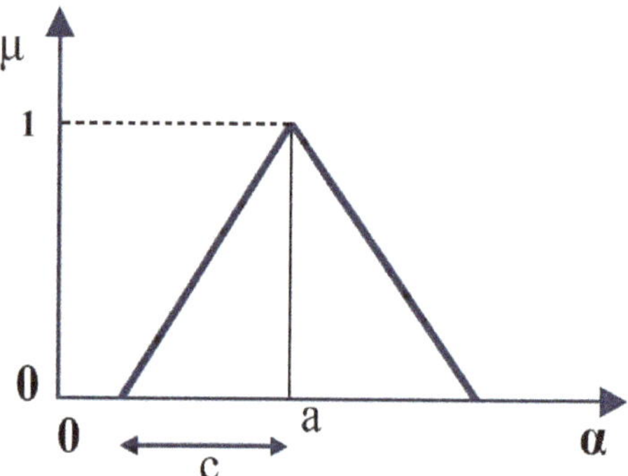

Figure 1. Fuzzy symmetric triangular number.

4.3. Implementation

For the modulation of the auxiliary variables X_1, X_2, X_3, X_4 and X_5, several parameters had to be calculated. The sedimentation rate in the unit stream power equation of Yang Equation (4) was determined by means of Zanke's [78] formula:

$$\omega = 11 \cdot v \cdot \left(\sqrt{1 + 0.01D^{*3}} - 1 \right) / d_{ch} \tag{16}$$

where v is the water kinematic viscosity (m^2/s); D^* is the Bonnefille number; and d_{ch} the characteristic grain diameter (m). The Bonnefille number, D^*, is given by:

$$D^* = \left(\rho' \cdot g / v^2 \right)^{1/3} \cdot d_{ch} \tag{17}$$

$$\rho' = (\rho_F - \rho_W) / \rho_W \tag{18}$$

In the above relation, ρ_F is the density of sediment (kg/m^3) and ρ_W is the density of water (kg/m^3). The kinematic viscosity, v, of water is given by the equation:

$$v = 1.78 \cdot 10^{-6} / (1 + 0.0337 \cdot T + 0.00022 \cdot T^2) \tag{19}$$

where T (°C) is the temperature of the water.

The shear velocity, V_*, was determined by means of the following formula:

$$V_* = \sqrt{ghS} \tag{20}$$

where g is the gravitational acceleration (m/s^2); h (m) is the flow depth; and S is the energy slope (m/m). In Equation (20), the hydraulic radius is replaced approximately by the flow depth. In the case of uniform flow, the energy slope equals the bed slope.

When the auxiliary variables X_1, X_2, X_3, X_4 and X_5 Equation (21) are introduced into the fuzzified version of the Yang's equation Equation (4), then Equation (22) results in:

$$\begin{cases} X1 = \log(\omega \cdot d50/\nu) \\ X2 = \log(V_*/\omega) \\ X3 = \log(V \cdot S/\omega - Vcr \cdot S/\omega) \\ X4 = \log(V \cdot S/\omega - Vcr \cdot S/\omega) \cdot \log(\omega \cdot D50/\nu) \\ X5 = \log(V \cdot S/\omega - Vcr \cdot S/\omega) \cdot \log(V_*/\omega) \end{cases} \tag{21}$$

$$\log \widetilde{C}_{F,j} = \widetilde{A}_0 + \widetilde{A}_1 \cdot X_{1,j} + \widetilde{A}_2 \cdot X_{2,j} + \widetilde{A}_3 \cdot X_{3,j} + \widetilde{A}_4 \cdot X_{4,j} + \widetilde{A}_5 \cdot X_{5,j} \tag{22}$$

The dependent variable is the logarithm of the concentration of the total load, C_F, which is produced as fuzzy symmetric triangular number, as well. By introducing the above auxiliary variables X_1 to X_5 for numeric data, the problem of non-linear fuzzy regression is reduced to a linear fuzzy regression problem. In the fuzzy linear regression model, the coefficients of the independent variables are fuzzy numbers that were determined using the Matlab program.

Furthermore, a simplified version of the Yang's Equation (4), which contains only the exerted unit stream power minus the critical unit stream power, is investigated:

$$\log \widetilde{C}_{F,j} = \widetilde{A}_0 + \widetilde{B} \cdot X_{3,j} \tag{23}$$

A criterion for the successfulness of this simplification will be the produced uncertainty. More analytically, if the uncertainty is increased significantly, this will indicate an irrational simplification (undertraining behavior).

5. Results

The comprehensive results of the calculation of the remainder parameters, as well as the results of both the classic multiple regression and fuzzy regression analyses, are presented in the following sections.

5.1. Determination of Yang's Formula Independent Variables

As demonstrated in Section 4.3, in an effort to determine the dimensionless independent variables of Equation (4), a set of supplementary hydraulic parameters, in addition to those displayed in Table 1, had to be calculated. Hence, the following parameters were computed, on the basis of available data: water kinematic viscosity, fall velocity, shear velocity, and the dimensionless critical velocity, V_{cr}/ω.

Fall velocity was deemed to be the most decisive parameter in Yang's formula, as it is the only parameter that appears in all independent variables. For this reason, a special attention was given to fall velocity, which was calculated by two widely used formulas for settling particles, those of Zanke [78] and Rubey [2]. Following the computation of kinematic viscosity and shear velocity, the dimensionless critical velocity, V_{cr}/ω, was calculated, for each set of data, by means of Equations (5) and (6). The ranges of values for all calculated parameters and variables, for each dataset, are provided in Table 2.

In Table 2, X_1, X_2, X_3, X_4, X_5 are the dimensionless variables $log(\omega \cdot d_{50}/\nu)$, $log(V_*/\omega)$, $log(VS/\omega - V_{cr}S/\omega)$, $log(\omega \cdot d_{50}/\nu) \cdot log(VS/\omega - V_{cr}S/\omega)$, $log(V_*/\omega) \cdot log(VS/\omega - V_{cr}S/\omega)$, respectively, C_t is the total measured sediment concentration obtained from the experimental data, and C_F is the total calculated sediment concentration, as obtained from the unit stream power formula, using Yang's coefficients and by replacing the independent variables with the calculated values of X_1, X_2, X_3, X_4, X_5.

Table 2. Calculated parameters and variables for unit stream power equation.

Kinematic Viscosity × 10^{-6}, ν (m²/s)	Fall Velocity, ω (m/s), Zanke (1977)	Shear Velocity, V_* (m/s)	Dimensionless Critical Velocity, V_{cr}/ω	x_1	x_2	x_3	x_4	x_5	Total Measured Sediment Concentration, C_t (ppm)	$\log C_t$	Total Calculated Sediment Concentration, C_F (ppm)	$\log C_F$	Number of Data, N
						Nomicos' data (1956) [61]							
0.88	0.02	0.04–0.05	3.46–4.0	0.53–0.54	0.27–0.43	−1.79–(−0.84)	−0.97–(−0.44)	−0.52–(−0.36)	300–5767	2.48–3.76	303–7387	2.48–3.87	12
						Vanoni and Brooks' data (1957) [62]							
0.85–1.04	0.015–0.017	0.03–0.05	3.87–4.77	0.29–0.45	0.29–0.46	−1.98–(−0.98)	−0.74–(−0.28)	−0.71–(−0.42)	37–3000	1.57–3.48	132–4419	2.12–3.65	14
						Kennedy's data (1961) [63]							
0.8–0.91	0.04–0.04	0.04–0.09	2.56–3.26	–	0.01–0.36	−1.53–(−0.43)	−1.56–(−0.44)	−0.22–(−0.01)	730–34,700	2.86–4.54	1070–27,619	3.03–4.44	14
0.86–0.91	0.08–0.09	0.05–0.1	2.09–2.42	1.72–1.75	−0.24–0.06	−1.73–(−0.42)	−2.97–(−0.73)	−0.02–0.42	1680–35,900	3.23–4.56	1075–29,057	3.03–4.46	14
0.85–0.94	0.036–0.038	0.04–0.12	2.4–3.34	0.96–1.02	0.02–0.52	−1.7–(−0.25)	−1.7–(−0.24)	−0.22–(−0.02)	490–58,500	2.69–4.77	619–40,833	2.79–4.61	13
						Stein's data (1965) [64]							
0.83–1.01	0.065–0.069	0.04–0.14	2.12–2.75	1.41–1.52	−0.19–0.32	−2.66–(−0.61)	−3.85–(−0.88)	−0.25–0.51	93–24,249	1.97–4.39	55–15,860	1.74–4.2	42
						Guy et al. data (1966) [65]							
1.02–1.23	0.01–0.02	0.04–0.14	2.62–4.94	0.09–0.55	0.24–0.81	−2.0–(−0.26)	−1.1–(−0.10)	−0.9–(−0.21)	29–47,300	1.46–4.68	122–38,227	2.09–4.58	29
1.05–1.3	0.02–0.04	0.04–0.14	2.67–3.82	0.51–0.85	0.1–0.79	−2.38–(−0.22)	−1.79–(−0.12)	−0.66–(−0.17)	12–35,800	1.08–4.55	50–42,848	1.7–4.63	18
1.07–1.3	0.1–0.04	0.04–0.13	2.65–4.22	0.13–0.99	−0.06–0.83	−2.51–(−0.24)	−2.49–(−0.14)	−0.88–0.14	12–42,400	1.08–4.63	43–41,340	1.64–4.62	33
1.01–1.35	0.01–0.08	0.03–0.11	2.26–4.66	0.19–1.56	−0.34–0.92	−3.72–(−0.37)	−4.92–(−0.22)	−1.04–1.25	10–15,100	1.0–4.18	1–28,845	0.15–4.46	34
0.97–1.1	0.05–0.12	0.03–0.13	2.05–2.54	1.5–2.16	−0.59–0.37	−3.13–(−0.63)	−6.7–(−1.19)	−0.45–1.84	15–10,200	1.18–4.01	28–15,613	1.45–4.19	32
0.74–1.43	0.02–0.06	0.04–0.18	2.29–3.86	0.57–1.46	−0.13–0.73	−2.33–(−0.17)	−3.33–(−0.15)	−0.59–0.3	55–49,300	1.74–4.69	105–45,297	2.02–4.66	29
1.0–1.02	0.02–0.05	0.04–0.13	2.34–4.24	0.71–1.27	0.05–0.5	−2.05–(−0.27)	−2.11–(−0.3)	−0.63–(−0.09)	47–18,400	1.67–4.27	176–37,507	2.25–4.57	12
0.92–1.02	0.01–0.05	0.03–0.12	2.82–5.95	0.09–1.19	−0.2–0.91	−2.46–0.07	−2.94–0.03	−0.82–0.49	12–22,500	1.08–4.35	64–93,556	1.81–4.97	14
0.91–1.28	0.03–0.08	0.03–0.11	2.15–3.54	0.82–1.74	−0.36–0.39	−1.99–(−0.21)	−2.77–(−0.3)	−0.39–0.647	23–17,700	1.362–4.25	314–48,987	2.5–4.69	50
0.9–1.09	0.06–0.09	0.03–0.2	2.05–3.17	1.35–1.79	−0.45–0.44	−3.02–(−0.37)	−4.66–(−0.55)	−0.23–1.37	17–50,000	1.23–4.7	17–26,180	1.24–4.42	35
						Williams' data (1967) [66]							
0.79–1.24	0.15–0.16	0.03–0.1	2.05–2.25	2.22–2.43	−0.7–(−0.2)	−3.33–(−1.07)	−7.75–(−2.53)	0.23–2.31	10–9223	1.0–3.97	35–8038	1.55–3.91	37
						Schneider's data (1971) [67]							
–	–	–	–	–	–	–	–	–	18–17,152	–	–	–	31

As mentioned above, the calculations were carried out two times, one by using the fall velocity obtained by the formula of Zanke [78], and one with the fall velocity from the formula of Rubey [2]. By means of comparison between the 432 values of $logC_t$ and $logC_F$, Nash-Sutcliffe Efficiencies (NSEs) of 0.79 and 0.72 were achieved with Zanke's and Rubey's formulas, respectively. Hence, in Table 2, only the results obtained by the use of Zanke's formula, for fall velocity, are presented.

As a first comment, and by observing both the range values of $logC_t$ and $logC_F$, and the NSE value of 0.79, it can be said that the approximation between the measured and calculated results, as well as the quality of data is deemed satisfactory.

5.2. Multiple Regression Analyses

5.2.1. Multiple Regression Analysis for the Reconstruction of the Unit Stream Power Formula

As a natural sequence, and to further test the successfulness of the results, the unit stream power formula Equation (4) was rebuilt. Basically, knowing both the dependent and independent variables, the coefficients of the equation were recalculated.

In the case that the conventional least square-based regression is used, the following relation is achieved:

$$\log C_F = -0.2215X_1 - 0.4369X_2 + 1.7105X_3 - 0.4271X_4 - 0.4742X_5 + 4.9998 \tag{24}$$

The coefficient of determination, R^2, is equal to:

$$R^2 = 1 - \frac{\sum\limits_{j=1}^{m}\left(\log C_{t_j} - \log C_{F_j}\right)^2}{\sum\limits_{j=1}^{m}\left(\log C_{t_j} - \overline{\log C_t}\right)^2} = 0.857 \tag{25}$$

where $\log C_{t_j}$, $\log C_{F_j}$, $\overline{\log C_t}$ are the measured jth value of the concentration, the calculated based on Equation (24) (crisp regression) and the mean value, respectively.

5.2.2. Fuzzy Regression Analysis

In the case that the aforementioned fuzzy regression is used, the following equation is produced:

$$\log \widetilde{C}_F = 0.1602X_1 + (0.2842, 0.1193)X_2 + (1.2643, 0.1394)X_3 - 0.1694X_4 \\ + (-0.2768, 0.1191)X_5 + (4.1880, 0.4014) \tag{26}$$

The first term in the parentheses expresses the central value and the second term the semi-width of the produced fuzzy coefficient.

The total amount of uncertainty, namely the sum of the semi-widths regarding the available data, is:

$$J = \left\{ 432c_0 + \sum_{j=1}^{432}\sum_{i=1}^{5} c_i\left|X_{ij}\right| \right\} = 289.069 \tag{27}$$

The projection of the produced logarithmic concentration Equation (26), with respect to several input variables, is presented in Figure 2.

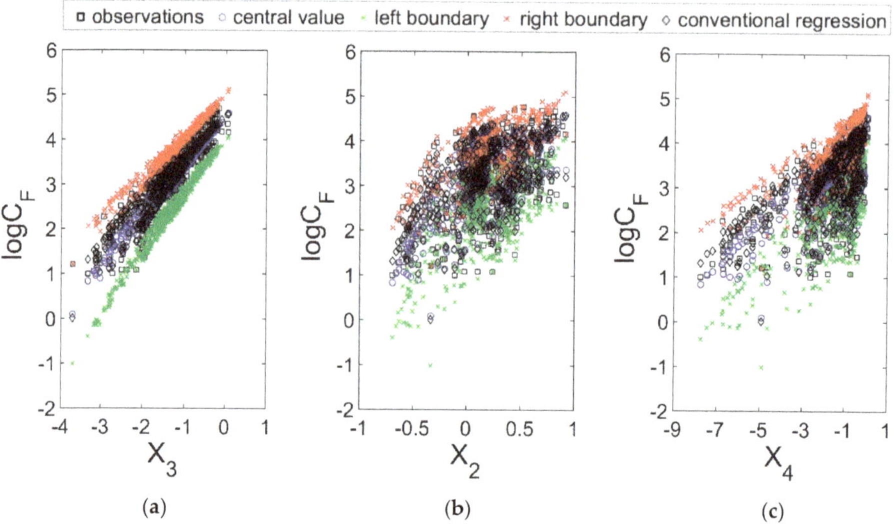

(a)   (b)   (c)

Figure 2. Projection of the achieved fuzzy relation regarding the log of the total sediment concentration with respect to (**a**) X_3, (**b**) X_2, (**c**) X_4.

As can be seen, all the available data are included within the produced fuzzy band. Furthermore, the use of only one input variable—in this case, variable X_3—is separately investigated (Figure 3). In case that the crisp linear regression is used with the X_3 as the only independent variable, the results are similar, and the squared correlation coefficient, r^2, is equal to 0.801. Hence, the linear dependence can be suggested. In this case, Equation (26) becomes:

$$\log \widetilde{C}_F = (1.1872,\ 0.2243)X_3 + (4.5954,\ 0.4948) \tag{28}$$

and the corresponding objective function obtains the following value:

$$J = \left\{ 432c_0 + \sum_{j=1}^{432} c_3|X_3| \right\} = 343.305 \tag{29}$$

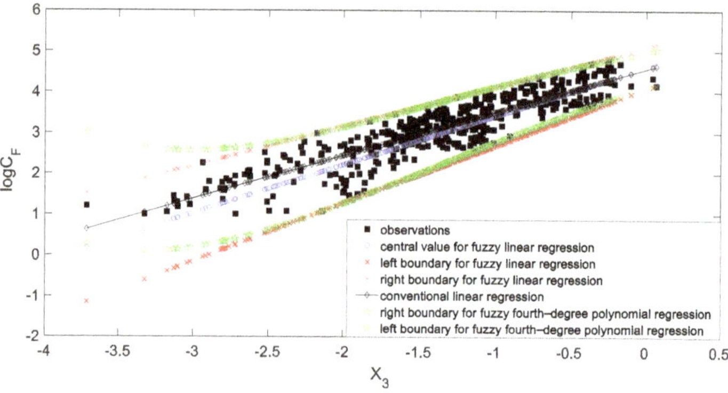

Figure 3. Fuzzy linear regression to achieve the log of the total sediment concentration with respect to X_3.

Based on the value of the objective function, J, it is obvious that this simplification (i.e., considering X_3 as the only independent variable) increases the uncertainty. However, the usefulness is that the emphasis is put on the subtraction of the critical unit stream power from the exerted unit stream power, as a main independent variable. Going in the opposite direction, if only the variable X_3 is removed, then the uncertainty of the produced fuzzy band is greater than the above value ($J = 396.75$).

An interesting perspective is that by adopting a polynomial form, in the above simplification, a small reduction of the uncertainty is achieved and hence, Equation (28) cannot be further improved. Indeed, a small reduction of the fuzzy band is achieved, if the fourth-degree polynomial regression is used.

In Figure 3, the observations against the results of the conventional linear regression, as well as the results of fuzzy linear and polynomial regressions, by using only X_3 as input variable, are depicted. As it can be observed from the figure, the data of the fuzzy fourth-degree polynomial regression and the data of the fuzzy linear regression almost overlap for the most part. However, the fourth-degree polynomial regression presents an "irrational behavior" in the area of low X_3 values, from a physical meaning point of view. To better explain this, as the difference "exerted unit stream power minus critical unit stream power" (here, represented by X_3) grows larger, a higher sediment transport, and therefore a higher sediment concentration is expected. Simply put, $logC_F$ and X_3 are similar amounts and the increase of one by the decrease of the other is not justified. The negligible reduction of the uncertainty, as well as the "irrational behavior" of the fuzzy fourth-degree polynomial regression, indicates the improperness of the polynomial models for these data. From the above it is concluded that the auxiliary variable X_3 is the most significant parameter parameter. The use of high polynomial extension to Equation (28) did not improve the results. Equation (26) results in significantly less uncertainty and should be preferred.

However, a fuzzy band with high spread will include all the data, but this will be a non-useful approach. Therefore, another suitability measure, JJ, is proposed, which is equal to the mean ratio of the total spread $\left(\log C_{F_j}{}^+ - \log C_{F_j}{}^-\right)$ to the central value ($\mu = 1$, with the index j), $\overline{\log C_{F_j}}$, and when it is applied for Equation (26), it leads to the following result [37]:

$$JJ = \frac{1}{N}\sum_{j=1}^{N}\frac{\left(\log C_{F_j}{}^+ - \log C_{F_j}{}^-\right)}{\overline{\log C_{F_j}}} = 0.5644 \tag{30}$$

where N is the number of data; in brief, the measure JJ expresses the mean uncertainty of the produced fuzzy band as a percentage of the central value. It is desirable to get low values for JJ [37]. At this point, it must be clarified that from a sediment transport point of view, the results can be characterized as sufficiently good. It should be noted that the measure JJ takes a better value compared with the corresponding JJ measure achieved by Kaffas et al. [79]. However, that study was based only on the experimental data of Guy et al. [65].

5.2.3. Validation

While the unit stream power formula was built upon only laboratory data, as stated in [80], Yang primarily built his dimensionless unit stream power equation to be used by engineers for the estimation of the total sediment concentration in both laboratory flumes and natural rivers. To make sure of the applicability of his unit stream power formula in natural streams, Yang validated it with total sediment concentrations and total suspended sediment loads from several natural rivers and streams [81–85]. The results revealed that Equation (4) is fairly accurate in predicting total sediment load or total bed-material load in the sand size range in natural rivers, as it is for laboratory flumes [55,80].

To test the applicability of the crisp and fuzzy regression formulas, presented in this study, data from three different sandy-bed rivers in Wisconsin, USA, taken from a US Geological Survey [86], were used. More specifically, total sediment concentration (bed load and suspended load) measurements, from Wisconsin River at Muscoda, Black River near Galesville, Chippewa River at Durand and Chippewa

River near Pepin, were used for the validation of Equations (24) and (26). The median particle diameters (d_{50}) were obtained from granulometric curves, which were constructed upon sieve analysis data, and are in a range between 0.38 mm and 0.88 mm. Along with the sediment data, basic hydraulic parameters, such as flow velocity, flow depth, energy slope and water temperature were available in the same survey. These data were used for the computation of the independent variables in Equations (24) and (26). The independent variables in any of the Equations (4), (24) or (26) represent the geometric and flow characteristics of the stream that they are applied for. A total of 55 sets of data were used for the validation of Equations (24) and (26).

Several well-known metrics, like the Root Mean Square Error (RMSE), the Mean Absolute Error (MAE), the Mean Bias Error (MBE), the Index of Agreement (d) and the NSE, were used to test the validity of the crisp multiple regression Equation (24). Though the comparison between observations and computations resulted in low statistical errors (RMSE = 0, MAE = 0.3, MBE = −0.124), and a fair Index of Agreement (d = 0.483), a negative Nash-Sutcliffe Efficiency (NSE = −1.207) (see in Appendix A) indicates that Equation (24) cannot be applied for the selected data sets. However, this was not received entirely as a surprise. Despite the validated suitability of Yang's formula for both laboratory flumes and natural rivers in the sand range [55,80], Yang and Stall, in their report "Unit stream power for sediment transport in natural rivers" [80], stress also the constraints of the unit stream power theory for natural rivers. According to them, these constraints can be reduced to particle size, temperature and water depth. Adding to these the stream sediment transport uncertainties, mentioned in Section 3, it is realized that the successfulness of Equation (24)—which is the crisp regression—is not guaranteed for natural streams.

This deficiency of the crisp regression is overcome by the multiple fuzzy regression Equation (26), which contains 96.36% of the observed data in the fuzzy band. This can be clearly seen in Figure 4; 53 out of 55 observations are included in the produced fuzzy band of Equation (26).

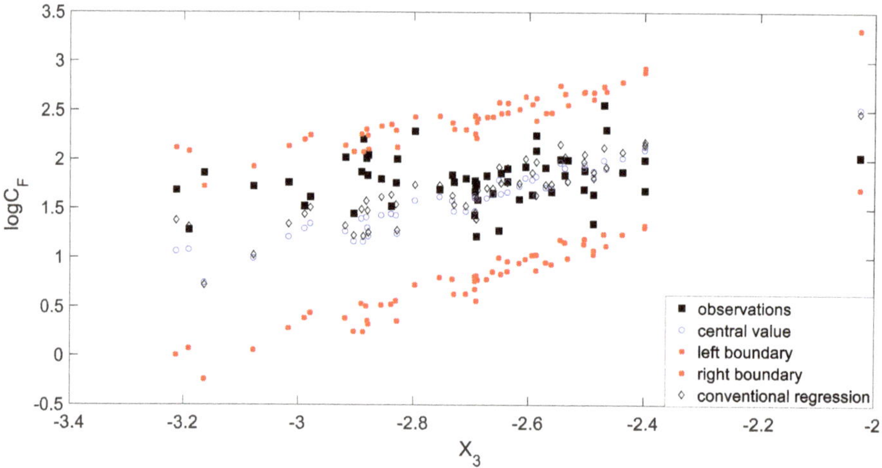

Figure 4. Multiple fuzzy regression of total sediment concentration in natural streams.

In order to check the performance of the proposed fuzzy curve upon the observations, the following validation measures are proposed:

$$E_1 = \left[\sum_{j=1}^{m} a_{R_j} \left(\log C_{t_j} - \log C_{F_j}{}^{+} \right)^2 + \sum_{j=1}^{m} a_{L_j} \left(\log C_{t_j} - \log C_{F_j}{}^{-} \right)^2 \right]^{\frac{1}{2}}$$

$$a_{R_j} = \begin{cases} 0 & if \quad \log C_{F_j}{}^{+} \geq \log C_{t_j} \\ 1 & if \quad \log C_{F_j}{}^{+} \leq \log C_{t_j} \end{cases} \qquad a_{L_j} = \begin{cases} 0 & if \quad \log C_{F_j}{}^{-} \leq \log C_{t_j} \\ 1 & if \quad \log C_{F_j}{}^{-} \geq \log C_{t_j} \end{cases}$$

(31)

This validation measure expresses the divergence of the produced fuzzy band to include all data. In other words, the squared penalty term, E_1, is activated if, and only if, the observed data are not included within the produced fuzzy band. This measure was initially proposed by Ishibuchi et al. [87] as a cost function to be minimized in the learning process, regarding a neural network with interval weights.

The second validation measure is to examine the number of points (observed data) which are outside the produced fuzzy band:

$$E_2 = \sum_{j=1}^{m} a_{R_j} + \sum_{j=1}^{m} a_{L_i} \tag{32}$$

Obviously, the application of the above validation measures on the training data (laboratory data) leads to the identical values $E_1 = E_2 = 0$. By applying the validation measures to the data for natural streams, the following values are achieved: $E_1 = 0.1868$, $E_2 = 2$. This means that only two points do not belong to the fuzzy band ($E_2 = 2$), but these points are not far from the produced fuzzy band, as suggested by the low value of the E_1 measure.

Ultimately, the success of the crisp regression Equation (24) is not guaranteed when applied for a dataset different than the one it was created from (in this case a dataset from natural rivers). Indeed, there is a large dispersion of the measurement data from the crisp curve Equation (24) and hence, the Nash-Sutcliffe Efficiency obtains a negative value (Figure 4). Contrarily, by applying the fuzzy curve of Equation (26) for a dataset different than that it was created from, it is concluded that rather all the data is included within the produced fuzzy band with a small divergence (Figure 4). Therefore, the fuzzy curve can be used in order to achieve a fuzzy estimation of the logarithmized total sediment concentration.

6. Conclusions

The objective of this research is to transform the arithmetic coefficients of the total sediment transport rate formula of Yang, into fuzzy numbers, and thus create a fuzzy relationship that will provide a fuzzy band of in-stream sediment concentration. A very large set of experimental data, in flumes, was used for the fuzzy regression analysis. The reason for selecting the fuzzy regression is that it provides a fuzzy band not only for the coefficients of the independent variables, but for the final result, as well, which is the total sediment concentration. This means that the resulting sediment concentration is not a crisp value, but a range of values, which stretch to a value equal to the semi-width on both sides of the central value. It is proved well, by the results, that this range of values deals efficiently with the uncertainties and the ambiguous nature of sediment transport processes. Apart from the measurement errors, the computational part, and specifically the physical simplifications, i.e., one-dimensional, uniform and steady flow, grain size distribution, etc., increase the uncertainty. An interesting perspective is that even if the validation data are observations from natural rivers, where significant uncertainty and simplifications take place, these are successfully captured by the proposed fuzzy band. The minimum advantage of the fuzzy band produced is that all the data must be included. However, a main criterion is the produced width of the fuzzy band. Based on this criterion, the authors concluded that the simplification of using only one variable should be avoided, and furthermore that a determinant variable is the subtraction of the critical unit stream power from the exerted unit stream power (X_3). Nevertheless, a simplification based on the X_3 (i.e., only the variable X_3 is taken into account) leads to a fuzzy linear curve that can be used to interpret the phenomenon. The produced fuzzy band compared with the central values indicates the good performance of the proposed fuzzy curve. In terms of elaboration of the original data utilized by Yang for the establishment of the unit stream power theory, this research goes the closest possible to what could be called "fuzzy twin" of Yang's stream sediment transport formula.

Author Contributions: Conceptualization, M.S. (Mike Spiliotis); Methodology, M.S. (Mike Spiliotis), K.K. and V.H.; Validation, M.S. (Mike Spiliotis), K.K. and M.S. (Matthaios Saridakis); Investigation, K.K.; Data curation, K.K.;

Writing—original draft preparation, K.K. and M.S. (Mike Spiliotis); Writing—review and editing, V.H., K.K., M.S. (Mike Spiliotis) and M.R.; Visualization, M.S. (Mike Spiliotis), M.S. (Matthaios Saridakis) and K.K.; Supervision, V.H. and M.S. (Mike Spiliotis). All authors have read and agreed to the published version of the manuscript.

Funding: K.K. and M.R. were supported by the project Sediplan–r (FESR1002) and financed by the European Regional Development Fund (ERDF) Investment for Growth and Jobs Programme 2014–2020.

Acknowledgments: The authors would like to express their thankfulness to the reviewers for their comments and suggestions, which helped to improve the quality of this study.

Conflicts of Interest: The authors declare no conflict of interest.

Appendix A

The Nash-Sutcliffe Efficiency (NSE), proposed by Nash and Sutcliffe in 1970, is defined as one minus the sum of the squared differences between the observed and predicted data normalized by the variance of the observed values [88]:

$$NSE = 1 - \frac{\sum\limits_{j=1}^{m} \left(y_j - \hat{y}_j\right)^2}{\sum\limits_{j=1}^{m} \left(y_j - \bar{y}\right)^2} \tag{A1}$$

where y_j are the observed values; $\hat{y}_j$ the predicted values; $\bar{y}$ the mean observed value; and $j = 1, \ldots, m$.

The use of NSE is not restricted solely in regression models, but extends for any application of hydrological modeling and, therefore, in its general use, it takes values between $-\infty$ and 1, i.e., $NSE \in (-\infty, +1]$. An efficiency lower than zero indicates that the mean value of the observed time series would have been a better predictor than the model [89]. In such cases, the model should be rejected.

At this point, the relation between the coefficient of determination, R^2, and the Nash-Sutcliffe Efficiency, NSE, should be clarified. In case of a multiple regression, R^2 and NSE are simply identical. However, the coefficient of determination has a value bounded between 0 and 1 [90], that is $R^2 \in [0, 1]$ [82].

An interesting perspective is the comparison of the correlation coefficient, r, with the Nash-Sutcliffe Efficiency. In case of a conventional linear regression model with only one independent variable, the squared value of the correlation coefficient, r^2, is equal to $R^2 = NSE$. The correlation coefficient, r, indicates the strength and the direction of a linear relationship with respect to the data, whilst it cannot imply causation.

In the present model and with regard to the training set, namely the laboratory data, in the case of the crisp multiple regression, the NSE is identical to the R^2, and obviously non-negative, i.e., $NSE = R^2 = 0.857$. However, when the crisp multiple regression equation is used upon the other validation measurements, from natural streams, then it takes negative values, i.e., $NSE = -1.207$.

Finally, if only one independent variable is used (in this case, X_3), again for the training data, then the NSE is identical to the R^2 and with the squared value of the correlation coefficient, r, $NSE = R^2 = r^2 = 0.801$.

References

1. Gilbert, G.K.; Murphy, E.C. *The Transportation of Debris by Running Water*; US Geological Survey Professional Paper 86; US Government Printing Office: Washington, DC, USA, 1914.
2. Rubey, W.W. Settling Velocities of Gravel, Sand and Silt Particles. *Am. J. Sci.* **1933**, *148*, 325–338. [CrossRef]
3. Anderson, A.G.; Johnson, J.W. A Distinction Between Bed Load and Suspended Load in Natural Streams. *Eos Trans. Am. Geophys. Union* **1940**, *21*, 628–633.
4. McCully, P. *Silenced Rivers: The Ecology and Politics of Large Dams*; Zed Books: London, UK, 1996.
5. Kaffas, K.; Hrissanthou, V.; Sevastas, S. Modeling Hydromorphological Processes in a Mountainous Basin Using a Composite Mathematical Model and ArcSWAT. *Catena* **2018**, *162*, 108–129. [CrossRef]

6.   Kaffas, K. Development of Mathematical Model for Calculating Continuous Hydrographs and Sediment Graphs in a Basin Due to Rainfall. Ph.D. Thesis, Democritus University of Thrace, Xanthi, Greece, 28 June 2017.

7.   Graf, W.; Leitner, P.; Hanetseder, I.; Ittner, L.D.; Dossi, F.; Hauer, C. Ecological degradation of a meandering river by local channelization effects: A case study in an Austrian lowland river. *Hydrobiologia* **2016**, *772*, 145–160. [CrossRef]

8.   Spalevic, V.; Lakicevic, M.; Radanovic, D.; Billi, P.; Barovic, G.; Vujacic, D.; Sestras, P.; Darvishan, A.K. Ecological-Economic (Eco-Eco) Modelling in the River Basins of Mountainous Regions: Impact of Land Cover Changes on Sediment Yield in the Velicka Rijeka, Montenegro. *Not. Bot. Horti Agrobot. Cluj-Napoca* **2017**, *45*, 602–610. [CrossRef]

9.   Samaras, A.G.; Koutitas, C.G. Modeling the impact of climate change on sediment transport and morphology in coupled watershed–coast systems: A case study using an integrated approach. *Int. J. Sed. Res.* **2014**, *29*, 304–315. [CrossRef]

10.   Kondolf, G.M. PROFILE: Hungry water: Effects of dams and gravel mining on river channels. *Environ. Manag.* **1997**, *21*, 533–551. [CrossRef]

11.   Pisaturo, G.R.; Righetti, M. Sediment flushing from reservoir and ecological impacts. *EPiC Ser. Eng.* **2018**, *3*, 1692–1697.

12.   Yagci, O.; Celik, M.F.; Kitsikoudis, V.; Kirca, V.S.O.; Hodoglu, C.; Valyrakis, M.; Duran, Z.; Kaya, S. Scour patterns around isolated vegetation elements. *Adv. Water Res.* **2016**, *97*, 251–265. [CrossRef]

13.   Summerfield, M.A.; Hulton, N.J. Natural controls of fluvial denudation rates in major world drainage basins. *J. Geophys. Res.* **1994**, *99*, 13871–13883. [CrossRef]

14.   Galy, A.; France-Lanord, C. Higher erosion rates in the Himalaya: Geochemical constraints on riverine fluxes. *Geology* **2001**, *29*, 23–26. [CrossRef]

15.   Lavé, J.; Avouac, J.P. Fluvial incision and tectonic uplift across the Himalayas of central Nepal. *J. Geophys. Res.* **2001**, *106*, 26561–26591. [CrossRef]

16.   Turowski, J.M.; Rickenmann, D.; Dadson, S.J. The partitioning of the total sediment load of a river into suspended load and bedload: A review of empirical data. *Sedimentology* **2010**, *57*, 1126–1146. [CrossRef]

17.   Meyer–Peter, E.; Müller, R. Formulas for bed-load transport. In Proceedings of the International Association for Hydraulic Research, 2nd Meeting, Stockholm, Sweden, 7 June 1948.

18.   Camenen, B.; Larson, M.A. General formula for non-cohesive bed load sediment transport. *Estuar. Coast. Shelf Sci.* **2005**, *63*, 249–260. [CrossRef]

19.   Lane, E.W.; Kalinske, A.A. Engineering calculations of suspended sediment. *Eos Trans. Am. Geophys. Union* **1941**, *22*, 603–607. [CrossRef]

20.   van Rijn, L.C. Sediment transport, part II: Suspended load transport. *J. Hydraul. Eng.* **1984**, *110*, 1613–1641. [CrossRef]

21.   Ackers, P.; White, W.R. Sediment transport: New approach and analysis. *ASCE J. Hydraul. Div.* **1973**, *99*, 204–254.

22.   Yang, C.T. Incipient motion and sediment transport. *ASCE J. Hydraul. Div.* **1973**, *99*, 1679–1704.

23.   Schriften, D.V.W.K. *Feststofftransport in Fließgewässern-Berechnungsverfahren für die Ingenieurpraxis*; Heft 87; Verlag Paul Parey: Hamburg/Berlin, Germany, 1988.

24.   Einstein, H.A. *The Bed-Load Function for Sediment Transportation in Open Channel Flows*; No. 1488-2016-124615; US Department of Agriculture Technical Bulletin: Washington, DC, USA, 1950.

25.   Toffaleti, F.B. Definitive computation of sand discharge in rivers. *ASCE J. Hydraul. Div.* **1969**, *95*, 225–248.

26.   Karim, M.F.; Holly, F.M.; Kennedy, J.F. *Bed Armoring Procedures in IALLUVIAL and Application to the Missouri River*; No. 269; Iowa Institute of Hydraulic Research, University of Iowa: Iowa City, IA, USA, 1983.

27.   Yang, C.T. Unit stream power and sediment transport. *ASCE J. Hydraul. Div.* **1972**, *98*, 1805–1826.

28.   Engelund, F.; Hansen, E. *A Monograph on Sediment Transport in Alluvial Streams*; Teknisk Forlag: Copenhagen, Denmark, 1967.

29.   Zanke, U. *Grundlagen der Sedimentbewegung*; Springer: Berlin/Heidelberg, Germany, 1982.

30.   van Rijn, L.C. Closure to "Sediment transport, part I: Bed load transport". *ASCE J. Hydraul. Eng.* **1984**, *110*, 1431–1456. [CrossRef]

31.   Yang, C.T. *Sediment Transport: Theory and Practice, reprint ed.*; Krieger: Malabar, FL, USA, 2003.

32.   Yang, C.T. Unit stream power equation for gravel. *ASCE J. Hydraul. Eng.* **1984**, *110*, 1783–1797. [CrossRef]

33. Aqil, M.; Kita, I.; Yano, A.; Nishiyama, S. Analysis and prediction of flow from local source in a river basin using a neuro-fuzzy modeling tool. *J. Env. Manag.* **2007**, *85*, 215–223. [CrossRef] [PubMed]

34. Zhu, S.; Heddam, S.; Nyarko, E.K.; Hadzima-Nyarko, M.; Piccolroaz, S.; Wu, S. Modeling daily water temperature for rivers: comparison between adaptive neuro-fuzzy inference systems and artificial neural networks models. *Environ. Sci. Pollut. Res.* **2019**, *26*, 402–420. [CrossRef] [PubMed]

35. Tayfur, G.; Ozdemir, S.; Singh, V.P. Fuzzy logic algorithm for runoff-induced sediment transport from bare soil surfaces. *Adv. Water Res.* **2003**, *26*, 1249–1256. [CrossRef]

36. Chachi, J.; Taheri, S.M.; Pazhand, H.R. Suspended load estimation using L_1—Fuzzy regression, L_2—Fuzzy regression and MARS—fuzzy regression models. *Hydrol. Sci. J.* **2016**, *61*, 1489–1502. [CrossRef]

37. Spiliotis, M.; Kitsikoudis, V.; Kirca, V.S.O.; Hrissanthou, V. Fuzzy threshold for the initiation of sediment motion. *App. Soft Comp.* **2018**, *72*, 312–320. [CrossRef]

38. Zanke, U.C.E. On the influence of turbulence on the initiation of sediment motion. *Int. J. Sed. Res.* **2003**, *18*, 17–31.

39. Spiliotis, M.; Kitsikoudis, V.; Hrissanthou, V. Assessment of bedload transport in gravel-bed rivers with a new fuzzy adaptive regression. *Eur. Water* **2017**, *57*, 237–244.

40. Kitsikoudis, V.; Spiliotis, M.; Hrissanthou, V. Fuzzy regression analysis for sediment incipient motion under turbulent flow conditions. *Environ. Proc.* **2016**, *3*, 663–679. [CrossRef]

41. Özger, M.; Kabataş, M.B. Sediment load prediction by combined fuzzy logic—Wavelet method. *J. Hydroinf.* **2015**, *17*, 930–942. [CrossRef]

42. Kişi, Ö. Evolutionary fuzzy models for river suspended sediment concentration estimation. *J. Hydrol.* **2009**, *372*, 68–79. [CrossRef]

43. Kişi, Ö.; Karahan, M.E.; Şen, Z. River suspended sediment modelling using a fuzzy logic approach. *Hydrol. Proc. Int. J.* **2006**, *20*, 4351–4362. [CrossRef]

44. Lohani, A.K.; Goel, N.K.; Bhatia, K.S. Deriving stage—Discharge—Sediment concentration relationships using fuzzy logic. *Hydrol. Sci. J.* **2007**, *52*, 793–807. [CrossRef]

45. Zadeh, L.A. Fuzzy sets. *Inf. Control* **1965**, *8*, 338–353. [CrossRef]

46. Firat, M.; Güngör, M. Monthly total sediment forecasting using adaptive neuro-fuzzy inference system. *Stoch. Environ. Res. Risk Assess.* **2010**, *24*, 259–270. [CrossRef]

47. Tanaka, H. Fuzzy data analysis by possibilistic linear models. *Fuzzy Sets Syst.* **1987**, *24*, 363–375. [CrossRef]

48. Kaffas, K.; Hrissanthou, V. Estimate of continuous sediment graphs in a basin, using a composite mathematical model. *Environ. Proc.* **2015**, *2*, 361–378. [CrossRef]

49. Kaffas, K.; Hrissanthou, V. Computation of hourly sediment discharges and annual sediment yields by means of two soil erosion models in a mountainous basin. *Int. J. River Basin Manag.* **2019**, *17*, 63–77. [CrossRef]

50. Nakato, T. Tests of selected sediment—Transport formulas. *J. Hydraul. Eng.* **1990**, *116*, 362–379. [CrossRef]

51. Baosheng, W.U.; van Maren, D.S.; Lingyun, L.I. Predictability of sediment transport in the Yellow River using selected transport formulas. *Int. J. Sed. Res.* **2008**, *23*, 283–298.

52. Moore, I.D.; Burch, G.J. Sediment transport capacity of sheet and rill flow: Application of unit stream power theory. *Water Res. Res.* **1986**, *22*, 1350–1360. [CrossRef]

53. Hui-Ming, S.; Yang, C.T. Estimating overland flow erosion capacity using unit stream power. *Int. J. Sed. Res.* **2009**, *24*, 46–62.

54. Hrissanthou, V.; Hartmann, S. Measurements of critical shear stress in sewers. *Water Res.* **1998**, *32*, 2035–2040. [CrossRef]

55. Yang, C.T. Unit stream power equations for total load. *J. Hydrol.* **1979**, *40*, 123–138. [CrossRef]

56. Salas, J.D.; Shin, H.S. Uncertainty analysis of reservoir sedimentation. *J. Hydraul. Eng.* **1999**, *125*, 339–350. [CrossRef]

57. Kleinhans, M.G. Flow discharge and sediment transport models for estimating a minimum timescale of hydrological activity and channel and delta formation on Mars. *J. Geophys. Res. Planets* **2005**, *110*. [CrossRef]

58. Fischer, S.; Pietroń, J.; Bring, A.; Thorslund, J.; Jarsjö, J. Present to future sediment transport of the Brahmaputra River: Reducing uncertainty in predictions and management. *Reg. Environ. Chang.* **2017**, *17*, 515–526. [CrossRef]

59. Beckers, F.; Noack, M.; Wieprecht, S. Uncertainty analysis of a 2D sediment transport model: An example of the Lower River Salzach. *J. Soils Sediments* **2017**, *18*, 3133–3144. [CrossRef]

60. Azamathulla, H.M.; Ghani, A.A.; Fei, S.Y. ANFIS-based approach for predicting sediment transport in clean sewer. *Appl. Soft Comp.* **2012**, *12*, 1227–1230. [CrossRef]
61. Nomicos, G.N. Effects of Sediment Load on the Velocity Field and Friction Factor of Turbulent Flow in an Open Channel. Ph.D. Thesis, California Institute of Technology, Pasadena, CA, USA, 1956.
62. Vanoni, V.A.; Brooks, N.H. *Laboratory Studies of the Roughness and Suspended Load of Alluvial Streams*; Report No. E-68; Sedimentation Laboratory, California Institute of Technology: Pasadena, CA, USA, 1957.
63. Kennedy, J.F. *Stationary Waves and Antidunes in Alluvial Channels*; Report No. KH-R-2; W.M. Keck Laboratory of Hydraulics and Water Resources, California Institute of Technology: Pasadena, CA, USA, 1961.
64. Stein, R.A. Laboratory studies of total load and apparent bed-load. *J. Geophys. Res.* **1965**, *70*, 1831–1842. [CrossRef]
65. Guy, H.P.; Simons, D.B.; Richardson, E.V. *Summary of Alluvial Channel Data from Flume Experiment, 1956–1961*; US Geological Survey Professional Paper: Washington, DC, USA, 1966.
66. Williams, G.P. *Flume Experiments on the Transport of a Coarse Sand*; US Geological Survey Professional Paper: Reston, VA, USA, 1967.
67. Schneider, V.R. *Personal Communication of Yang. Data Collected from an 8-Foot Wide Flume at Colorado State University*; Colorado State University: Fort Collins, CO, USA, 1971.
68. Simons, D.B.; Richardson, E.V.; Albertson, M.L. *Flume studies using medium sand (0.45 mm)*; US Geological Survey Water-Supply Paper: Washington, DC, USA, 1961.
69. Inter-Agency Committee on Water Resources, Subcommittee on Sedimentation. *Some Fundamentals of Particle Size Analysis*; St. Anthony Falls Hydraulic Laboratory: Minneapolis, MN, USA, 1957.
70. Tsakiris, G.; Spiliotis, M. Uncertainty in the analysis of urban water supply and distribution systems. *J. Hydroinf.* **2017**, *19*, 823–837. [CrossRef]
71. Papadopoulos, C.; Spiliotis, M.; Angelidis, P.; Papadopoulos, B. A hybrid fuzzy frequency factor based methodology for analyzing the hydrological drought. *J. Desal. Water Treat.* **2019**, *167*, 385–397. [CrossRef]
72. Klir, G.; Yuan, B. *Fuzzy Sets and Fuzzy Logic: Theory and Applications*; Prentice Hall: Upper Saddle River, NJ, USA, 1995; p. 574.
73. Buckley, J.; Eslami, E. *An Introduction to Fuzzy Logic and Fuzzy Sets*; Springer: Berlin, Germany, 2002; p. 285.
74. Hanss, M. *Applied Fuzzy Arithmetic, an Introduction with Engineering Applications*; Springer: Berlin, Germany, 2005; p. 256.
75. Spiliotis, M.; Angelidis, P.; Papadopoulos, B. A hybrid probabilistic bi-sector fuzzy regression based methodology for normal distributed hydrological variable. *Evol. Syst.* **2019**, 1–14. [CrossRef]
76. Tsakiris, G.; Tigkas, D.; Spiliotis, M. Assessment of interconnection between two adjacent watersheds using deterministic and fuzzy approaches. *Eur. Water* **2006**, *15*, 15–22.
77. Spiliotis, M.; Hrissanthou, V. Fuzzy and crisp regression analysis between sediment transport rates and stream discharge in the case of two basins in northeastern Greece. In *Conventional and Fuzzy Regression: Theory and Engineering Applications*, 1st ed.; Hrissanthou, V., Spiliotis, M., Eds.; Nova Science Publishers: New York, NY, USA, 2018; pp. 1–49.
78. Zanke, U. *Berechung der Sinkgeschwindigkeiten von Sedimenten*; Mitt. des Franzius-Instituts für Wasserbau, Technical University of Hannover: Hannover, Germany, 1977.
79. Kaffas, K.; Saridakis, M.; Tsangaratos, P.; Spiliotis, M.; Hrissanthou, V. Application of Yang formula for calculating total sediment transport rate with fuzzy regression. In Proceedings of the 14th Conference of the Hellenic Hydrotechnical Association, Volos, Greece, 16–17 March 2019. (In Greek).
80. Yang, C.T.; Stall, J.B. *Unit Stream Power for Sediment Transport in Natural Rivers*; WRC Research Report 88; Water Resources Centre, University of Illinois: Urbana-Champaign Water Resources Center, IL, USA, 1974.
81. Einstein, H.A. *Bed-Load Transportation in Mountain Creek*; United States Department of Agriculture, Soil Conservation Service: Washington, DC, USA, 1944.
82. Colby, B.R.; Hembree, C.H. *Computation of Total Sediment Discharge, Niobrara River near Cody, Nebraska*; Water Supply Paper 1357; US Geological Survey: Washington, DC, USA, 1955.
83. Hubbell, D.W.; Matejka, D.Q. *Investigations of Sediment Transportation, Middle Loup River at Dunning, Nebraska*; Water Supply Paper 1376; US Geological Survey: Washington, DC, USA, 1959.
84. Nordin, C.F. *Aspects of Flow Resistance and Sediment Transport, Rio Grande River near Bernalillo, New Mexico*; Water Supply Paper 1398-H; US Geological Survey: Washington, DC, USA, 1964.

85. Jordan, P.R. *Fluvial Sediment of the Mississippi River at St. Louis, Missouri*; Water Supply Paper 1802; US Geological Survey: Washington, DC, USA, 1965.

86. Williams, G.P.; Rosgen, D.L. *Measured Total Sediment Loads (Suspended Loads and Bedloads) for 93 United States Streams*; Open File Report; US Geological Survey: Reston, VA, USA, 1989.

87. Ishibuchi, H.; Tanaka, H.; Okada, H. An architecture of neural networks with interval weights and its application to fuzzy regression analysis. *Fuzzy Sets Syst.* **1993**, *57*, 27–39. [CrossRef]

88. Nash, J.E.; Sutcliffe, J.V. River flow forecasting through conceptual models: Part 1. A discussion of principles. *J. Hydrol.* **1970**, *10*, 282–290. [CrossRef]

89. Krause, P.; Boyle, D.P.; Bäse, F. Comparison of different efficiency criteria for hydrological model assessment. *Adv. Geosci.* **2005**, *5*, 89–97. [CrossRef]

90. Mays, L.; Tung, Y.K. Water distribution systems. In *Hydrosystems Engineering and Management*; McGraw-Hill: New York, NY, USA, 1992; pp. 354–386.

Article

Evolution Pattern of Tailings Flow from Dam Failure and the Buffering Effect of Debris Blocking Dams

Guangjin Wang [1,2,3], **Sen Tian** [3,*], **Bin Hu** [4,*], **Zhifa Xu** [2,5], **Jie Chen** [3] and **Xiangyun Kong** [2]

1   State Key Laboratory of Geomechanics and Geotechnical Engineering, Institute of Rock and Soil Mechanics, Chinese Academy of Sciences, Wuhan 430071, China; wangguangjin2005@kust.edu.cn
2   Yunnan Key Laboratory of Sino-German Blue Mining and Utilization of Special Underground Space, Faculty of Land Resources Engineeirng, Kunming University of Science and Technology, Kunming 650093, China; zhifaxu@kust.edu.cn (Z.X.); kxyun0527@kust.edu.cn (X.K.)
3   State Key Laboratory of Coal Mine Disaster Dynamics and Control, School of Resources and Safety Engineering, Chongqing University, Chongqing 400044, China; jiechen023@cqu.edu.cn
4   School of Resources and Environmental Engineering, Wuhan University of Science and Technology, Wuhan 430081, China
5   Guangxi Branch of CSCEC Strait Construction and Development Co., Ltd., Nanning 530000, China
*   Correspondence: sentian@cqu.edu.cn (S.T.); hbin74@wust.edu.cn (B.H.)

Received: 31 August 2019; Accepted: 10 November 2019; Published: 14 November 2019

Abstract: Tailings ponds are the indispensable facilities in the mine production and operation. Once the dam is destabilized and damaged, it will pose a serious threat on the life and property of the downstream population and could also potentially cause an environmental disaster. With an engineering background, this paper dynamically and numerically simulates the evolution process of tailings flow from dam failure and the influence scope of any resulting disaster in context. The evolution characteristics of leaked tailings flow are analyzed at various downstream riverbed slopes and debris blocking dam settings. In addition, parameters such as flow rate, impact force and deposition range of leaked tailings flow at downstream arrival are studied, as well as their correlations. The results indicate that the flat terrains upstream and downstream of passage zone show a relatively larger area of inundation by tailings flow. Both the maximum and final downstream inundated ranges increase with the elevating slope of downstream riverbed, and the leaked tailings are deposited mainly in the nearby villages in front of the dam and the flat terrains of the downstream passage zone. Additionally, rational establishment of debris blocking dams on the downstream side is effective in diminishing the damage of tailings flow to the downstream section. This study can also provide an important basis for the quantitative evaluation of post-disaster influence scope for tailings pond as well as for the design of dam body.

Keywords: tailings pond; leaked tailings flow; dam failure; impact force; deposition range; debris blocking dam

---

1. Introduction

Tailings ponds are the artificial hazard sources with high potential energy. In the event of an accident, the resulting tailings flow might bring huge losses to the life and property of downstream populations while also polluting the ecological environment severely [1–3]. Tailings are the kinds of solid wastes generated from mineral separation. Tailings are also a special kind of soil, generally structured with single granule and beehive. Due to the poor comprehensive utilization of tailings, the majority of them are stored in the tailings ponds aside from the cases of beneficiation, mine filling and building material recovery [4–6]. With the continuous exploration by mine companies as well as the growing demand for mineral products, the dam height and storage volume of tailings ponds have been gradually increasing [7]. Meanwhile, the risk of dam failure in the tailings ponds also heightens greatly in the face of extreme conditions such as earthquakes, rainstorms and failure of flood drainage systems. Cases of severe accidents caused by tailings ponds project failure are not uncommon [8–11]. For example, two dam burst accidents occurred in Minas Gerais, Brazil, in January 2019 and November 2015, leading to at least 232 deaths and 19 deaths, respectively [7,12]. In September 2010, a tin tailings pond in China's Guangdonge collapsed, causing at least 18 deaths and direct economic losses of approximately 460 million yuan [13].

Evolution of tailings flow resulting from tailings pond leakage is a complex mechanical process, which involves multiple interdisciplinary fields including fluid mechanics, geological hazards and sediment transport mechanics [14–16]. So far, numerous scholars have estimated the flow characteristics of tailings pond dam failure such as total tailings leakage, dam breach width and maximum tailings flux through theoretical derivation, statistical analysis and other approaches and also formed relevant theoretical formulas [17–21]. With the rise of computer simulation software, plentiful fluid dynamics software has been applied to the evolution pattern research of tailings flow resulting from tailings pond leakage [22–25]. The fluid rheological equation embedded in FLO-2D, a US Federal Emergency Management Agency (FEMA)-approved flood and mudflow hazard simulation software [26,27], can be used for the simulation of the rheological state during evolution of leaked tailings flow.

This study explores the evolution of leaked tailings flow following dam failure and its effects on the downstream structures and personnel during evolution. Using the FLO-2D simulation software (FLO-2D Software Inc, Nutrioso, AZ, USA), the evolution process of tailings flow is studied under different downstream riverbed slopes and debris blocking dam construction parameters. At the same time, the evolution pattern, sedimentary characteristics and energy loss of tailings flow during evolution are analyzed. In addition, the effects of downstream riverbed slope and debris blocking dam construction parameters on the tailings flow characteristics are also investigated.

2. Model and Computational Method

2.1. Simulation Model and Boundary Conditions

Digital Elevation Model (DEM) is created by collecting the digital elevation data within the scope of studied tailings pond combined with in situ investigation. The studied tailings pond dam has a crest elevation of 2005 m (the height above sea level), a fill dam height of 30 m, an initial dam height of 32 m, a total dam height of 62 m, a whole storage capacity of 3.996 million m^3, and an effective storage capacity of 3.197 million m^3, and it is located in Huili Country of Liangshan, Sichuan Province of China. The tailings pond adopts upstream damming, in which the height of each sub-dam is 5 m, the ratio of sub-dam to outer slope is 1:3.5, and the total slope ratio is 1:4.233. The objective of the sub-dams located upstream of the main dam is the deposition of tailings upstream of the sub-dams and, consequently, the increase of the storage capacity of the tailings ponds and the decrease of the streambed slope. Besides the storage capacity, maximum pond water level at overtopping, basic mechanical parameters of dam tailings and flow hydrograph range (Figure 1) of the tailings pond are determined based on the data collected in situ and the information provided by mining companies.

Dam failure computation models are built through in situ investigation at three downstream riverbed slopes, i.e., original terrain slope (average slope ratio of 4.2% along the valley bottom), elevation by 5% (average slope ratio of 9.2% along the valley bottom) and elevation by 10% (average slope ratio of 14.2% along the valley bottom), which can be found in Figure 2.

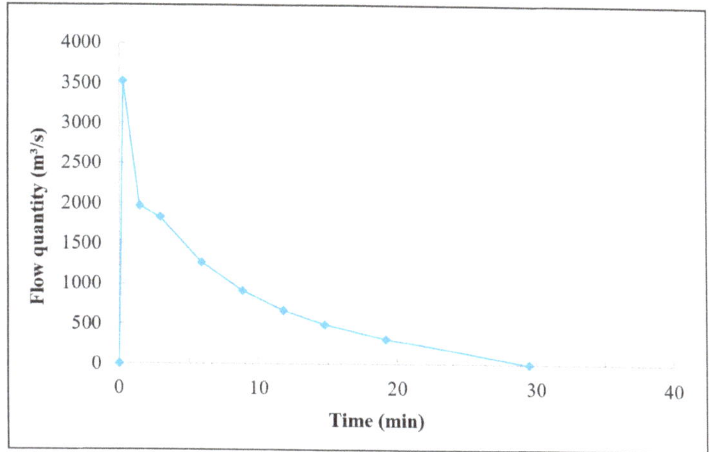

Figure 1. The flow hydrograph for FLO-2D model.

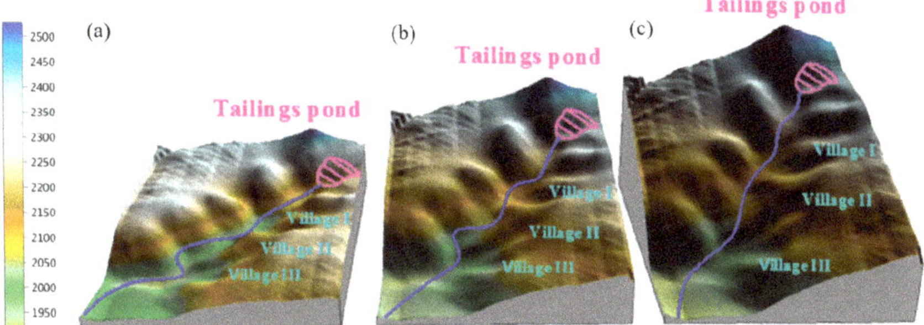

Figure 2. The 3D computation models for dam failure. (**a**) Original terrain slope; (**b**) elevation by 5%; (**c**) elevation by 10%.

2.2. Computational Parameters

After vertical integration, the controlling equations can be reduced to a 2D format [28], which is described by the Navier-Stokes equations. The simplified controlling equations are shown below.

The continuity equation:

$$\frac{\partial h}{\partial t} + \frac{\partial(uh)}{\partial x} + \frac{\partial(vh)}{\partial y} = I \tag{1}$$

The equations of motion:

$$S_{fx} = S_{ox} - \frac{\partial h}{\partial x} - \frac{\partial u}{g\partial t} - u\frac{\partial u}{\partial x} - v\frac{\partial y}{g\partial y} \tag{2}$$

$$S_{fy} = S_{oy} - \frac{\partial h}{\partial y} - \frac{\partial v}{g\partial x} - u\frac{\partial v}{\partial x} - v\frac{\partial v}{g\partial y} \tag{3}$$

where h denotes the draining tailing flow depth (m), I is the drop in the water surface per unit distance within the simulated range and is called the hydraulic gradient (%), u and v refer to the flow rate in the horizontal and vertical directions (m/s), respectively; S_{fx} and S_{fy} are the differences in the unit distance of the frictional resistance in the x and y directions, respectively, and are called the frictional slope (%), S_{ox} and S_{oy} are the differences in elevation within a unit distance in the x and y directions, respectively, and are called the riverbed slope (%).

When the flood or tailings flow is simulated by the FLO-2D, it can be performed in the dynamic wave mode or the diffused wave mode. According to the similarity criterion, Equation (1) is a mass conservation equation, while Equation (2) and Equation (3) are the momentum conservation equations.

Additionally, the evolution process of leaked tailings flow can be described by the rheological equation for high sediment concentration, which is proposed by O'Brien [29], as shown below.

$$S_f = S_y + S_v + S_{td} = \frac{\tau_y}{\gamma_m h} + \frac{K\eta u}{8\gamma_m h^2} + \frac{n^2 u^2}{h^{4/3}} \tag{4}$$

where S_f, S_y, S_v and S_{td} represent frictional slope, yield slope, viscous slope and turbulence-distribution slope, respectively. τ_y denotes the yield stress of the fluid during the flow, γ_m denotes the fluid specific gravity, η is the viscous coefficient of fluid, K is the laminar flow resistance coefficient, and n refers to the Manning coefficient representing the roughness of ground surface.

Yield stress in the current study refers to the Bingham yield stress, which is reflected primarily as the internal stress pattern of viscous tailings flow and present in the form of viscous force [30–32]. The viscous force is a resistance produced by the interaction between shear and tensile stresses of fluid [33,34]. As a result, the Bingham viscosity coefficient and viscous force are closely correlated in the present work, and the increase in fluid volume concentration leads to exponential increases in the Bingham yield stress and Bingham viscosity coefficient [35,36].

The relationship between the yield stress and the volume concentration is shown in Equation (5).

$$\tau_y = \alpha_1 e^{\beta_1 C_V} \tag{5}$$

and the relational expression between the Bingham viscous coefficient and the volume concentration is presented in Equation (6).

$$\eta = \alpha_2 e^{\beta_2 C_V} \tag{6}$$

where volume concentration (C_V) indicates the percentage of soil and debris like aggregate and gravel in the leaked tailings flow over the entire tailings flow volume. α_1, β_1, α_2 and β_2 are the empirical coefficients of the yield stress and viscous force, which can be derived experimentally. On this basis, Table 1 provides the relevant parameters used in the present numerical simulation.

Table 1. The FLO-2D simulation parameters.

Simulation Parameters		Values
Fluid relative density γ_m (g/cm^3)		1.8
Retardation coefficient of laminar flow K		2285
Manning coefficient in passage zone		0.05
Parameters of yield stress coefficients	α_1	0.128
	β_1	12
Parameters of viscous force coefficients	α_2	0.0473
	β_2	21

3. Results and Discussion

3.1. Various Downstream Riverbed Slope Conditions

3.1.1. Flow Depth of Leaked Tailings

This study chooses the passage zone downstream of the tailings pond as the simulation target. As presented in Figure 3, the leaked tailings all flow downstream along the valley bottom (lowest point) within the passage zone after collapse of tailings pond in three different slope conditions.

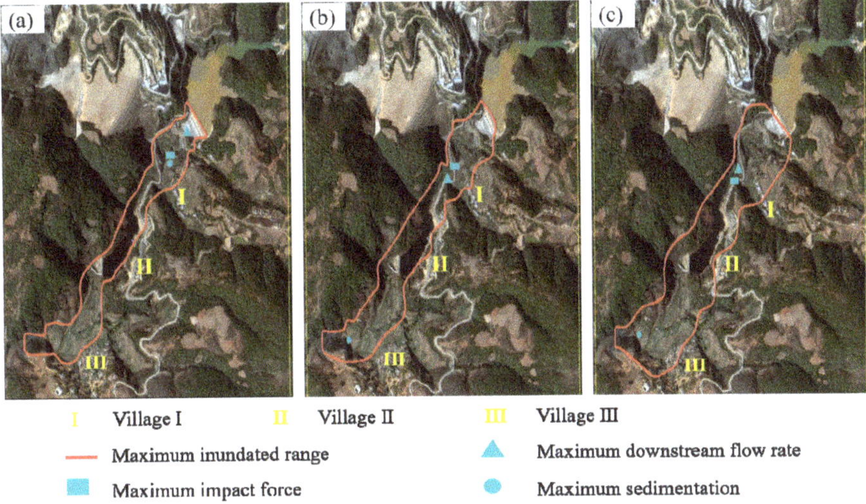

Figure 3. The maximum influence scope of leaked tailings flow under different slope conditions. (**a**) Original terrain slope; (**b**) elevation by 5%; (**c**) elevation by 10%.

Due to the presence of fluted or relatively flat terrains upstream and downstream of the passage zone, the inundated areas by leaked tailings flow are larger on the upstream and downstream sections than the midstream section, where the terrain changes more drastically. The inundated ranges increase with the increasing slope of downstream riverbed, which maximize at 2.8×10^5 m^2, 3.7×10^5 m^2 and 4.9×10^5 m^2, respectively. In the earlier stage of dam failure, the volume of water in the leaked tailings flow is considerably higher than the volume of tailings. As the dam breach opening develops further, the content of tailings increases. In the meanwhile, the impact height increases gradually with the descending terrain, and the influence scope on downstream section accordingly increases. The elevated content of tailings in the leaked tailings flow leads to a corresponding increase in the fluid viscosity. In the later stage of dam failure, the water content in leaked tailings flow declines gradually, and the drag force of tailings flow decreases. By contrast, the frictional resistance increases to ultimately attain equilibrium of forces and maximization of flow rate. With further increase in the frictional resistance, the flow rate of leaked tailings flow declines to zero to achieve the maximum inundated area. This result is consistent with the energy storage-dissipation-deposition process during physical motion [37].

Figure 4 presents a diagram of maximum impact height of leaked tailings flow along the valley bottom downstream of dam in the earlier stage. In the original terrain scenario, the tailings in leaked tailings flow are deposited substantially within the valleys or flat terrains on the upstream 150–650 m section and the downstream 1200–1800 m section after the failure of tailings pond, while deposited slightly within the areas with sharp slope changes on the midstream section. After terrain elevation by 5% and 10% separately on the basis of original topography, significant reduction of tailings storage capacity is noted for both the upstream and downstream sections with the rising terrain, as well as substantial decline in the amount and depth of deposition.

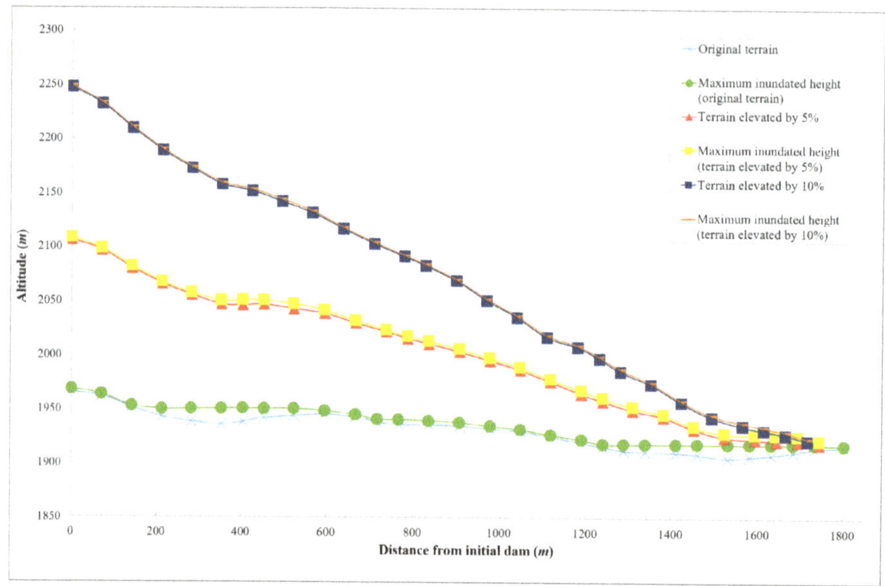

Figure 4. The maximum impact height of leaked tailings flow along the valley bottom.

Maximum depth of tailings deposition decreases gradually with the elevating slope of downstream riverbed. Maximum impact height is found near Village I for the original terrain (14.43 m). After terrain elevation by 5% (4.97 m), the maximum impact height is found at the valley downstream of Village III; and after terrain elevation by 10% (3.30 m), the maximum impact height is found also at the valley downstream of Village III. Deepest sedimentation point of tailings is developed downstream, and the deposition thickness also decreases accordingly.

3.1.2. Flow Rate of Leaked Tailings

As shown in Figure 5, the maximum velocities of tailings flow are all located downstream of initial dam, and the maximum evolution velocities of tailings flow increase with the elevating terrain, which are 10 m/s, 10.9 m/s and 13.5 m/s, respectively. In the original terrain scenario, the terrain and altitude difference decrease sharply within the areas downstream of initial dam and upstream of Village III after dam failure, thus enabling fast evolution of the leaked tailings flow. At the downstream section of Village III, however, the altitude difference changes little and the evolution distance is long. Therefore, a gradual decline of flow rate is noted with the increasing evolution distance. According to Figure 5, after elevating terrain by 5% and 10%, the altitude difference changes rather greatly for a scope from the initial dam to the 1000 m downstream section, and the flow rate of leaked tailings continues to heighten. The velocity tends to stabilize when the downstream altitude difference narrows down.

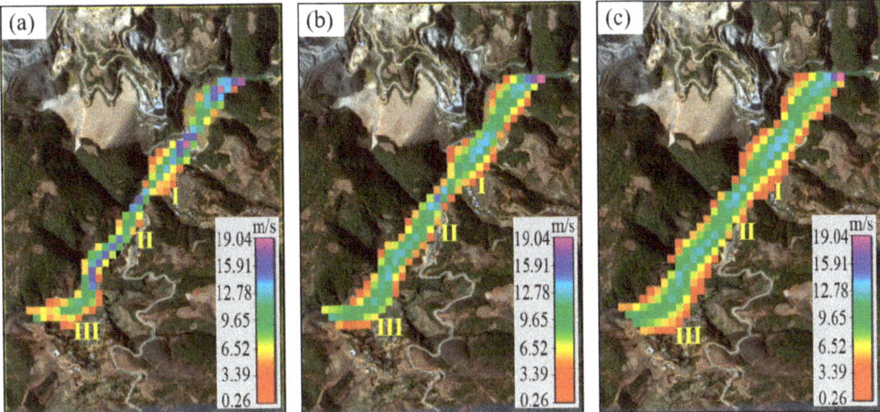

Figure 5. The maximum velocity distribution of leaked tailings flow under different slope conditions. (**a**) Original terrain slope; (**b**) elevation by 5%; (**c**) elevation by 10%.

3.1.3. Impact Force of Leaked Tailings Flow

This study investigated the effect of the impact during the evolution process of leaked tailings flow on its downstream constructs, and the exerting range of impact force is the region that the discharged sediments flow through. The impact force of leaked tailings flow on the downstream constructs is calculated using Equation (7) through Equation (9) [27,37–40]. Multiple factors, including the impact angle, flow rate and density of tailings flow, can affect the computational results by varying degrees [38–40].

$$P_i = k\rho_f v^2 \tag{7}$$

$$k = 1.261 e^{C_w} \tag{8}$$

$$F = P_i h \tag{9}$$

where P_i denotes the impact pressure imposed by leaked tailings flow on other objects within the flow region; ρ_f refers to the fluid density of leaked tailings flow; h denotes the maximum depth of leaked tailings flow; k is the impact coefficient, and its value range from 1.261 to 3.427 [26,27]; v is the flow rate of leaked tailings at arbitrary time; C_w represents the weight concentration; e is a constant; and F is the impact force per unit width.

According to Equation (9), increases in the flow rate and depth lead to enhanced impact force per unit area.

From Figures 3 and 6, it can be seen that the locations of maximum impact forces in three different conditions are all adjacent to the maximum evolution velocity location. In the original terrain condition, the maximum impact force is located downstream from the initial dam with the value of 706 kN/m. With the elevation of slope, the maximum impact force increases first and then decreases, which is 883 kN/m at an elevation by 5% and is 495 kN/m at an elevation by 10%, showing a value approximately 70% of the original terrain scenario.

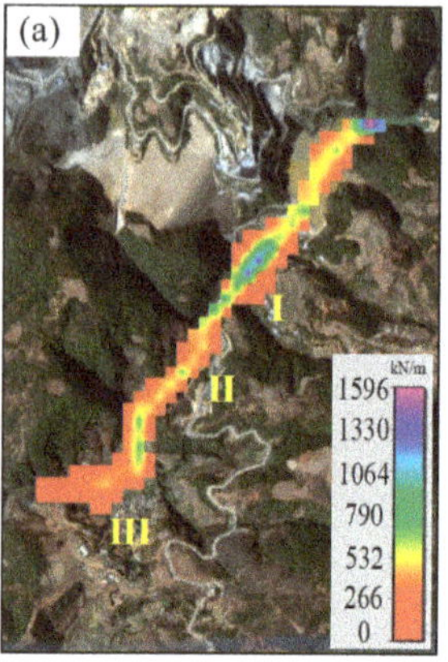 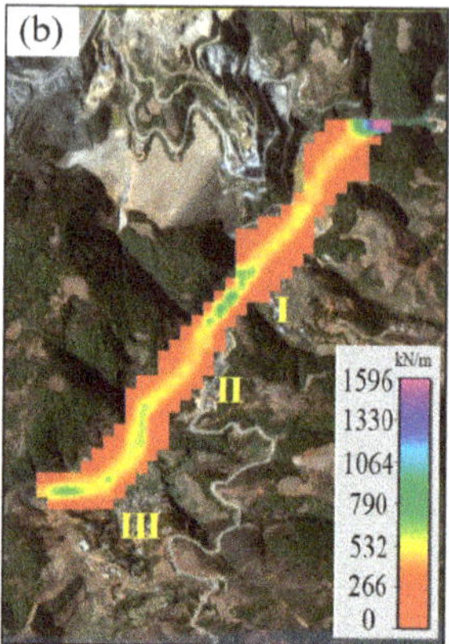 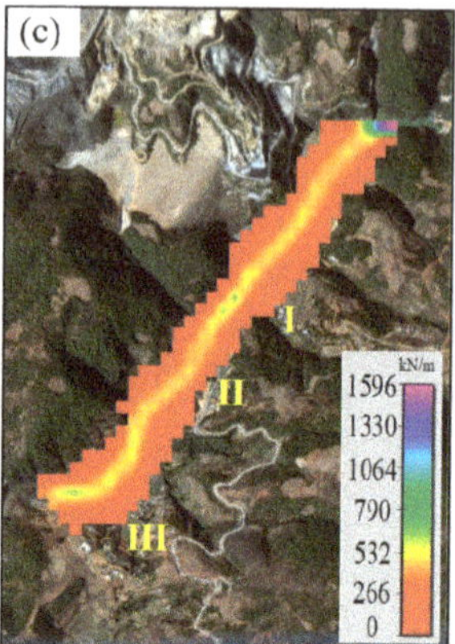

Figure 6. The maximum impact force of leaked tailings flow downstream under different slope conditions. (**a**) Original terrain slope; (**b**) elevation by 5%; (**c**) elevation by 10%.

3.2. Effects of Debris Blocking Dam on Tailings Flow

Debris blocking dams are set up at three characteristic locations downstream of initial dam in this study (Dam 1: valley mouth upstream of Village III, 1000 m away from the initial dam; Dam 2: valley mouth upstream of Village II, 600 m away from the initial dam; and Dam 3: valley mouth downstream of Village I, 400 m away from the initial dam). The height of these dams is 10 m.

According to the data collected in situ and the information provided by mining companies, there are villages downstream of these three locations, where the constructions of the debris blocking dams at these characteristic locations downstream are more conducive to reducing losses. Additionally, these three places are located at the top of the valley, the terrain on both sides is higher and the cost of dam constructions is relatively low.

3.2.1. Flow Depth of Leaked Tailings

As presented in Figure 7, the leaked tailings flow evolves downstream along the topographically lowest point in the passage zone in all three different conditions. The tailings inundated area is larger for the upstream and downstream sections of passage zone than for the midstream section. Under three dam distance conditions, the maximum inundated ranges are 2.7×10^5 m^2, 2.5×10^5 m^2 and 2.37×10^5 m^2, respectively. Consequently, the downstream inundated area shrinks with the narrowing distance between debris blocking and initial dams.

From Figure 8, it can be found that the deposition thicknesses of tailings crossing over the debris blocking dams decrease slightly compared with the case without such dams. The maximum sedimentation thicknesses are all found in front of the debris blocking dams, which increase with the shortening distance between initial and debris blocking dams and are all located in the valley near Village I. After flowing over the debris blocking dams, the tailings are deposited primarily in the valley near Village III, and the inundated height increases with the increasing of the distance between the debris blocking dam and the initial dam.

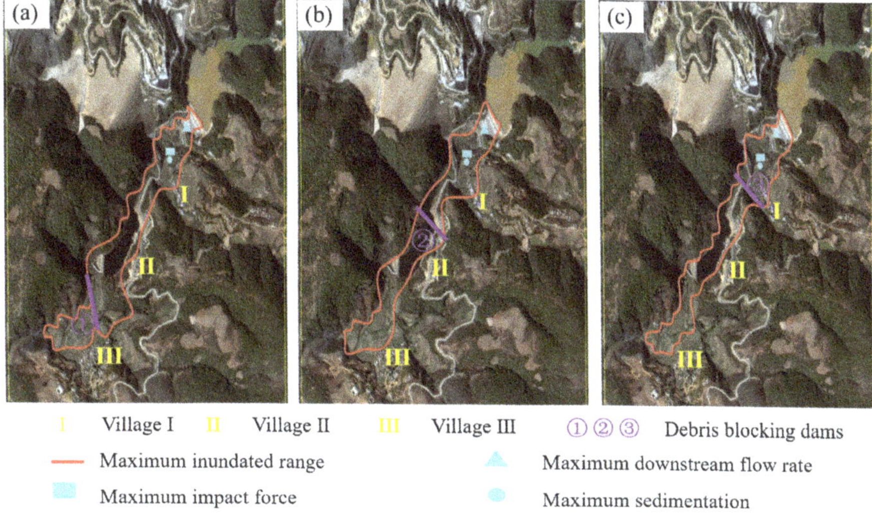

Figure 7. The maximum influence scope of leaked tailings flow with debris blocking dams at different distances. (**a**) 1000 m away from initial dam; (**b**) 600 m away from initial dam; (**c**) 400 m away from initial dam.

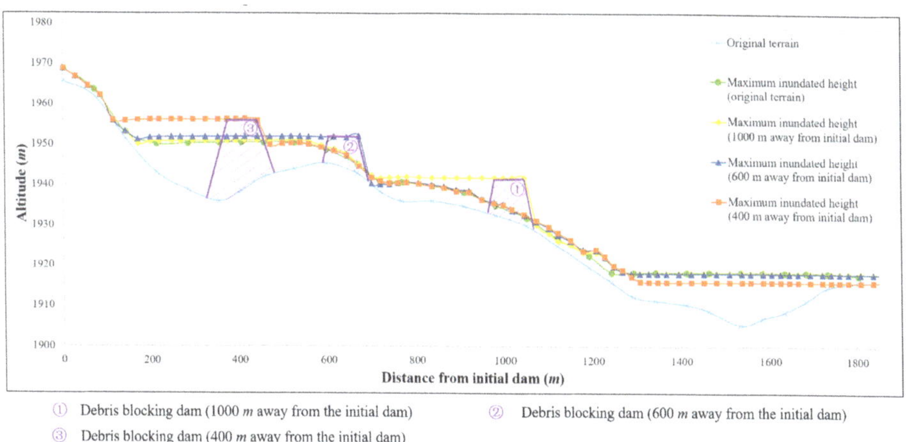

① Debris blocking dam (1000 *m* away from the initial dam)   ② Debris blocking dam (600 *m* away from the initial dam)
③ Debris blocking dam (400 *m* away from the initial dam)

Figure 8. Schematic diagram of the maximum impact height along the valley bottom.

3.2.2. Flow Rate of Leaked Tailings

Figure 9 depicts the flow rate distribution of leaked tailings after establishment of debris blocking dams at different distances downstream of initial dam. The flow rates are 10.3 m/s in all three cases before crossing the dams, and the maximum flow rates are all located at the valley near Village I. Meanwhile, the maximum flow rates downstream of the debris blocking dams are all found at the valley mouth near the upstream side of Village III. After setting up the debris blocking dams, the maximum downstream flow rates also decrease accordingly.

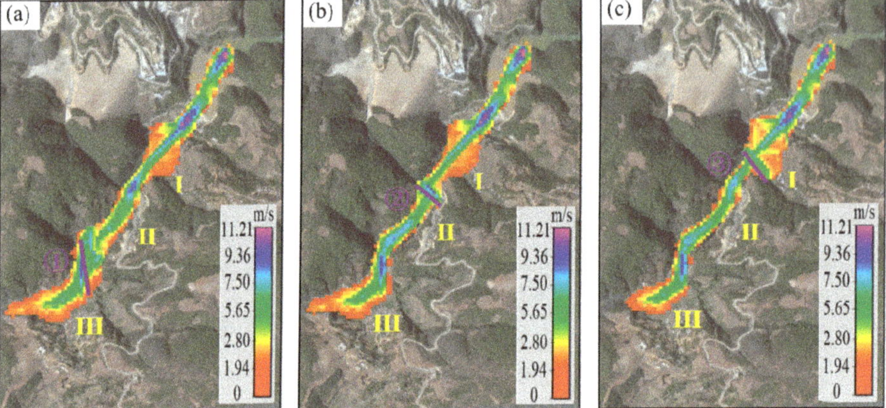

Figure 9. The maximum flow velocity distribution of leaked tailings flow with debris blocking dams at different distances. (**a**) 1000 m away from initial dam; (**b**) 600 m away from initial dam; (**c**) 400 m away from initial dam.

3.2.3. Impact Force of Leaked Tailings Flow

As shown in Figures 7 and 10, the locations of maximum impact forces after dam failure are all at the valley near Village I, whose values are all 1000 kN/m. Moreover, marked enhancement of impact force is observed in the vicinities of debris blocking dams. After crossing these dams, the impact forces of tailings flow decrease due to the decline in flow rate and depth. The locations of maximum impact forces are all at the valley mouth near the upstream side of Village III, which are adjacent to the maximum downstream flow rate locations.

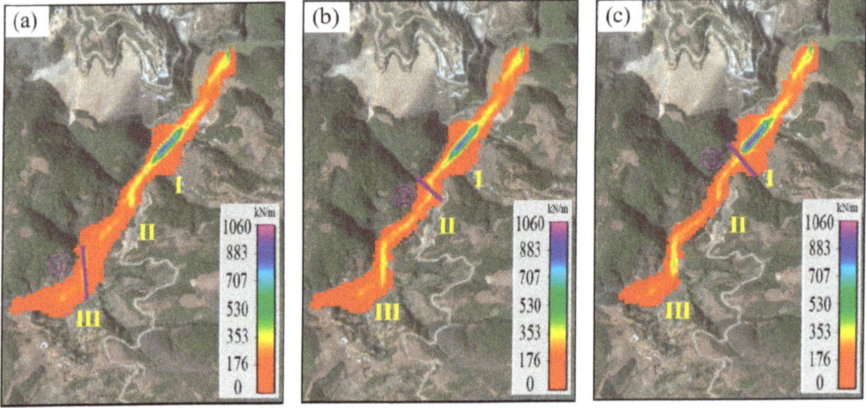

Figure 10. The maximum impact force of leaked tailings flow downstream with debris blocking dams at different distances. (**a**) 1000 m away from initial dam; (**b**) 600 m away from initial dam; (**c**) 400 m away from initial dam.

4. Conclusions

In this study, different conditions of downstream riverbed slopes and debris blocking dam construction are analyzed for the tailings pond. The evolution characteristics and deposition laws of leaked tailings flow after dam failure are studied through simulation. During the evolution process, the overtopping flows carry the tailing particles towards the downstream, among the drag forces on the tailing particles, the friction is greater than the resistance, the tailing particles therefore migrate downstream rapidly. When the friction of the drag forces is equal to its resistance, the flow rate of tailing particles reaches a maximum. With the decrease in water content in the discharged sediments, the uplift force is gradually reduced, the resistance is increased, and the flow rate of tailing particles is decreased. The tailing particles gradually deposit until the flow rate reaches the isolation rate, and they finally deposit completely.

The results demonstrate larger inundated area of tailings flow on the upstream and downstream sections than the midstream, where there are drastic terrain changes. With the elevation of downstream riverbed slope, the inundated area increases, while the maximum depth of tailings deposition decreases gradually. In addition, the deepest sedimentation point of tailings is developed downstream, and the deposition thickness also decreases accordingly. The maximum evolution velocity increases with the elevating terrain, and the maximum impact forces are all located adjacent to the locations of maximum evolution velocity or maximum flow depth.

Setting up debris blocking dams at different distances (characteristic locations) downstream of the initial dam leads to a decline in the maximum inundated range of leaked tailings flow with the shortening distance of these dams from the initial dam. After blockage by the dams, the energy storage-dissipation-deposition process needs to be repeated again. Both the flow rate and depth decrease after crossing the debris blocking dams, and the downstream inundated area is accordingly reduced. Thus, the downstream inundated area ultimately decreases with the shortening distance between initial and debris blocking dams.

Author Contributions: G.W. and S.T. designed the study, participated in data collection and analysis, carried out the statistical analyses, wrote the graph of the data and drafted the manuscript. Z.X. and X.K. participated in data collection, statistical analysis and drafted the manuscript. J.C. and B.H. participated in statistical analyses, and helped to interpret the data and draft the manuscript. All authors gave final approval for publication.

Funding: This research is supported by the Program for the National Key Research and Development Plan (2017YFC0804600), Open Research Fund of State Key Laboratory of Geomechanics and Geotechnical Engineering, Institute of Rock and Soil Mechanics, Chinese Academy of Sciences (Z018017), Changjiang Scholars and Innovative Research Team in University (IRT_17R112), National Natural Science Foundation of China (41602307; U1802243; 51904040), Open Issue with Key Laboratory of Mine Geological Hazards Mechanism and Control and Department of Natural Resources of Shaanxi Province (KF2018-09).

Acknowledgments: The authors would like to acknowledge the colleagues from the State Key Laboratory of Coal Mine Disaster Dynamics and Control for their perspectives and suggestions related to data collection and statistical analysis.

Conflicts of Interest: The authors declare no competing interests.

References

1. Yin, G.; Li, G.; Wei, Z.; Wan, L.; Shui, G.; Jing, X. Stability analysis of a copper tailings dam via laboratory model tests: A Chinese case study. *Miner. Eng.* **2011**, *24*, 122–130. [CrossRef]
2. Mura, J.C.; Gama, F.F.; Paradella, W.R.; Negrão, P.; Carneiro, S.; De Oliveira, C.G.; Brandão, W.S. Monitoring the Vulnerability of the Dam and Dikes in Germano Iron Mining Area after the Collapse of the Tailings Dam of Fundão (Mariana-MG, Brazil) Using DInSAR Techniques with TerraSAR-X Data. *Remote Sens.* **2018**, *10*, 1507. [CrossRef]
3. Sun, E.; Zhang, X.; Li, Z. The internet of things (IOT) and cloud computing (CC) based tailings dam monitoring and pre-alarm system in mines. *Saf. Sci.* **2012**, *50*, 811–815. [CrossRef]
4. Azam, S. Tailings dam failure: A review of the last one hundred years. *Geotech. News* **2010**, *28*, 50–53.
5. Zhou, X.; Cheng, H. Stability analysis of three-dimensional seismic landslides using the rigorous limit equilibrium method. *Eng. Geol.* **2014**, *174*, 87–102. [CrossRef]
6. Parbhakar-Fox, A.; Glen, J.; Raimondo, B. A Geometallurgical Approach to Tailings Management: An Example from the Savage River Fe-Ore Mine, Western Tasmania. *Minerals* **2018**, *8*, 454. [CrossRef]
7. Wang, K.; Yang, P.; Hudson-Edwards, K.A.; Lyu, W.; Yang, C.; Jing, X. Integration of DSM and SPH to Model Tailings Dam Failure Run-Out Slurry Routing Across 3D Real Terrain. *Water* **2018**, *10*, 1087. [CrossRef]

8. Jia, T.; Wang, R.; Fan, X.; Chai, B. A Comparative Study of Fungal Community Structure, Diversity and Richness between the Soil and the Phyllosphere of Native Grass Species in a Copper Tailings Dam in Shanxi Province, China. *Appl. Sci.* **2018**, *8*, 1297. [CrossRef]

9. Wang, X.; Zhan, H.; Wang, J.; Li, P. The Stability of Tailings Dams under Dry-Wet Cycles: A Case Study in Luonan, China. *Water* **2018**, *10*, 1048. [CrossRef]

10. Ismet, C. Guest editorial-Special issue on dynamic failures in underground mines. *Int. J. Min. Sci. Technol.* **2018**, *28*, 719–720. [CrossRef]

11. Aly, A.; Keizo, U.; Yang, Q. Assessment of 3D Slope Stability Analysis Methods Based on 3D Simplified Janbu and Hovland Methods. *Int. J. Geomech.* **2012**, *12*, 81–89.

12. Santamarina, C.J.; Torres-Cruz, L.A.; Bachus, R.C. Why coal ash and tailings dam disasters occur. *Science* **2019**, *364*, 526–528. [CrossRef] [PubMed]

13. Wu, Z.; Mei, G. Statistical analysis of tailings pond accidents and cause analysis of dam failure. *China Saf. Sci. J.* **2014**, *24*, 70–76. (In Chinese)

14. Wei, Z.; Yin, G.; Wang, J.; Wan, L.; Li, G. Design, construction and management of tailings storage facilities for surface disposal in China: case studies of failures. *Waste Manage Res.* **2013**, *31*, 106–112. [CrossRef] [PubMed]

15. Rico, M.; Benito, G.; Díez-Herrero, A. Floods from tailings dam failures. *J. Hazard. Mater.* **2008**, *154*, 79–87. [CrossRef]

16. Peng, K.; Zhou, J.; Zou, Q.; Zhang, J.; Wu, F. Effects of stress lower limit during cyclic loading and unloading on deformation characteristics of sandstones. *Constr. Build. Mater.* **2019**, *217*, 202–215. [CrossRef]

17. Roussel, N.; Stefani, C.; Leroy, R. From mini-cone test to Abrams cone test: Measurement of cement-based materials yield stress using slump tests. *Cem. Concr. Res.* **2005**, *35*, 817–822. [CrossRef]

18. Hu, J.; Liu, C.; Li, Q.; Shi, X. Molecular simulation of thermal energy storage of mixed CO_2/IRMOF-1 nanoparticle nanofluid. *Int. J. Heat Mass Tran.* **2018**, *125*, 1345–1348. [CrossRef]

19. Tian, S.; Chen, J. Multi-hierarchical fuzzy judgment and nested dominance relation of the rough set theory-based environmental risk evaluation for tailings reservoirs. *J. Cent. South Univ.* **2015**, *22*, 4797–4806. [CrossRef]

20. Deng, D.; Li, L.; Wang, J.; Zhao, L. Limit equilibrium method for rock slope stability analysis by using the Generalized Hoek–Brown criterion. *Int. J. Rock Mech. Min.* **2016**, *89*, 176–184.

21. Jing, X.; Chen, Y.; Xie, D.; Williams, D.J.; Wu, S.; Wang, W.; Yin, T. The Effect of Grain Size on the Hydrodynamics of Mudflow Surge from a Tailings Dam-Break. *Appl. Sci.* **2019**, *9*, 2474. [CrossRef]

22. Zheng, B.; Zhang, D.; Liu, W.; Yang, Y.; Yang, H. Use of Basalt Fiber-Reinforced Tailings for Improving the Stability of Tailings Dam. *Materials* **2019**, *12*, 1306. [CrossRef] [PubMed]

23. Dutto, P.; Stickle, M.M.; Pastor, M.; Manzanal, D.; Yague, A.; Moussavi Tayyebi, S.; Lin, C.; Elizalde, M.D. Modelling of fluidised geomaterials: The case of the Aberfan and the Gypsum tailings impoundment flowslides. *Materials* **2017**, *10*, 562. [CrossRef] [PubMed]

24. Marsooli, R.; Wu, W. 3-D finite-volume model of dam-break flow over uneven beds based on VOF method. *Adv. Water Resour.* **2014**, *70*, 104–117. [CrossRef]

25. Zhang, L.; Tian, S.; Peng, T. Molecular Simulations of Sputtering Preparation and Transformation of Surface Properties of Au/Cu Alloy Coatings Under Different Incident Energies. *Metals* **2019**, *9*, 259. [CrossRef]

26. Federal Emergency Management Agency. Numerical Models Meeting Minimum Requirement of National Flood Insurance Program 2010. Available online: http://www.fema.gov (accessed on 10 January 2019).

27. FLO-2D Software Inc. *FLO-2D Users Manual*; Version 2018, 06; FLO-2D Software Inc.: Nutrioso, AZ, USA, 2018.

28. Hübl, J.; Steinwendtner, H. Two-dimensional simulation of two viscous debris flows in Austria. *Phys. Chem. Earth* **2001**, *26*, 639–644. [CrossRef]

29. Du, X.; Sun, S.; Zhao, Z.; Qin, L. Research on Engineering Control Effect on Hongchun Gully Debris Based on FLO-2D Model. *Earth Environ.* **2016**, *44*, 376–381. (In Chinese)

30. Pashias, N. A fifty cent rheometer for yield stress measurement. *J. Rheol.* **1996**, *40*, 1179. [CrossRef]

31. Alhasan, Z.; Jandora, J.; Riha, J. Comparison of specific sediment transport rates obtained from empirical formulae and dam breaching experiments. *Environ. Fluid Mech.* **2016**, *16*, 1–23. [CrossRef]

32. Gawu, S.K.; Fourie, A.B. Assessment of the modified slump test as a measure of the yield stress. *Can. Geotech. J.* **2004**, *41*, 39–47. [CrossRef]

33. Hu, L.; Zhang, Z.; Li, Q.; Guo, X. Sequential dam break simulation and risk analysis of earth-rock dams of cascade reservoirs. *J. Hydroelectr. Eng.* **2018**, *37*, 65–73.

34. Tian, S.; Chen, J.; Dong, L. Rock strength interval analysis using theory of testing blind data and interval estimation. *J. Cent. South Univ.* **2017**, *24*, 168–177. [CrossRef]
35. Mizani, S. Rheology of Thickened Gold Tailings for Surface Deposition. Ph.D. Thesis, Carleton University, Ottawa, ON, Canada, 2010.
36. Zhou, Y.; Li, Q.; Wang, Q. Energy Storage Analysis of UIO-66 and Water Mixed Nanofluids: An Experimental and Theoretical Study. *Energies* **2019**, *12*, 2521. [CrossRef]
37. Zhang, P.; Ma, J.; Shu, H.; Wang, G. Numerical simulation of erosion and deposition debris flow based on FLO-2D model. *J. Lanzhou Univ. Nat. Sci.* **2014**, *50*, 363–368. (In Chinese)
38. Okuda, S.; Okunishi, K.; Suwa, H. Observation of Debris Flow at Kamikamihori Valley of Mt. Yakedake. In Proceedings of the 3rd Meeting of IGU Commission on Field Experiment in Geomorphology, Kyoto, Japan, 24–30 August 1980; pp. 116–139.
39. Li, J.; Luo, D. The Formation and Characteristics of Mudflow and Flood in the Mountain Area of the Dachao River and Its Prevention. *Z. Geomorphol. N.F.* **1981**, *25*, 470–484.
40. Nam, D.H.; Kim, M.I.; Kang, D.H.; Kim, B.S. Debris Flow Damage Assessment by Considering Debris Flow Direction and Direction Angle of Structure in South Korea. *Water* **2019**, *11*, 328. [CrossRef]

MDPI

St. Alban-Anlage 66

4052 Basel

Switzerland

Tel. +41 61 683 77 34

Fax +41 61 302 89 18

www.mdpi.com

Water Editorial Office

E-mail: water@mdpi.com

www.mdpi.com/journal/water